Technology's Past *Vol. 2*

More Heroes of Invention and Innovation

By Dennis Karwatka

Tech Directions Books
Prakken Publications, Inc.

Copyright © 1999 Prakken Publications, Inc.
P.O. Box 8623, Ann Arbor, MI 48107-8623

Library of Congress Catalog Card Number: 95-72938
ISBN 0-911168-96-6

Printed in the United States of America

Cover photos, clockwise from top left: a wooden clock mechanism used in Eli Terry's early wall clock (page 47); Michael Faraday's original iron-ring transformer (page 66); Isambard Kingdom Brunel, a builder of bridges, tunnels, and the world's first all-metal, screw-propeller-driven ship (page 85); Marie Curie, in her lab, sometime after 1906 (page 182); Alexander Gustave Eiffel's tower, under construction in Paris in 1888 (page 137); a U.S. postage stamp that commemorates Benjamin Banneker, astronomer, almanac publisher, and surveyor who helped lay out Washington, D.C. (page 20); Amelia Earhart, preparing to fly Juan de la Cierva's autogiro aircraft (page 209); one of the world's first production automobiles offered for sale to the public, built by Karl Benz in 1888 (page 149).

This book is dedicated to Josiah Willard Gibbs, America's greatest native-born, theoretical scientist. A courteous and productive nineteenth-century investigator, he rarely mentioned his international successes. He simply encouraged people to do their very best. Gibbs's life is an honorable model for everyone.

Contents

Special Section: Technology's Past Overviews

Recommended Reference

Introduction

Investigating the lives of people who helped create our modern world is an exciting activity filled with all the twists and turns of an unfamiliar roller coaster ride. This book will introduce you to the colorful personal lives of some technical pioneers and the events that led to their successes.

More than six dozen international technologists are in this upbeat book of short biographies. The subjects are partly based on the "Technology's Past" column in Tech Directions, a monthly magazine for teachers and students of industrial technology.

When reading about an inventor, avoid thinking of that person as "the" inventor of a device. Most inventions are improvements rather than entirely new innovations. Isaac Singer did not invent "the" sewing machine, but he improved the basic design and did more than anyone else to introduce it to the public. William Burroughs did not invent "the" calculator, but he made the first with a keyboard. Karl Benz did not invent "the" automobile, but he was the first to offer one for sale to the general public.

Photographs

Every biography has an image of its subject. Most came from the Smithsonian Institution in Washington, D.C., the Science Museum in London, the Deutsches Museum in Munich, or from a company. A few are renditions sketched by artist Tim Harmon. Although many technologists were in their 20s or 30s when they did their best work, the images usually show them at an older age, after they became better known.

Many additional supporting images were photographed by the author at technical sites in America, Great Britain, and Germany. Some are of better quality than others but we hope each gives you a more complete appreciation of the individual's technical achievement. We tried to include as many photographs as space would allow but the number is not consistent among the biographies. The work of some individuals, like Ada Lovelace in early computer programming, did not lend itself to hardware images. Also, technical museums often emphasize the accomplishments of better-known inventors. There are many displays associated with James Watt's steam engine but fewer connected with Erastus Bigelow's power loom.

The people from technology's past included some giants and many whose names may be less familiar. But everyone's contribution was important, and you will find no distinctions here: Each person's biography receives the same amount of space.

References

A sweeping book such as this one required many references. More than 280 were used to make it both technically accurate and pleasantly readable. Important starting points were specialty encyclopedias and dictionaries. They are listed at the back of the book in general order of usefulness to the author. In addition, each biography lists unique references, again in order of usefulness.

Acknowledgments

The acknowledgments section is where an author names those who assisted in a book's development. Such sections are well worth reading. They show the wide range of individuals who worked behind the scenes.

This book required assistance from many different quarters. These are the people the author wishes to formally thank for their help:

Librarians
- Paulette Hayes, Assistant Librarian, Fayette County Library, Connersville, Indiana
- Barbara Heathcote, Local Studies Librarian, City Library, Newcastle upon Tyne, England
- Irene McCabe, Librarian, Royal Institution, London
- Venita Paul, Librarian, Science Museum, London
- Peter Van Dulken, Historical Patent Information Officer, British Library, London

Educational Researchers
- Vicente Cano, Professor of Romance Languages, Morehead State University, Morehead, Kentucky
- Alan Hague, past Department Head of Manufacturing Engineering, Nottingham-Trent University, Nottingham, England
- James Lewis, Project Coordinator, Middlesex University, London
- Russ Moniz, Computer Analyst, Tiverton, Rhode Island
- Rodney Stanley, Associate Professor of Industrial Education, Morehead State University, Morehead, Kentucky
- Ray Watson, Vicar, Parish Church of St. Mary Magdalene, Hucknall, England

Museum Researchers
- Ed Battison, past Director, American Precision Museum, Windsor, Vermont
- Graham Bradshaw, Curator, Newcastle Discovery Museum, Newcastle, England
- Walter Rathjen, Director, Deutsches Museum, Munich
- Joyce Stoffers, Managing Director, American Clock and Watch Museum, Bristol, Connecticut
- Michael Wright, Curator of Machine Tools, Science Museum, London

Company and Business Personnel
- Molly M. Gant, Corporate Relations, Thermo King Corporation
- Dan Guzewich, Editorial Assistant, Carrier Corporation
- Richard James, Corporate Affairs Manager, J. I. Case Corporation
- Sue Langer, Public Relations Administrator, Rolls-Royce Corporation
- Michael Lombardi, Historian, Boeing Company
- Lori McJavish, Chrysler Corporation
- Jack Morris, Chairman, Business Design Centre, London
- Gerbrand Poster, Director of Merchandising, Bigelow Corporation
- Deeann Siler, Advertising Coordinator, Roots Corporation
- Pat Zerbe, Public Relations Manager, Beech Aircraft Corporation

Over the past dozen years or so, the author's researches into the history of technology have been well supported by Morehead State University. Several people were most helpful in securing that support:
- Carole Morella, Director of Research Grants
- Robert Hayes, Department Head of Industrial Education and Technology
- Charles Derrickson, past Dean of the College of Applied Sciences and Technology
- Gerald DeMoss, Dean of the College of Science and Technology
- John C. Philley, past Executive Vice President of Academic Affairs
- C. Nelson Grote, past President
- Ronald Eaglin, President

The editors and professional staff at Tech Directions Books / Prakken Publications take a roughly worded manuscript and convert it to smoothly flowing text. They have done an equally impressive job with attractively presenting the images. Insights provided by Christine Ecarius, Sharon Miller, Pam Moore, and others have resulted in a technically rich, pleasantly readable, and well illustrated book. Another person who deserves acknowledgment is Susanne Peckham, editor of Tech Directions Books. A continually supportive person, she spearheaded the entire project. Without her, this book would have never been published.

Dennis Karwatka
Morehead, Kentucky
1999

Johann Gutenberg

People communicate with words in several different ways. Visual information from the Internet is one example. Listening to the radio, telephone, or instructional lectures are others. Although possessing the advantage of immediacy, these media can't match the heritage or staying power of the printed page. Little printing today uses individually positioned pieces of metal type. Computerized techniques are now the preferred method. But using hand-set type is a hobby for some people. They make limited print runs for personal enjoyment. Like any hobbyist, these people have a greater interest in quality than quantity. They might print small books, hand bind them, and present them as gifts. Such beautiful publications often use very old presses and are sometimes considered to be an art form.

Books sold before the mid-1400s were costly and rare. Each was individually handwritten using quill pens or printed from hand-carved wooden pages. In either case, the books were too expensive for average people to buy. A high-quality handwritten Bible cost about 80 guilders in fifteenth-century Germany. For comparison, a stone house cost 80 to 100 guilders. The development of reusable metal printing type used with a printing press led to a drop in book prices and an increase in level of education. It started a revolution in mass communication. The first book produced using such methods was an impressive two-volume large-format Bible printed by Johann Gutenberg in 1456. Of the 200 copies that Gutenberg printed, 47 still exist.

Gutenberg was born in an industrial city on the Rhine River not far from Frankfurt. He was the youngest child in his family and may have had two siblings. He received training as a goldsmith or silversmith and

Born:
1397 (?), in Mainz, Germany

Died:
February 3, 1468, in Mainz, Germany

earned his living at that trade. His parents were apparently well-to-do but lost their fortune and social status following a revolt by the lower-class citizens of Mainz. To escape the political upheaval, Gutenberg's family moved to Strasbourg, France, about 100 miles to the south, in 1411. The younger Gutenberg spent many years there and developed an interest in the possibility of mechanical printing. Goldsmiths marked their work with a metal punch and that activity may have fostered Gutenberg's interest in printing with replaceable, or *movable*, type.

Like other printers at the time, Gutenberg began by obtaining a block of wood as large as a page. He carved rows of letters that made words, inked the block, and pressed a sheet of paper onto to it. The process worked fairly well but was too labor intensive to be practical. He tried carving individual letters from wood. He personally carved many hundreds and combined the letters into words. Tightening them within a metal frame and pressing paper against the inked letters produced a readable image. However, that method was also labor intensive and the wooden letters wore out

The Gutenberg Bible at the Gutenberg Museum in Mainz is displayed under security glass in a darkened vault. Flash photography is not allowed. This is one volume of a two-volume Bible that was purchased in 1978 for about $2 million.

more quickly than Gutenberg had expected. That resulted in smudged words and unevenly printed letters.

Gutenberg returned to Mainz in 1448 and began work on a method that used interchangeable metal type. To purchase the necessary supplies, he borrowed 800 guilders from wealthy lawyer Johann Fust (1400 (?)-1466). That was a very large amount of money, since a master craftsman at the time typically earned 20 to 30 guilders a year. To make individual letters, Gutenberg engraved each character in relief on the end of a steel punch. He used a heavy hammer blow to strike the image into a piece of softer copper and formed a matrix. He then filed the copper matrix to fit in his two-part mold. The mold was an ingenious invention and the most important part of his printing system. Gutenberg poured in molten lead with a little tin and antimony to improve flowability and hardness. The lead alloy solidified within seconds, Gutenberg split the mold, and a cast letter fell out. The mold was adjustable for different letter widths, wider for a "w" and narrower for an "i". He could easily make as many identical metal letters as he wanted. Considering the alphabet, punctuation marks, and special characters, he needed about 150 different Latin characters for his Bible. On a good day, Gutenberg and an assistant could cast 600 lead characters. Each one had to be filed to a uniform height and stored in cases.

Gutenberg also developed an oil-based

Gutenberg's innovative adjustable two-piece mold could cast letters of different widths, such as "m" and "l". This was the most important feature of his printing system.

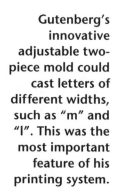

ink for use with metal type. The water-based ink used with quill pens tended to form into droplets and would not properly spread over the surface of the metal letters. Borrowing an idea from artists, Gutenberg mixed ground carbon with boiled linseed oil and produced a suitable ink. He also had to deal with hard-surfaced papers made for quill pens. He solved the problem by slightly moistening the sheets before pressing the inked type onto their surface.

Gutenberg came from a grape-growing region and adapted his printing press from one used to squeeze grapes for wine. Turning a large screw with a heavy wooden handle forced two horizontal plates together. The lower plate held the inked type in a clamped frame. The upper plate pressed paper against the type, forming an image on the paper. After printing several copies of one page, the type was removed and reset for another page. A photo on the next page shows a reconstruction of Gutenberg's printing press at the Gutenberg Museum in Mainz. His printing rate was about 30 pages per hour and the method remained essentially unchanged for more than 300 years. Gutenberg used six presses that probably cost him 40 guilders each.

As was common in his day, Gutenberg wanted his first major effort to have a religious connection. He and his assistant Peter Schoeffer (1425 (?)-1502) assumed the four-year task of setting all the type for a two-volume 1,286 page Bible. It was printed in Latin with 42 lines per column. It eventually required 3 million hand-set characters. Gutenberg purchased a high-quality handwritten Bible to use as a master. His printed book turned out beautifully. Although almost impossible for a nonscholar to read because it is written in Latin with an ancient type face, few modern books can

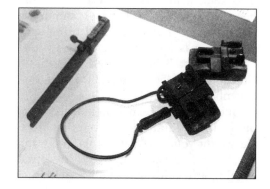

This replica of Gutenberg's printing press and print shop are at the Gutenberg Museum in Mainz. Demonstrations are regularly given for visiting school groups. With little mechanical advantage to the screw on the press, it took a great deal of effort to press the inked type against the paper.

rival its impressive appearance. Each sheet was elegantly type set, had decorative marks at the beginning of sentences, and was printed on the best Italian paper or parchment. Gutenberg printed each sheet so that the front and back were precisely aligned. He used tiny needles in the paper as positioning guides. Bookbinders put the sheets together for him.

The Gutenberg Bible was the first substantial publication to come from a printing press. One on display at the Gutenberg Museum was purchased in 1978 for more than $2 million. It is called the *Shuckburgh Bible*. Original copies in America are at the Library of Congress, Yale and Harvard Universities, the New York Public Library, and the Huntington Museum in California. It is hard to identify other books Gutenberg may have printed because he never put his name on his work.

The introduction of reusable type and a printing press provided a spark for other potential printers throughout the world. Within the next 45 years, about 15 million books had been printed. Gutenberg's innovation had changed human civilization forever. Books and magazines could be printed at low cost. That created an emphasis on literacy and led to the establishment of major universities and specialized fields of knowledge.

Gutenberg's indebtedness to Fust eventually amounted to more than 2,000 guilders, an incredible amount of money. As was his legal right, Fust foreclosed on the loan. Gutenberg could not come up with the repayment and Fust acquired all his tools and supplies. Gutenberg was almost reduced to poverty. Late in life, he received an adequate pension from the German government and died at about the age of 71.

For all his abilities as a professional printer, Gutenberg left no written material about his personal life. He has been a very difficult person for historians to research. About 30 documents from his lifetime mention his name, but most are legal papers. Gutenberg was continually plagued with financial problems of his own making. Only two documents make any mention of his printing profession, and those comment on it only in passing. It has been impossible to gain any insight into his personality or motivation. But there is little doubt about Gutenberg's technical successes. Secondary sources clearly verify his accomplishments. Although his abilities were little appreciated in his own time, he found a unique niche in history. He printed the only ancient book immediately known to much of the general public, the Gutenberg Bible.

References

The Gutenberg Bible by Martin Davies, British Library Press, 1996.

Stories of Great Craftsmen by S. H. Glenister, Books for Libraries Press, 1939 (reprinted 1970).

Printers and Printing by David Pottinger, Harvard University Press, 1941 (reprinted 1971).

Leonardo da Vinci

Born:
April 15, 1452, in
Vinci, Italy

Died:
May 2, 1519, in
Amboise, France

People working in the field of technology occasionally move from one specialty to another. James Buchanan Eads (1820-1887) invented a diving bell to salvage boats sunk in the Mississippi River. But he is better known for building the world's first steel bridge, which spans the Mississippi River at St. Louis. Eads had no earlier bridge-building experience. It is more unusual to find someone who is equally talented in technology and literature. Michael Pupin (1858-1935) was one such person. He developed improved telephone communication equipment in the early 1900s. He also won the 1924 Pulitzer Prize for his book *From Immigrant to Inventor*. But it is very hard to identify a person who has built several world-recognized careers in such diverse fields.

There is at least one: Leonardo da Vinci, the painter of the 1497 *The Last Supper* in Milan and the 1504 *Mona Lisa* in Paris. Many know that da Vinci worked with mechanisms at some points in his life. But only in the twentieth century have historians come to appreciate how much effort he put into science and technology. His artistic accomplishments went on display immediately, and for 500 years he has been known as a great painter. That reputation has overshadowed his other work. His brilliant technical efforts were often hidden in notebooks that he wrote in code and rarely shared with others. Leonardo left thousands of pages of manuscripts to his friend Francesco da Melzi, who hoarded them for 50 years. The papers were then dispersed haphazardly by da Melzi's son, Orazio. The full extent of Leonardo's work did not begin to appear until about 1880. By then, much of what he recorded had been independently discovered by others. Because of the misguided concerns of da Melzi, and the foolish actions of his son,

Courtesy Deutsches Museum, München

the world lost the opportunity for early gains in canal construction, aeronautics, architecture, and many other technical areas.

Leonardo was born near Florence in central Italy. His parents were not married and he was raised by his father who was a government official and a good provider. Leonardo learned the basics of reading and writing at an early age. He was a pleasant child who had many friends. He had shown some skill at sculpting, so his father apprenticed him to a leading painter and sculptor in Florence. Starting at age 15, Leonardo received training not only in art but in a variety of technical crafts like furniture making and bronze casting. Over his six-year obligation period, he learned his skill well and stayed on as an assistant. He opened his own studio in 1477. By 1482, he wanted new challenges in a new city.

At the time, it was not unusual for small wars to erupt between cities and countries.

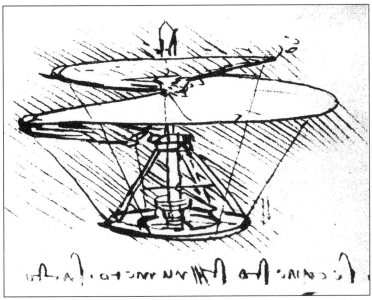

It was during his 17 years in Milan that Leonardo established his reputation with unique mechanisms and water-powered equipment. He built machinery to cut screw threads and nails. He also invented a water-powered saw. He constructed a 70-foot-tall wood-framed hydraulic pump that lifted stream water into the Duke of Milan's castle. Leonardo was the first to construct models of rivers. By studying the flow, he could anticipate certain conditions during unusually wet or dry weather. One of his related projects was to irrigate large agricultural fields outside of Milan. This afforded him the opportunity to put together mechanisms that helped people. He constructed a whole new range of mechanical cranes and shovels. The use of gears and levers allowed him to multiply forces and easily dig out the canals.

In Leonardo's time, many engineers and technologists were convinced that the work done by a water-wheel could be indefinitely increased. While that is theoretically true for forces, technical people in the fifteenth century often confused *work* and *force*. Their belief led to the idea of perpetual-motion machines. Uncountable talent has been wasted on that truly unattainable goal. Leonardo was the first to state publicly that perpetual motion was impossible. Using the stilted terminology from his era, Leonardo wrote in 1494, "Oh specu-

Polished brass helicopter model patterned after the one da Vinci sketched. Made in 1988, one of the plaques states, "Leonardo da Vinci—Helicopter Design—1480."

Leonardo had that in mind when he wrote a letter to Lodovico Sforza, the Duke of Milan. He offered his services as a designer of novel weapons. His letter mentioned new types of guns, collapsible bridges, and instruments for destroying ships. In reality, Leonardo had no experience with such things. Sforza was impressed, however, offered him a job, and Leonardo moved to Milan, the richest city in Italy. He was a tireless designer of unique engines of war. The word *engine* meant device, and the job title of "engineer" was just coming into use. An engineer in Leonardo's time designed military equipment like bridges and structures. Today, we would call these people "civil engineers."

Leonardo built and tested many different devices. They included armored vehicles, rapid-firing crossbows, and one-man pocket battleships. He also built experimental rope ladders for scaling walls, giant catapults, and a mobile crane for lifting heavy building blocks. Leonardo was endlessly ingenious coming up with mechanical methods to achieve his objective. He used gears, chains, ratchets, and whatever else he could think of. He also suggested items for personal protection like parachutes and bullet-proof vests. His squared-off linen parachute was shaped like an Egyptian pyramid and about 12 feet on each side. It is believed to have been successfully tested from a tower built for the purpose. Leonardo's mind was so active that his innovations make up a seemingly endless list.

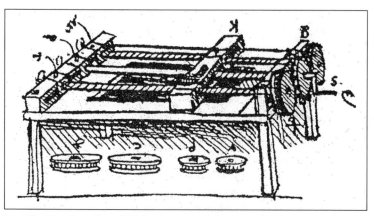

Da Vinci's drawing of his screw-cutting machine

A model at London's Science Museum of Da Vinci's screw-cutting machine taken from a drawing. The small wooden gears are for different thread characteristics.

lators about perpetual motion, how many vain chimeras [foolish goals] have you created in the like quest? Go and take your place with the seekers after gold [those who think they can change lead into gold]!"

Leonardo also had an interest in human anatomy, which he approached in a professional and dispassionate manner. To some, the body was a temple and should be treated as such. Leonardo saw it as a structure for a life system and he wanted to know how everything operated. Considering his knowledge of human anatomy, it is unusual that he would design human-powered airplanes instead of gliders. Leonardo was fascinated by flight and left 500 sketches and 35,000 words devoted to flying machines.

One of his designs was a helicopter with a large air screw, like the one shown in the photograph here. The surfaces were to be covered with linen and starched for stiffness. Another flying machine—a favorite project— used flapping wings. An experimental flapping-wing airplane is called an *ornithopter*. Lying flat, the pilot used both arm and leg power. Leonardo believed a pilot could develop enough force to lift 850 pounds, a wildly optimistic figure. Because of the low power-to-weight ratio of the human body, it is all but impossible for such aircraft to fly. Like others, Leonardo was influenced by birds and fixated on flapping-wing flight. It is not known whether he attempted to build a full-size working model. Had Leonardo focused on gliding instead, he could have easily been the first to fly. All birds glide as well as flap their wings. The first person to lift from the ground in a heavier-than-air flying machine was German glider pilot Otto Lilienthal (1848-1896) in 1891. (See page 151.)

The French were victorious in a small war

that erupted with the city-state of Milan in 1499. The Duke had built few of the defense mechanisms Leonardo had developed and Leonardo returned to Florence. The citizens remembered him for his skillful painting rather than his technical work. He made his living primarily as an artist. Then it was back to Milan in 1506 to work for King Louis XII of France who appointed Leonardo his "painter and engineer."

French King Francis I invited Leonardo to France in 1516. He gave Leonardo the title of "first painter, architect, and mechanic to the king." He was provided with a country home. One of his favorite pastimes was playing the lute, but he left few clues about his personal life. Unmarried and with no immediate family, he died at 67.

Leonardo lived so long ago that he has remained a shadowy figure in technology. Historians were still organizing his writings as recently as 1967. His notebooks, drawings, plans, and diaries far exceeded the output of any of his colleagues. Much of it was lost over the centuries, but some is slowly being recovered. In the late 1990s, one of his 20 surviving notebooks was auctioned for $30.8 million. The full scope of Leonardo's technical triumphs may not be known for many more years.

References

Leonardo da Vinci by Richard Friedenthal, Viking Press, 1959.

Leonardo da Vinci, Fine Arts Department, International Business Machines, 1951.

Biographical Dictionary of the History of Technology, edited by Lance Day and Ian McNeil, Routledge Publishers, 1996.

Leonardo da Vinci, A&E Television Biography Series, 1997.

Galileo Galilei

Born:
February 15, 1564, in Pisa, Italy

Died:
January 8, 1642, in Arcetri, Italy

The sun rises in the east every day and sets in the west. To almost everyone in the sixteenth century, that proved that the sun revolved around the earth. Those same people were certain that a 10-pound rock fell faster than a 5-pound rock. Neither premise is correct, but they were common points of view within the educated classes at the time. Such facts were even taught at the universities. Government and church officials wielded great political power and could imprison anyone who professed otherwise. University studies in science and technology were driven by philosophy, not by observation and measurement. If something appeared to be true, then it was considered true. The sixteenth century was an uncomfortable time for anyone who wanted to extend the world's technical reach.

The turning point occurred in Italy around 1600. A strong-willed, self-assured, and totally confident scientist began using modern scientific methods. Data obtained from his astronomical telescope conclusively proved that the earth revolved around the sun. He dropped rocks from the Leaning Tower of Pisa to show that heavy and light weights fall at the same rate. He invented the pendulum for measuring time and the thermometer for measuring temperature. He was the first to investigate the strength of materials. Galileo Galilei was a forceful personality who came up against powerful but small-minded leaders. His work was pivotal to the advancement and acceptance of scientific experimentation.

Galileo was the first child of Vincenzio Galilei and his wife Giulia. He is commonly known by his first name. He had at least two sisters and one brother. Galileo's father was a talented musician and textile merchant who provided an above-average standard of living. His mother had a distant relationship with her children. Formal schooling was uncommon in the 1500s and Galileo was educated by monks in a nearby monastery. They taught him Latin, Greek, and art. His father taught him to play the lute. His education was good enough to gain acceptance to the University of Pisa when he was 17.

Galileo began his university career with the study of medicine. To deal with the international makeup of students who spoke different native languages, most instruction was in Latin. Galileo happened to pass by a mathematics class that was being taught in Italian. He listened and enjoyed what he heard. He soon became fascinated by the predictive ability of mathematics and physical science. Today for example, mathematics allows people to predict under what conditions a bridge might fail. Galileo was only a teenager and already thinking about experimentation.

Galileo's cantilever beam drawing in his 1638 book *Dialogues Concerning Two New Sciences*. Unlike beam drawings in modern textbooks, Galileo's is refreshingly embellished with vegetation and an unusual rock wall, all in three-dimensional perspective.

While attending church services at Pisa Cathedral in 1582, Galileo noticed wind currents causing a chandelier to swing. Whether the arc of swing was large or small, the amount of time of each swing took was the same. Galileo used his own pulse beat to make crude measurements. He later tested his theory and found that the period of a pendulum swing depends only upon its length. Galileo built a small timing device incorporating a pendulum and an escapement. It did not record time as a clock does. It measured time intervals like a stopwatch. Physicians found it useful for evaluating heart beats. Galileo designed a clock controlled by a pendulum. His son Vincenzio constructed it after Galileo's death.

Galileo studied physical sciences and developed a hydrostatic balance. It used water and allowed people to determine the density of irregularly shaped objects. The invention brought him his first taste of fame. Galileo also wrote a scientific paper on detailed aspects of geometry. Those successes helped him get a teaching position at the University of Pisa in 1589. He was an excellent instructor and students flocked to his classes. They liked his refreshing approach to experimentation. Galileo often dropped differing weights from the Leaning Tower of Pisa, having his students note which ones hit the ground first. They always struck at the same instant. As long as air resistance is not a con-

sideration, the weight of an object has no bearing on how rapidly it falls. In a related experiment, Galileo rolled weights down inclines and came to the same conclusion.

Young, dynamic, and searching for a technical niche, Galileo invented a thermometer in 1593. It was among the first measuring devices used in science. It had a glass bulb attached to a thin glass tube partly filled with colored water. The air in the glass bulb expanded or contracted as the temperature changed. The effect caused the water to rise or fall. Although temperature standards had yet to be developed, this device was the forerunner of improved temperature-measuring instruments. Gabriel Fahrenheit (1686-1736) from Gdansk, Poland, introduced his alcohol thermometer in 1709.

One of Galileo's other interests was analyzing the strength of beams. A drawing in his 1638 book *Dialogues Concerning Two New Sciences* shows a cantilever beam with a load hanging from the end. It was the first technical drawing of a structural test. Galileo determined that the beam's shape was important to safely carry the load. And he explained that strength was a function of cross-sectional area. He was the first to start closing the gap between science and technology. The "two new sciences" Galileo introduced in his book were *materials testing* and *dynamics*, the science of motion. Galileo wrote his book long after his initial observations.

Since Galileo's position at the University of Pisa did not pay well, the death of his father forced him to look for work elsewhere. As eldest son, Galileo had responsibility for the care of his mother and siblings. He found a better salary at the Uni-

This reproduction of Galileo's laboratory is displayed at the Deutsches Museum in Munich. It represents the period between 1610 and 1623.

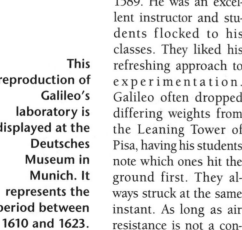

versity of Padua near Venice and moved there in 1592. He said that the 18 years he spent in Padua were the happiest of his life. He had a nice house, pleasant surroundings, and the respect of his colleagues. Galileo never married, but Marina Gamba moved in with him when he was 35. This was a common arrangement at the time and not at all unusual. The couple had two daughters and one son before separating in 1610.

The telescope had been invented in 1608 by Hans Lippershey (1570-1619) of Holland. Lippershey's was only three power and he used it to view ships, animals, people, and other similar subjects. Galileo heard about the telescope in 1609 and immediately made his own. He had great skills at fabrication and made a telescope of 32 power. It was about 24 inches long and 1-3/4 inches in diameter. It had a concave eyepiece and a convex object lens. The design came to be known as a *Galilean telescope.*

Galileo turned his telescope upward toward the planets and stars. He saw ragged mountains on the moon, spots on the sun, stars too faint to see with the unaided eye, and four moons of Jupiter. Jupiter's moons and the phases of Venus provided irrefutable astronomical evidence that the sun was at the center of the solar system. Galileo published his results in a 1610 book titled *Starry Messenger.* Galileo was a brilliant writer and the new ideas in his book made him famous. But the findings contradicted traditional religious thought, which put the earth at the center of the universe. Galileo

ran headlong into leaders unwilling to accept the clear evidence provided by careful scientific observation.

Galileo's published works and teaching methods triggered negative comments from many religious leaders. He was ordered in

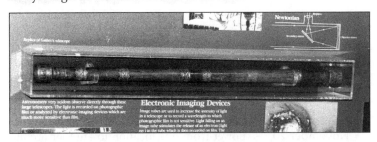

1616 not to "hold, teach, or defend in any manner whatsoever" that the sun was at the center of the solar system. But since the sun is at the center of the solar system, Galileo found it impossible to comply. He rarely started arguments, but he never backed down when he knew he was right. He stood trial in 1632 and was forced into house arrest for the last 10 years of his life. He spent his time conducting experiments, writing books, and corresponding with colleagues. Galileo died at the age of 77.

Often abrasive and always outspoken, Galileo had some personal enemies. But he had none in the fields of science and technology. His methods were correct and all technically competent people knew it. Galileo's greatest contribution was his use of experimentation to prove or disprove scientific theories. The Scientific Revolution started with him. It was a period when reason overcame tradition and the world began to move forward. Galileo's forceful personality and his conflict with religious leaders help make him the most romantic figure in all of science.

This is a replica of Galileo's telescope. The original was approximately 2 feet long and 1-3/4 inches in diameter at the eyepiece.

References

Galileo at Work: His Scientific Biography by Stillman Drake, University of Chicago Press, 1978.

Galileo and the Scientific Revolution by Laura Fermi and Gilberto Bernardini, Basic Books Publishers, 1961.

Dialogues Concerning Two New Sciences by Galileo Galilei, Northwestern University Press, 1946.

To Engineer is Human by Henry Petroski, Barnes and Noble Books, 1994.

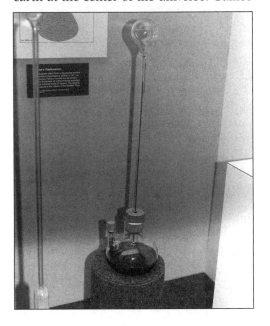

This is a replica of Galileo's original thermometer, the first instrument intended to measure temperature.

Otto von Guericke

Born:
November 20, 1602, in Magdeburg, Germany

Died:
May 11, 1686, in Hamburg, Germany

Two common methods of transferring power from one location to another are *mechanical* and *electrical*. A bicycle chain is an example of a mechanical device. It transmits power from the crank sprocket to the rear wheel. Electrical power travels by wires from a power station to a factory, business, school, or home. But there is a third method: *fluid* power. The fluid is usually air (a gas) or hydraulic oil (a liquid). If a car has power brakes, for example, air pressure from the engine amplifies the force of a person's foot on the brake pedal. Hydraulic oil transfers that force to each wheel through small tubes. Brake parts push against the wheel's rotation and the car stops. Fluid power that uses a gas is called *pneumatics*. The word comes from the Greek *pneuma*, meaning "wind," and *atics*, meaning "use of." Fluid power that uses a liquid is called *hydraulics*. The word comes from the Greek *hydro*, which means "water."

Pneumatics began with atmospheric air. Many early technologists considered removing air from a container an impossibility. Aristotle swayed them in about 340 B.C. when he said, "Nature abhors a vacuum." People tended to do less creative work in pneumatics until seventeenth-century investigators had some successes. No one showed the potential of vacuum pressure more dramatically than Germany's flamboyant Otto von Guericke. In 1657, he put two copper hemispheres together and removed the air inside them with a hand pump. Sixteen horses, eight attached to each hemisphere, could not pull them apart.

Guericke was born in Magdeburg, about 80 miles west of Berlin. His ancestors had lived in the north-central Germany city for three centuries. He received his primary education in local city schools. Guericke's wealthy aristocratic parents saw that he received an

Courtesy Deutsches Museum, München

excellent education at universities in Leipzig and Helmstedt. Although he studied mathematics and engineering, his major subject was law. Following graduation, Guericke traveled through France and England for nine months as young noblemen often did at the time. He returned home and spent most of his professional life in public service. He was elected to serve as a magistrate at the age of 23 and had responsibility for architectural affairs. He married Margarethe Alemann during the same year. The word "von" in Guericke's name was a title of nobility granted in 1666 by Leopold I.

Europe in the 1600s was a continent in turmoil. Germany had internal hostilities and also battled with Sweden, France, Spain, and other countries in a religious conflict called the Thirty Years War (1618-1648). General Johan Tserclaes, the Count of Tilly (1559-1632), destroyed the city of Magdeburg in 1631 in a savage act of the war. Guericke lost everything. He went to work as a military engineer in Sweden but returned to Magdeburg in 1632. He retained his position as magistrate and played a key role in the the city's reconstruction. He also functioned as a representative between Magdeburg and the occupying powers. Diplomacy consumed most of Guericke's time from 1642 on and he was elected mayor in 1646. After the war ended in 1648, he represented his city at the signing of the peace treaty. Because of Guericke's honorable approach to all his political dealings, the citizens admired him. They kept him as mayor for 30 years and he served in politics for more than 50 years.

From his university studies, Guericke developed an interest in atmospheric air. He wondered if it contained anything measurable. His investigations were always a hobby and done in his leisure time. Guericke's political activities sent him to many international meetings. He used these opportunities to exchange technical ideas. His German contacts included the mathematician Gottfried Leibniz (1646-1716) and inventor Gaspar Schott (1608-1666). Guericke was encouraged by the work of Italy's Evangelista Torricelli (1608-1647). Torricelli invented a mercury-filled, glass-tube barometer in 1640. He produced a vacuum over the column of mercury as it registered atmospheric pressure, about 30 inches. But it was Guericke who first used the barometer to predict the weather. His forecasts were based on systematic observations over many years. He made a special barometer in which a column of mercury moved the arm of a mannequin to indicate rising or falling pressure. Guericke suggested a network of weather stations to report on weather conditions and barometric pressures. Modern weather forecasters measure barometric pressure as "29.92 inches [of mercury]," for example. Nominal atmospheric pressure also can be specified as 14.7 psi.

Guericke constructed his vacuum pump in 1647. Using a cylinder and a piston with a pair of one-way leather flapper valves, it worked like a modern tire pump in reverse. He tested it by evacuating air from a clear glass cylinder. A ring-

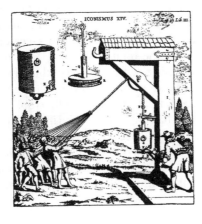

ing bell could not be heard in the vacuum nor would candles burn. Ever a showman and wanting the recognition that many politicians desire, Guericke colorfully demonstrated the capabilities of his vacuum pump at Rattisbon in 1654. He had a cylinder made and attached a heavy rope to its close-fitting piston. While several men pulled on the rope, Guericke used his pump to evacuate air from the cylinder and the piston slowly moved toward the other end. The men were powerless to stop its motion. Although the piston's size is uncertain, it may have had a diameter of about 18 inches. If Guericke's crude pump removed only three-quarters of the air in the cylinder, it would have produced a vacuum pressure of about 11 psi (3/4 x 14.7 psi). That pressure difference, acting over an 18-inch piston size, would have resulted in a large force of about 2,800 pounds.

A period painting in front of the museum display of the Magdeburg hemispheres. It shows two teams of horses that cannot separate hemispheres that have had air removed from them.

But the Magdeburg hemispheres were Guericke's most dramatic demonstration. Emperor Ferdinand III visited Magdeburg in 1657 and Guericke wanted to acknowledge the event in some notable way. He connected two 20-inch copper hemispheres together along a well-finished greased flange, and placed them on his pump. He removed as much air as his crude device would allow. It was enough so that atmospheric pressure held the hemispheres together. Even having eight horses pull on each side could not separate the hemispheres. Afterward, Guericke opened an air valve, admitting atmospheric pressure inside. The hemispheres fell apart. He duplicated the experiment in Berlin in 1663 for royal observers

A drawing from Guericke's 1672 book shows men pulling against an evacuated cylinder. The man at the right uses a transfer vessel that has had the air removed by a vacuum pump.

Guericke's 1657 Magdeburg hemispheres and vacuum pump. To remove air, the combined hemispheres were held together and placed on top of the pump where the glass globe appears.

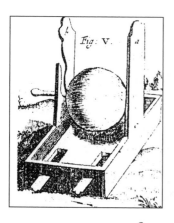

This rotating sulfur ball was the first static electricity generator. The drawing appeared in Guericke's 1672 book, which described the ball as "about the size of a child's head."

and many other times over the next several years. Munich's Deutsches Museum displays the original Guericke hemispheres and vacuum pump.

Guericke was the first to notice the elasticity of air. Shock absorbers on modern storm doors make use of that principle. Guericke often receives credit for inventing the vacuum pump, although he had competitors in England and France. He published a book in 1672 titled *New Experiments Concerning Empty Space.*

Guericke did not stop with pneumatics. He made a ball of sulfur in 1660 that could be rotated with a hand crank. He wanted to simulate the planet Earth and its magnetic properties. Holding his hand on the rotating sphere created sparks and static electricity. In a dark room, this produced a luminous glow. Although Guericke did not fully understand the phenomenon, he had created the first electrostatic generator. His discovery initiated a full century of electrical experimentation.

Guericke retired from public life and

moved to Hamburg in 1681 to live with his son. There, he worked on developing a complete physical explanation of the universe with magnetism as the primary force. Albert Einstein (1879-1955) tried a similar explanation with nuclear forces almost 300 years later. Neither succeeded. Guericke died at 83.

Guericke left two technical legacies. His sulfur ball started people thinking about static electricity. Improved electrical friction devices reached their peak with the work of Benjamin Franklin (1706-1790), America's greatest eighteenth-century scientist. But perhaps of greater importance, Guericke showed the potential of fluid power. He was among the first to question the 2,000-year-old wisdom of Aristotle, who incorrectly stated that a vacuum could not exist. Once Guericke opened the door to investigating fluid power, the possibilities seemed limitless. Fluid power now provides water to homes, lifts elevators, and operates equipment in automotive service centers. In modern American factories, about 75 percent of industrial machine tools use fluid power as a prime mover. So do amusement park rides, industrial robots, and aircraft and space flight simulators. The list is almost endless. And it all started with sixteen horses pulling on copper hemispheres more than 300 years ago.

References

Dictionary of Scientific Biography edited by Charles Coulston Gillispie, Charles Scribner's Sons Publishers, 1972.

Biographical Dictionary of the History of Technology edited by Lance Day and Ian McNeil, Routledge Publishers, 1996.

Chambers Concise Dictionary of Scientists by David Millar and others, W&R Chambers Publishers, 1990.

Asimov's Biographical Encyclopedia of Science and Technology by Isaac Asimov, Doubleday and Co. Publishers, 1964.

From Compass to Computer by W. A. Atherton, San Francisco Press, 1984.

The Timetables of Technology edited by Bryan Bunch and Alexander Hellemans, Simon and Schuster Publishers, 1993.

Magdeburg University Website. Available: http://comserv.urz.uni-magdeburg.de

Thomas Newcomen

A story has a radio talk-show host telephoning a person and asking, "For $100, who invented the steam engine?" The person does not understand the question and replies, "What?" The radio host says, "Absolutely correct. James Watt invented the steam engine and you win $100." Not only is that fictionalized pun not especially funny, it's also not true. James Watt (1736-1819) did remarkable work on the steam engine, but he did not invent it. The first working steam engine operated near Birmingham, England, in 1712, 24 years before Watt's birth.

Coal mine owners encouraged interest in early steam engines. Coal was an important British export and many mines existed in central and northern England. But as mines went deeper and deeper, ground water seeped into them. Horse-operated lift pumps proved only partly effective in dealing with the problem. Scientists throughout the world wanted to design a steam engine that could pump water from the mines. None succeeded. But the blacksmith Thomas Newcomen did. He built a large, room-sized engine to operate mine pumps and has gone down in history as the steam engine's inventor.

Newcomen was born in Dartmouth on the south coast of England to a father who was a merchant and minister. Newcomen's mother died when he was three. No one knows his precise birth date, but church records confirm his baptism on February 24, 1664. Newcomen's father helped bring educator John Flavel to Dartmouth in 1656. The younger Newcomen received his earliest formal education in Flavel's one-room school and felt encouraged to pursue the blacksmith trade. Blacksmithing was a technically demanding profession and attracted only the most intelligent young people. Newcomen

Rendering by Tim Harmon

An artist's interpretation of Newcomen. No confirmed image of him exists.

Born:
1663 or 1664, in Dartmouth, England

Died:
August 5, 1729, in London, England

served an apprenticeship in nearby Exeter. He returned home to open a business in 1685 in partnership with his close friend John Calley (1663-1717). Newcomen and Calley made many items out of iron, such as tools, hinges, nails, and chains. They also worked with specialty metals and opened a retail shop, much like a hardware store. Newcomen married Hannah Waymouth in 1705. She routinely handled the shop's retail sales. The couple had three children.

As part of his business activities, Newcomen visited tin mines in the Dartmouth area to sell tools. Miners there, too, were having trouble with water seepage and discussed the problem with Newcomen. That was probably when he heard about the steam-powered pump invented by Thomas Savery (1650-1715).

Savery was the most prolific inventor of his time. He held seven patents. One was issued in 1698 for a pump that had no ma-

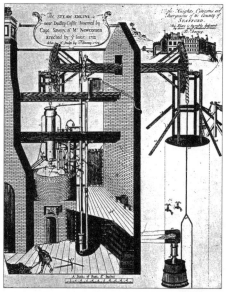

The oldest drawing (1719) of Newcomen's first engine at Dudley Castle, near Birmingham in 1712. Thomas Savery's name appears as co-inventor because he held an earlier related patent. Savery actually had nothing to do with inventing Newcomen's engine.

Bronze and stone monument dedicated to Thomas Newcomen in his hometown of Dartmouth, England. The plaque says in part, "This memorial is erected to the memory of Thomas Newcomen . . . the first to conceive the idea of working a piston by steam." It includes a drawing of Newcomen's engine.

jor moving parts. Reduced to its simplest concept, Savery's pump was a pipe with a large bulge in it. One end of the pipe was positioned under the water in the mine and the other end extended outside. The pump system also included a series of simple valves and pipe branches. A separate boiler sent steam to the bulge. Pouring water over the bulge would condense the steam, producing a partial vacuum. The vacuum would pull the water from the mine up a small distance in the pipe. By repeating the procedure over and over, water would be slowly removed from the mine. At least that was the idea. Savery demonstrated his engine to King William III and the Royal Society but no Savery pump was ever installed in a mine.

The idea of pistons moving up and down in pump cylinders led Newcomen to think about a similar approach for an engine. He made an experimental seven-inch-diameter brass cylinder in 1698. He cast the metal himself, but no equipment existed that could bore it to a true size. Newcomen and Calley spent hours hand finishing the inside to get it as smooth and circular as possible. The piston was less of a problem because a leather flap around the edge provided a reasonable seal.

The technique Newcomen developed for operating his test engine was the best he could come up with at the time. But he used it on all his future engines and many worked for more than a century. He positioned his brass test cylinder vertically. The piston had a rod con-

nected to it that pointed straight up. Newcomen had a series of hand-operated valves and pipes connected at various locations. Steam from a separate boiler went to the bottom of the cylinder, pushing the piston up. A jet of water then squirted into the cylinder, condensed the steam, and produced a partial vacuum. Atmospheric pressure forced the piston down to its original position. The process produced a continuously repeated up-and-down motion. The engine was called an *atmospheric steam engine.*

After 14 long years of experimentation, with no experts to guide him or technical books to consult, Newcomen was ready. He sold shares of stock to obtain financing and built the world's first full-size steam engine in 1712. It was a huge complicated 5 hp stationary beam engine unlike anything seen today. It was built for the Conygree Coal Works at Dudley Castle near Birmingham. The eight-foot-tall brass cylinder used a piston that had a diameter of 21 inches. Like most Newcomen engines, it operated at a leisurely six cycles per minute. Each stroke took five seconds. The piston connected to

This 1760 Newcomen engine is the oldest steam engine in the world. It pumped water from a coal mine in England. A haystack boiler is at the extreme left. The cylinder is partly visible behind the left side of the masonry support and the pump piston is in a pit at the right. The engine is displayed at the Henry Ford Museum in Dearborn, Michigan.

one end of a strong overhead wooden beam or giant lever, about 28 feet long. Large bicycle-type chains allowed for the arcing motion of the beam. As the piston moved up and down, the beam transferred the motion to a pump at the other end. The pump removed water from the coal mine at about 3,780 gallons per hour. The project cost £1,000 at a time when a skilled worker earned £1 per week. In spite of its expense, the engine was an immediate success and was soon followed by many others.

Newcomen's mechanical beam engine was totally different from Savery's unsuccessful vacuum-pressure pump. But the laws were such that it was assumed to infringe on Savery's patent, which an act of Parliament had extended to 1733. Newcomen was forced to accept Savery as a partner, though it is unlikely the two ever met.

When Savery's patent expired in 1733, more than 100 Newcomen steam engines were operating throughout the world. Besides England, they were also used in Belgium, France, Germany, Hungary, and Sweden. This was remarkable considering that steam engines were a new and expensive technology. The largest Newcomen engine ever built was for the Walker Coal Mines. Its 74-inch-diameter cylinder had such a large volume that it required four boilers to operate. James Watt got his start in steam engine technology with a small Newcomen engine model. He repaired it for the University of Glasgow in 1763. It was the first steam engine Watt had ever seen.

Newcomen was a religious man. He sold his share of the business to Calley to work full time as a minister. He died unexpectedly during a visit to London in 1729. Newcomen never wrote a biography and historians have found it hard to piece together the story of his life. Only two of his letters have survived. No images or descriptions of his appearance exist. The image used at the beginning of this chapter is an artist's interpretation.

The oldest steam engine in the world is not on display in England but in America. It is at the Henry Ford Museum in Dearborn, Michigan. The engine was built in 1760 and used until 1827 to pump water from the Cannel Coal Works near Ashton-under-Lyne. The 20 hp Newcomen engine has a 28-inch-diameter piston with a stroke of 72 inches. It survived outdoors in a neglected state for more than 100 years before Henry Ford had it disassembled and shipped to his museum in 1930. The Henry Ford Museum displays the world's largest collection of steam engines.

Unlike many technical predecessors who were scientists or noblemen, Thomas Newcomen was a practical tradesman. The scientific world was reluctant to recognize and acknowledge his achievements. They found it incredible that someone with little formal education could succeed where they had not. That was part of the reason that Newcomen had to include Savery as co-inventor of his engine, when Savery had nothing to do with it. Both their designs used steam, but in completely different ways. Savery was not a bad or disagreeable person, but he was politically well connected and took advantage of his position. Newcomen blazed the trail for many self-taught inventors who followed. He showed others that technical success does not come from social position but from insight, hard work, and confidence in one's ability.

Five-foot-diameter Newcomen engine piston. The piston cracked during use, was buried in 1781, then dug up and recently put on display. A young British primary-school student poses by the piston for size comparison.

References

The Steam Engine of Thomas Newcomen by L. T. C. Rolt and J. S. Allen, Moorland Publishing Company, 1977.

Thomas Newcomen, revised edition, by H. W. Dickinson, Newcomen Society for the Study of the History of Engineering and Technology, 1989.

Reconstructed hillside building on Lower Street in Dartmouth, where Newcomen had his workshop in the shop on the left and his home in what is now the shop on the right.

Josiah Wedgwood

Born:
(baptized) July
12, 1730, in
Burslem, England

Died:
January 3, 1795, in
Etruria, England

Some products of the modern world are purely functional. Bridges on interstate highways are an example. They safely carry traffic, but most do not have the graceful beauty of a suspension or cantilever bridge. Wristwatches for sporting activities are not as elegant as those worn during important social events. Tables and chairs in many fast-food restaurants lack the visual appeal of those found in homes. None of these examples are necessarily undesirable. But another dimension is added when an item is both functional and attractively designed.

Once a product's function has been satisfied, designers can work on improving its form. For example, all remote controls for electronic equipment function quite well. But some are easier to use because the designers took more time considering their form. The function of cameras was determined decades ago. They record images on photographic film. But their form varies widely. There are compact cameras, single-lens-reflex cameras, Advanced Photo System cameras, and others.

The form/function idea is not new. People in the eighteenth century, for example, had pottery kitchenware that functioned satisfactorily. Then, a potter of the period developed a method for making kitchenware more attractive. His products were so beautiful that the demand outstripped the supply and Josiah Wedgwood became wealthy. His name is now synonymous with expensive pottery and china.

Wedgwood was born in a small community that is now a northern suburb of Stoke-on-Trent in central England. He was the youngest of 12 children in a family whose history in the pottery trade extended back to 1612. Brothers, sisters, cousins, uncles, aunts, and other Wedgwoods

From *The Life of Josiah Wedgwood* by Eliza Meteyard, Hurst and Blackett Publishers, 1865

worked in various businesses in the closely knit region. Wedgwood received a limited amount of formal education until he was nine. His father died that year and the youngster had to go to work. He continued to put time into self-education, though, and did particularly well with writing and mathematics. Wedgwood contracted smallpox in 1741. He survived, but his right knee was so weakened that he could not use the foot-operated potter's wheel.

At 15, he was apprenticed to his oldest brother, Thomas, who had inherited the modest family pottery business. Because of his knee problem, Wedgwood could not do much physical work. He spent his time learning business practices, pottery design, and the chemistry of various clays and glazes. Much of the region's industry went to producing functional but unattractive mottled-black pots, jugs, and mugs made from local clay and glazes. Wedgwood tried to convince his brother to experiment with new glazes and to use modern manufacturing methods. The idea of tampering with age-old methods upset his brother. Wedgwood left the family business in 1752.

First, Wedgwood formed a partnership with John Harrison and then with Thomas Whieldon in 1754. Whieldon was an outstanding potter and he and Wedgwood worked together for five years. Wedgwood continued experimenting and developed several new glazes. He saved enough money to open his own independent business in 1759. He rented a section in a small factory called the Ivy House. Once on his own, Wedgwood refined his glazes, improved his raw clay with various additives, and kept careful notes. His best-selling early pottery had a marbled effect that he obtained by using different colored clays.

Wedgwood's most important technical innovations involved his division of labor and use of powered machinery. In all previous pottery production, one person worked on a single piece from beginning to end. Wedgwood divided tasks into categories such as mixing clay, operating the potter's wheel, and applying glaze. This eliminated the apprenticeship program but allowed workers to become highly skilled in one area of production. The result was a less-expensive, higher-quality product. Workers resisted at first but saw potential for greater pay. Wedgwood was also the first to introduce steam-powered equipment to his industry. He bought an engine from James Watt (1735-1819) to crush raw materials and to turn potter's wheels, lathes, and other machines.

In the early 1760s, Wedgwood perfected a method for uniformly coloring kitchenware. Most white pottery intended for sale as a set usually had varying shades of color. Sometimes the difference in a single set ran from yellowish cream to pearl white. Wedgwood's consistently colored, high-quality plates came to the attention of Queen Charlotte of England, wife of King George III (1738-1820). She paid Wedgwood the high honor of commissioning him to make a tea service for her.

It was a soft cream color and later versions had painted or enameled designs. Aware of the publicity value of pottery made for royalty, Wedgwood requested permission to call the design Queen's Ware. He received it in 1765 and orders soon came from all

From *The Life of Josiah Wedgwood* by Eliza Meteyard, Hurst and Blackett Publishers, 1865

A view of the former Wedgwood factory alongside the Trent and Mersey Canal at Etruria. The factory was opened by Wedgwood in June 1769.

over. One of the largest was for 952 pieces from Empress Catharine the Great of Russia (1729-1796).

England's transportation system could not support expanding industrial production. The pottery district was particularly isolated. Clay from England's south coast had to follow a circuitous route. Finished pottery went to Hull on the east coast or Liverpool on the west by horse-drawn wagon. Rough roads caused many pieces to break in transit. Wedgwood helped spearhead the construction of a canal through his district. The British Parliament appointed him treasurer of the project. The 94-mile Trent and Mersey Canal was completed in 1777 and provided a smooth and efficient passage to either coast.

Wedgwood married a third cousin, Sarah Wedgwood, in 1764. The couple had eight children and were devoted parents. Their first child, Susannah, became the mother of Charles Darwin (1809-1882), the internationally famous naturalist. Wedgwood purchased a 350-acre estate just south of Stoke-on-Trent. The family grew up there with all the trappings that wealth could pro-

Jasperware has raised white figures along the outside of the container. Its basic color is most often blue, but it also appears in green, yellow, lilac, and black.

vide. Wedgwood also built a new factory and village on the grounds and called the area Etruria, after an ancient country in the region of central Italy.

Over the years, Wedgwood's right knee continued to cause problems. One incident resulted in a good consequence. Wedgwood was in Liverpool and was forced to remain in bed for several weeks. During that time, he met Thomas Bentley (1730-1780) who became his business partner and best friend. He had a less pleasant event in 1768. Doctors had to amputate his leg above the knee. After recuperation, Wedgwood continued an active life, using an artificial leg.

Chemical investigations gained Wedgwood membership in the Royal Society. The first presentation he made to the group in 1782 concerned his invention of a new *pyrometer*, a high-temperature thermometer. It was based on clay's property of contracting to different sizes when heated to different temperatures. Carefully made samples went into the kiln until they fully hardened. The samples were then placed in a slightly tapered groove that had temperature markings. Where they wedged in place showed the temperature reading. Nearly 20 years passed before anyone invented a more accurate method.

Of all Wedgwood's accomplishments, his development of *jasperware* is probably the best remembered. He mixed spar (a crystalline nonmetallic mineral), flint, barium carbonate, and clay. It produced a lustrous blue product that he called *jasper*, after a variety of colored quartz. Wedgwood made more than 5,000 experiments before he could consistently manufacture jasperware. Finished products had detailed, raised fig-

ures around the outside. Wedgwood had artisans make molds that resembled shallow cookie cutters. Workers forced clay into the molds, then removed the clay, and carefully positioned it on the unfired jasperware. When removed from the kiln, the teapot, bowl, or other object was a beautiful blue with raised white figures around the outside. Wedgwood took his design from ancient Greek vase paintings, so many of the raised-clay figures depicted that era.

Wedgwood became active in national politics and represented potters in a trade treaty with the newly independent United States of America. He also worked on treaties with Ireland and France. When he died at 64, he was one of England's wealthiest men.

Wedgwood's technical innovations had significant effects on table manners and how people ate. Early dinners were often served in a single container and people removed what they wanted with knives or ladles. When Wedgwood made inexpensive plates, many could afford individual dinner settings. As table manners improved and tea and coffee gained in popularity, there was a demand for creamers, sugar bowls, cups, saucers, and other items. Wedgwood was not content to make ordinary pottery. He blended art with industry by concentrating on elegance of form while paying attention to detail. Members of his family are still involved in the business.

References

Josiah Wedgwood by Richard Tames, Shire Publications, 1984.

The Story of Wedgwood by Alison Kelly, Faber and Faber Publishers, 1975.

Stories of Great Craftsmen by S. H. Glenister, Books for Libraries Press.

Extraordinary Origins of Everyday Things by Charles Panati, Harper and Row Publishers, 1987.

Wedgwood's pyrometer for determining the temperature inside a pottery kiln was a ceramic measuring device. Kiln-fired samples were inserted into the grooved gauge, which had temperature markings. This demonstration set was presented to King George III by Wedgwood in 1786.

Benjamin Banneker

Rendering by Tim Harmon

Born:
November 9, 1731, in Ellicott City, Maryland

Died:
October 9, 1806, in Ellicott City, Maryland

Eighteenth-century America was clearly a farming nation, with more than 70 percent of workers employed in agricultural occupations. Although the Industrial Revolution was beginning to make inroads, most people made their living as members of independent farming families. So it is not surprising that many early scientists and technologists had their roots in farming. Thomas Jefferson (1743-1826) thought the country should be centered around the lives of independent farmers. He was involved with agriculture his entire life. Even while active in politics, he found time to invent an improved plow and thresher.

It was during the Jefferson era that America's first black scientist gained public attention. A successful farmer, Benjamin Banneker took time to develop an interest in astronomy. That led in 1792 to his publishing a farmer's almanac with tables derived from astronomical observations. Banneker also had surveying skills that he used to help lay out the city of Washington, D.C. He corresponded with Jefferson and was the first African-American to have such a relationship with a high-ranking federal official.

Banneker's grandmother, Molly Welsh, was born in England and traveled to Maryland in 1683 as an indentured servant. She saved some money, and when her seven-year obligation was completed, Welsh bought a small farm and two slaves to help her. It was a time when owning people as slaves was a common and accepted practice. Welsh freed both of her slaves and married one, a man named Banna Ka. The couple altered their last name to Bannaky. Their eldest daughter married a free-born black man in 1730. The couple lived on the family farm near Baltimore and assumed the Bannaky name. They raised three daughters and a son, Benjamin. The son later modified his name to Benjamin Banneker.

During his lifetime, all Banneker's family members were "free blacks," a term used in the eighteenth and nineteenth centuries to describe legal status. There was no public school in the area so the grandmother, Molly Welsh Bannaky, taught all her grandchildren to read. A teacher from the Society of Friends (the Quakers) later spent a short time in the small community. He set up a school for the boys, where Banneker learned to write and do simple mathematics. He enjoyed math and continued educating himself for the rest of his life.

Banneker's father saved some money and purchased a 100-acre farm in 1737. The farm was known as Bannaky Springs because of fresh-water springs on the property. Banneker made ditches and small dams to control the water for irrigation. His carefully planned work helped the farm

The United States Post Office issued this commemorative stamp in 1980. It depicts Banneker as he might have appeared during the 1791 Washington, D.C., survey.

Banneker used a tripod-mounted compass almost identical to this one. These compasses were particularly popular in America because there were few established landmarks from which to take angle sightings. This one was made in 1760 by Thomas Streathfield in England.

flourish even in dry periods. The family grew some food crops, but the major cash crop was tobacco. Tobacco was so important in the early economy of the American colonies that it often was used in place of money. The family earned a modest living and was as well off as most others along the upper reaches of the Patapsco River.

Banneker's interest in science and technology was probably sparked when he was 21. At that time, a man named Joseph Levi, probably either a neighbor or traveling salesman, showed the young man a pocket watch. Such timepieces were uncommon in 1752. The mechanism fascinated Banneker and he made a large wooden model of it. The evidence is imprecise, but Banneker appears to have hand carved the gears of a clock that kept time for 20 years or more. He read widely and was more intelligent than many people realized. But other than for building the clock, Banneker generally led an unremarkable life for many years. He inherited the family farm when his father died in 1759 and shared it with his mother. All his sisters had married and left the area.

Banneker's life had three important turning points. The first was when his grandmother taught him to read. The second was the rudimentary education he received from the unknown Quaker teacher. The third occurred more than 30 years later when the Ellicott family chose to construct a nearby flour mill. The mechanical aspects of the 100-foot by 36-foot mill had a distant relationship to the wooden clock from Banneker's past. The mill fascinated Banneker and he often visited it during construction. He made friends with the workers and with George Ellicott, a surveyor by trade. At the time of their meeting, Banneker was 47 and Ellicott was only 19. But the men were drawn together by similar interests and they developed a lifelong friendship.

Ellicott shared his books with Banneker

and introduced him to the field of astronomy. He also lent Banneker scientific equipment such as a pedestal telescope and a set of drafting instruments. Banneker had stared at the nighttime heaven for years and was captivated by its order and repeatability. The books and telescope allowed him to put everything in perspective, and he became ever more deeply involved with astronomy. It soon dominated his life.

Ellicott saw Banneker advancing in his astronomical studies and encouraged him to publish an almanac. Almanacs were useful to people from all walks of life. They included a yearly calendar that identified holy days and festivals. They helped farmers by providing long-range weather forecasts, phases of the moon to plan for evening work, and recommended planting and harvesting dates. Sailors used the star charts to plot their position at sea. People who did not own clocks could even estimate the time with an almanac, which specified sunrise and sunset times.

Banneker was spending so much time with astronomy that he had neglected his farming responsibilities. To help him out, Ellicott purchased Banneker's farm, allowed Banneker to continue living there, and paid him an annual stipend. At night, Banneker typically wrapped himself in a blanket and studied the heavens until dawn. During the day, he slept or worked on mathematical problems based on his observations. For the next 20 years, Banneker devoted almost all his time to studying astronomy.

Then, national events began to occur that would briefly sweep Banneker away from his astronomical observations. The U.S. Congress of 1790 passed a bill to establish a national capital on the Potomac River. President George Washington (1732-1799) selected the Frenchman Pierre Charles L'Enfant (1754-1825) to head the project. He

also appointed Major Andrew Ellicott to the survey commission. A cousin of George Ellicott, Andrew Ellicott knew Banneker's ability with spatial measuring equipment and asked him to help with the field work. Although Banneker was 60 years old, he quickly agreed.

Banneker was one of six people who assisted in laying out the city of Washington,

Banneker would have used drawing instruments almost exactly like these. These were made in the early eighteenth century by instrument maker Richard Glynne in England.

D.C. He maintained surveying notes, made calculations, analyzed data, and established surveying base points. Banneker personally helped select sites for the Capitol, White House, and Treasury Building. After a dispute with federal officials, L'Enfant quit, taking all working drawings with him. But Banneker had worked closely with the data, and he remembered the plans well enough to satisfactorily redraw them. His work allowed the survey of Washington, D.C., to continue on schedule. He viewed the project as the greatest adventure of his life.

Encouraged by his membership in the federal surveying party, Banneker wrote a 12-page letter in 1791 to Jefferson, who then served as Secretary of State. Banneker expressed concern that black Americans were not being given a chance to prove their capabilities. Jefferson sent a reply 11 days later that included these words: "Nature has given our black brethren talents equal to those of other colors of men."

Banneker continued his work with astronomical observations and completed his first almanac in 1792. It was printed and sold by Joseph Crukshank of Philadelphia. Jefferson was so impressed with the alma-

nac that he sent a copy to the Academy of Sciences in Paris. Banneker produced a new almanac each year until 1797. Almanacs were as popular in Banneker's day as novels are today. They were typically the only nonreligious publications available outside of large cities.

Banneker never married. Other people described him as a very likable person. One wrote that he was "very noble in appearance . . . a perfect gentleman . . . kind, generous, respectable, humane, dignified, and pleasing to talk with." He often played a violin and smoked a pipe for relaxation. He died at the age of 74, following his usual Sunday walk.

Early American technology remains hard to separate from science. The two disciplines were once closely related. Banneker's name often appears in history books, which describe him as a self-taught scientist or mathematician. He has been affectionately called the Sable (black) Astronomer. His additional accomplishments in surveying, drafting, and printing also contribute to Banneker's ranking as the country's first great black technologist.

References

The Life of Benjamin Banneker by Silvio A. Bedini, Charles Scribner's Sons Publishers, 1972.

Benjamin Banneker, Scientist and Mathematician by Kevin Conley, Chelsea House Publishers, 1989.

The Real McCoy: African-American Invention and Innovation 1619-1930, by Portia P. James, Smithsonian Institution Press, 1989.

Some of Banneker's almanacs have been reprinted. This unclear copy was reprinted from his 1793 *Almanack and Ephemeris*. An *ephemeris* is a table of star and planet positions.

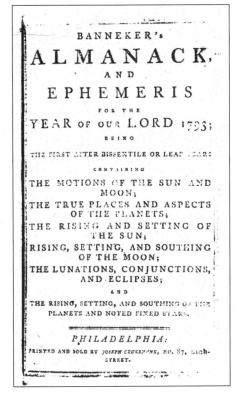

Richard Arkwright

Born:
December 23,
1732, in
Preston, England

Died:
August 3, 1792,
in Cromford,
England

It would be hard to name an item that was not manufactured or processed in a factory. Manufacturing is among America's most important economic activities. According to the Bureau of Labor Statistics, it produces more income than all retail and wholesale stores combined. Manufacturing also earns more than all financial institutions combined. The United States is the world's largest producer of manufactured products. All of its 375,000 factories can trace their roots to a single factory built in 1768.

The factory system of manufacture mechanized production under one roof. It eventually replaced the craft trades associated with cottage industries. The word *factory* comes from the word *factor*. It means "all the elements of a situation." Through hard work, organizational ability, and unbounded confidence, Richard Arkwright built the world's first factory in central England. He brought all aspects of cotton spinning, or thread making, into one building.

The youngest of 13 children, Arkwright was born in a small town on England's west coast. His formal education was slight, and he later credited an uncle with teaching him to read and write. Arkwright became a barber's apprentice in his early teens and later worked for a wig maker. He married Margaret Biggins in 1761 and used her moderate inheritance to open a business dying hair for wigs. Arkwright often obtained his raw materials from young women attending local fairs. Because he traveled so much, the work put him in contact with many cotton spinners and weavers who worked out of their homes.

As the British population increased during the mid-1700s, so did the demand for cotton thread for weaving. Several people were at work on textile inventions and

The Science Museum/Science & Society Picture Library

Arkwright heard discussions about mechanizing the process. He rented a small shop in which to build a machine that would automatically twist cotton fibers into yarn and thread. It is uncertain where he found financing for this venture. Clockmaker John Kay (1704-1764) was his paid assistant. Kay had earlier invented an improved shuttle for weaving and provided his technical insights. By the late 1760s, Arkwright devoted all his time to the difficulties of mechanizing cotton spinning.

Spinning a raw fiber into thread involved two separate operations: drawing and twisting. Arkwright's invention used two pairs of grooved horizontal wooden rollers to draw fibers from raw cotton. One roller in each pair was covered with leather, which allowed it to grip the cotton. The second pair rotated slightly faster than the first pair. This caused the fiber to stretch and improved the thread's uniformity. Vertical rotating spindles twisted the cotton to give it strength, while winding the thread on a spool. Arkwright's 1769 patent model had four spindles, although later factories would have thousands. His technique with rollers and spindles was not new—others had similar patents issued as early as 1738. However, Arkwright was the first to combine several steps in a system using a central source of power.

Arkwright realized that the efficiency of his system would result in lost work for some people. He built his first factory 20 miles north of Nottingham. This central-England city had some involvement with textile work, and it was more stable socially than others. Arkwright located his factory in an isolated rural area called Cromford on the River Derwent. Fed by hot springs, the river did not freeze in the winter. Arkwright started with 48 spindles. He received financial help from industrialist Jedediah Strutt (1726-1797). (One of Strutt's more famous apprentices was Samuel Slater (1768-1835) who established America's first factory in 1790.)

To provide for his workers, Arkwright built a company town at Cromford. He had roads, bridges, and a canal constructed to link with large cities like Nottingham and Manchester. He also built housing, stores, a church, school, hotel, and other buildings. His large manor house, under construction several years later, was not completed at the time of his death.

Arkwright's Cromford works is generally regarded as the world's first true factory. Powered by a water wheel that developed 5 to 10 horsepower, the factory manufactured an excellent product. Arkwright's was the first cotton uniformly strong enough to be used as the lengthwise threads in woven fabric. Cotton fabrics were popular because of their absorbency, washability, and smoothness against the skin. Arkwright's cotton fabric was used for long cotton stockings, which were fashionable in his day.

Arkwright had not initially planned to weave cloth, but weavers refused to purchase his thread because it had been made with a mechanical device. So Arkwright decided to expand his factory and ultimately became wealthy because of it. Many international visitors toured the factory and each usually purchased one or two dozen pairs of Arkwright's excellent stockings. Earlier handmade thread had resulted in cloth of variable quality. Arkwright's cotton was so much in demand that his factories ran 24 hours a day.

Arkwright's almost fanatical desire to accumulate wealth was well known. He typically worked from 5 A.M. until 9 P.M. He had a four-horse coach to carry him swiftly from one factory to another. A large and aggressive man, Arkwright became the richest cotton spinner in the country. He once made irresponsible plans to purchase the world's entire output of field cotton, process it, and use the proceeds to pay off the British national debt. Strutt was a partner with Arkwright in some factories. He did what he could to keep Arkwright's abrasive personality from causing problems with suppliers, customers, and government officials.

Arkwright was so powerful that he alone set prices for an entire industry. His dictatorial manner and negative personality played a role in causing some of the

This large factory on the River Derwent was constructed by Arkwright near Cromford, England. He called it the Masson Mill.

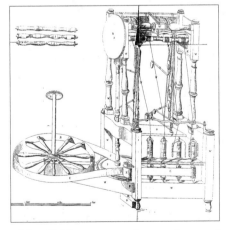

This 1769 patent shows four of Arkwright's spindles. Wooden replicas of this historically significant patent are on display in many technical museums throughout the world.

textile worker riots that occurred in the late 1700s. People destroyed factories that were taking away their livelihood. One was Arkwright's at Birkacre, the country's largest. By 1779, the conflicts had generally sub-

Arkwright had apartment-type housing constructed for his factory workers. The late-1700s building on North Street is still occupied by families who own parts of it as condominiums.

sided and Arkwright's factories had several thousand spindles tended by about 300 workers.

Since waterwheels powered most early factories, thread-making machines were called *water frames*. Arkwright's later factories used steam engines and he was involved with a total of about 20 that made thread, yarn, or cloth. Over 30,000 British workers in 1785 were using his patented machinery, and 5,000 of those worked for Arkwright himself. Sparked by Arkwright's work, textiles helped make Great Britain the world's dominant nineteenth-century manufacturer. Even 50 years after Arkwright's death, textiles accounted for 40 percent of Britain's exports.

Arkwright's factories were expensive to build, and it is not completely clear how he obtained his early financing. Constructing the town buildings in Cromford, for example, required large sums of money. Arkwright did not keep a diary nor write a biography. He was twice married, and had a son and daughter. A few years before he died in 1792, he destroyed all documents relating to his life. Nobody knows why. His son Richard tried to write his father's biography but was unable to locate any proper information.

A generally disagreeable person, Arkwright received criticism for taking the ideas of others and exploiting them. His patents were contested by nine textile com-

petitors. After a four-year battle, court judgments went against Arkwright and his patents were legally rescinded in 1785. But by that time, Arkwright was well established and the decision did him little financial damage.

For all his personal failings, Arkwright was an innovator. He made new technical ideas work in commercial operations. It took a strong-willed person to construct successful factories in the midst of an uneducated population that felt suspicious of anything new. Steam-engine pioneer James Watt (1736-1819) wrote, ". . . whoever invented the spinning machine, Arkwright certainly had the merit of performing the most difficult part, which was making it useful."

Arkwright had this thread-spinning model made in 1775 to show how his manufacturing process worked. It's a modified version of his 1769 patent and was used as a basis for his factory machines.

References

Stories of Great Craftsmen by S. H. Glenister, Books for Libraries Press, reprinted 1970.

The History of Invention by Trevor I. Williams, Facts on File Publications, 1987.

Works of Man by Ronald W. Clark, Viking Penguin, Inc, 1985.

Making of the Modern World edited by Neil Cossons, John Murray Publishers, 1992.

James Watt

As the world's population increased during the eighteenth century, people making hand-crafted items could not keep up with the demand. Clothing, farm implements, and even food production fell short of what was required. Quietly and almost without warning, the factory system of manufacture began to take hold. That period in history is known as the Industrial Revolution and it started in about 1760. It was not a "revolution" in the traditional sense of the word, but meant a rapid change. Countries went from farm-based economies to ones based on manufacturing. The Industrial Revolution started in Great Britain because it was the world's most important trading country at the time. Britain had many trained technologists and a banking system with enough money to support their efforts in building factories.

If any one invention can be identified with the Industrial Revolution, it would be James Watt's steam engine. Watt's improvements to existing steam engines made them practical as replacements for troublesome and underpowered waterwheels. People no longer had to locate factories near rivers, which often froze in the winter or had reduced flows in the autumn. Factories could be built near sources of raw material or transportation terminals like roads, canals, or seaports. With powerful Watt steam engines, factories made more items in less time. They could operate at the same rate, any day of the year. These advantages reduced the cost of manufactured goods and made more products available to more people.

Watt was born in a seaport community near Glasgow, Scotland. He grew up in a technically minded family. His father was a master ship builder, his grandfather taught mathematics, and his uncle was a surveyor. Watt was a slender unhealthy child who suffered

Born:
January 19, 1736, in Greenock, Scotland

Died:
August 19, 1819, in Heathfield, England

from headaches and could not attend school regularly. His parents assumed responsibility for his early education. At 13, he discovered geometry, which fascinated him. His focus on that subject was fortunate, because much of the design of his future engines would be based on geometry. Watt also learned to use tools in his father's workshop where he had a workbench and a small forge.

At 18, Watt wanted to be independent. He arranged for an apprenticeship in London with optical equipment manufacturer James Short. The trip by horseback took a rugged 12 days. During his time in London, Watt's health worsened from the smoke produced by coal-burning stoves. He endured London for a year, then returned to Glasgow to find work. Because his apprenticeship lasted only one year, instead of the required two, the only job he could find was a low-level position at the University of Glasgow. But luck was on his side. Alumnus Alexander MacFarlane died in Jamaica and bequeathed his large collection of astronomical instruments to the

"Old Bess, " one of Watt's earliest engines, built in 1777. The engine was installed experimentally at Watt's Birmingham factory to pump water. The 33-inch piston is at the left and the lift pump at the right. The engine is displayed in an incomplete form.

university. Many were damaged in transit and Watt was assigned to repair and calibrate them. His work was excellent and he soon developed an admirable professional reputation. Around this time, an event occurred that placed him on the path to world recognition.

The first piston-type steam engine was installed by Thomas Newcomen (1663-1729) near Birmingham in 1712. (See pages 13-15.) It powered a lift pump to remove the ground water that seeped into a coal mine. It was a huge room-sized beam engine. Its single large piston moved vertically at six strokes per minute. A shaft from the piston connected to one end of a strong horizontal wooden beam. The other end of the beam connected to a lift pump that drew water from the mine. Watt saw his first Newcomen engine in 1764. Not full size, it was a small model used for class demonstrations. The engine did not work properly and the teachers asked for Watt's help.

That was Watt's introduction to steam engine technology and legend has it that he blurted out, "But they're doing it all wrong." He was referring to the way the cylinder was alternately heated and cooled. The engine used low-pressure steam to push a piston up in a cylinder. Cold water sprayed into the cylinder then condensed the steam and atmospheric pressure forced the piston back down. This heating and cooling of the cylinder was the source of the inefficiency. Each fresh charge of steam wasted most of its energy as it reheated the cylinder.

Watt worked on developing a separate condenser, a device that

This model of a Newcomen engine is the one Watt repaired in 1763 while employed by the University of Glasgow. It is displayed at the university's Hunterian Museum.

would condense steam outside the engine and keep the cylinder hot. He wrote in 1814, "Steam would rush into a vacuum. If communication were made between the cylinder and an exhaustive vessel, it would rush into it and might be there condensed without cooling the cylinder." Watt used a large anatomical brass syringe as a test condenser with a small model engine. Pulling down on the piston in the syringe provided the necessary vacuum to empty steam from the engine's cylinder. This was the greatest single improvement ever made to the steam engine and it reduced coal consumption by 70 percent. Watt took out a British patent in 1769.

A Watt-type pumping engine, built in 1813. It raised 500 gallons of water per minute from a depth of 120 feet. The piston diameter is 36 inches with a stroke of 8 feet.

Watt later invented the double-acting piston, which applied steam pressure first to one side of the piston and then to the other. Patented in 1782, it effectively doubled an engine's power. Watt also patented the flyball governor in 1787 to control engine speed.

Watt's work was sponsored in part by John Roebuck (1718-1794), a coal mine investor. Then, Roebuck had some financial reversals and foundry owner, Matthew Boulton (1728-1809) purchased an interest in Watt's engine.

The two men struck up an immediate and lasting friendship. With Watt providing the technical ability and Boulton the financial expertise, the men established the successful Boulton and Watt engine factory in Birmingham in 1775.

Boulton and Watt did not sell completed engines. Instead, they provided important parts for their patented single-cylinder engines, along with drawings and licenses. Most of the parts were enormous in size. Boulton and Watt's first commercial steam engine had a cylinder diameter of 50 inches and a stroke of about 15 feet. It operated at 3 to 5 psi, pumping water from a coal mine in Tipton. The oldest operating Watt engine still in existence is the 1779 Smethwick engine at the Museum of Science and Industry in Birmingham, England. It has a 32-inch-diameter cylinder, a stroke of 6 feet, and produces 10.5 hp.

Boulton and Watt offered two types of engines, pumping and rotative. Pumping engines removed water from deep mines, but rotative engines were meant for factory use. Their power came from a large rotating flywheel. A belt around the flywheel connected to a shaft running along the ceiling of a factory building. Each machine tool had its own belt that connected to the shaft. Boulton and Watt engines typically developed 5 to 10 horsepower, a unit established by Watt after experimenting with many horses. He determined that the average horse could lift 550 pounds a distance of one foot in one second. His engines were rated accordingly. It is still a universal standard, 550 ft-#/sec = 1 hp.

The easiest way to convert up-and-down motion to rotary motion was with a crankshaft. Although Watt had been working on just such a design, the crankshaft and connecting rod arrangement had been patented by Matthew Washbrough in 1779. Washbrough took out the patent based on information leaked by one of Watt's employees. Watt found a way to get around the problem by inventing the sun-and-planet gear train, now used in many modern automatic trans-

missions. Most historians think the crankshaft had been in the public domain long enough that it was nonpatentable. Had Watt legally contested the Wash-brough patent, he most likely would have won.

Watt's partnership with Boulton ended in 1800 when his patent on the separate condenser expired. Over a span of 25 years, Watt and Boulton's company had licensed 183 pumping and 268 rotative engines. That represented about 40 percent of all the engines in the world. Watt was prosperous, respected, and successful when he retired that year.

More intelligent than many people realized, Watt read German, Italian, and French. He kept current on technical progress in all major countries. His closest friends included Antoine Lavoisier (1743-1794), who defined chemical elements; Joseph Priestly (1733-1804), who discovered oxygen; Adam Smith (1723-1790), a founder of modern economics; and Josiah Wedgwood (1730-1795) who invented a new type of pottery. But Watt never lost touch with his roots. After he retired, he set up an attic room with workbenches and tools so that he could continue to keep active with technology throughout the rest of his life.

References

Making of the Modern World edited by Neil Cossons, John Murray Publishers, 1992.

Beam Engines by T. E. Crowley, Shire Publications, 1986.

James Watt 1736-1819, Inverclyde District Council, 1991.

Watt's original experimental condenser. The small piston provided a partial vacuum that removed spent steam from a cylinder. Watt drilled the long rod to provide a water drain.

Sun-and-planet gear train on a rotative Watt engine. Patented by Watt in 1781 to avoid conflicting with another person's crankshaft patent. It allowed the flywheel to turn twice as fast.

Alessandro Giuseppe Antonio Anastasio Volta

Born:
February 18, 1745,
in Como, Italy

Died:
March 5, 1827, in
Como, Italy

Electricity isn't as new a technology as some might think. The Greek scientist Thales (636 (?) - 546 (?) B.C.) made a basic electrical observation more than 2,500 years ago. He rubbed a piece of amber, fossilized tree resin, with a cloth and saw that it attracted small bits of straw. Thales was the first person to purposely generate static electricity. The word *electricity* is from the Latin *electrum*, which means "amber." In 1646, British physician Thomas Browne (1605-1682) was the first person to use the word.

Scientists in the 1700s built electrostatic generators to produce high voltages. The generators resembled a large plastic dinner plate, set vertically, and rotated with a crank. Metal braid rubbed against the plate and electricity collected on the surface of metal globes. It did not flow along a wire. That is why the energy was called *static electricity*. *Static* means "motionless." Bringing the globes close together produced a momentary spark.

Electrostatic generators were sometimes used for entertainment. Incompetent individuals occasionally generated voltages that seriously shocked some people. Many considered electricity only a laboratory curiosity until Alessandro Volta made the first battery in 1800. Volta's was the first device to produce electrical energy that could continuously flow down a wire. It was a stack of metal and cardboard disks called a *voltaic pile*. Benjamin Franklin (1706-1790) coined the word *battery* several years earlier. *Battery* refers to a group of items arranged together. Franklin connected several electrical storage devices called *Leyden jars*.

Volta was born near the Swiss border in northern Italy. His hometown was on a beautiful lake that attracted many European

Pioneers of Electrical Communication by Rollo Appleyard, 1930

vacationers. His family was of local nobility. Though once financially comfortable, Volta's father had overspent the family fortune. They had fallen on hard times by the time Volta was born. He was the youngest of seven children. Other relatives from the extended family helped raise him.

Volta showed little early evidence of becoming a world-recognized scientist. He did not talk until he was four and his parents feared that he might have a mental deficiency. But he proved to be one of the brighter students in school. His father died when Volta was seven and relatives paid for his education through the age of 16. He showed special ability with drama, foreign languages, and chemistry. Aside from Italian, Volta spoke English, French, and Latin, and he could read Dutch and Spanish. Volta was particularly impressed with the work of Joseph Priestly (1733-1804). Priestly was a British-American chemist who discovered oxygen. A two-volume book Priestly wrote,

History of Electricity, inspired the young Volta to follow a technical career.

Volta's uncles wanted him to become a lawyer, a profession well represented on his mother's side of the family. But he chose to study electricity at the Royal Seminary in Como. After completing his studies sometime around 1763, Volta remained in Como for several years and began to establish a technical reputation. His early experiments included rubbing different metals with cloth and evaluating their electrical charges. He also described an improved electrostatic generator. Volta published his results over the years and they helped him gain a position in 1775 as a physics teacher at the high school in Como.

For some time, Volta had been working on a clever electrical device that he finally completed in 1775. He called it an *electrophorus.* The device used two metal plates, each about six inches in diameter. One was coated with ebonite and remained on a table. Ebonite is a dark hard rubber similar to plastic. The other plate was uncoated and had an insulating handle extending from its middle. Rubbing the ebonite with a cloth produced a negative charge. It attracted the positive charges to one side of the metal plate when the two plates were brought close together. A small wire drained negative charges from the other side and left the metal plate with extra positive charges. Continuing the process over and over built up a substantial positive charge on the metal plate. The amount of charge could get quite high and the invention brought Volta publicity and fame. The electrophorus has no practical use in the modern world other than to demonstrate static electricity in educational settings.

The University at Pavia, near Milan, offered Volta a professorship in 1779. He accepted and stayed for the next 40 years. To keep up with the latest developments, Volta traveled extensively. One lengthy trip in 1781 took him to Switzerland, France, Germany, Belgium, Holland, and England. In recognition of his contributions to the fields of electricity and chemistry, the British Royal Society awarded Volta its coveted Copley Medal in 1794. The medal was a precursor to the Nobel Prize.

Volta worked primarily with static electricity until he heard of observations made by Luigi Galvani (1737-1798) in Bologna. In 1791, Galvani announced the existence of animal electricity. When two probes made from different metals touched a frog leg, the muscle twitched. Galvani speculated that electricity was stored in the leg. His theory had a large following, but not everyone agreed with it. Volta strongly disagreed and began his own investigations.

Volta was convinced that the presence of two different metals was all that mattered. His experiments resulted in his battery in 1800. It consisted of a series of cups partly filled with vinegar. Volta connected them with U-shaped strips of metal. One end of each metal strip was copper and the other end was zinc. Volta put his fingers in the two end cups and felt a slight tingle. The tingle decreased as the number of cups was reduced. It stopped when a metal strip was removed. Volta knew he had produced electricity.

To make the battery more compact, Volta used round disks of copper, zinc, and cardboard soaked in vinegar and stacked one on top of the other. The battery generated about one volt for each set of disks. Volta referred to "30, 40, 60, or more pieces of copper, or rather silver, applied each to a piece of tin, or zinc, which is much better. . . ." The *voltaic pile* was the first source of continuous current electricity and its effect was immediate.

This unlabeled electrostatic generator is a nineteenth-century Wimshurst machine, named for James Wimshurst (1832-1903). Electrical charge from the spinning disk is stored in the two slender Leyden jars on either side. When enough charge builds up, a spark jumps between the two metal globes. The photograph in the background shows a modern demonstration unit safely dissipating electricity into the air though the girl's hair.

Within a few weeks, William Nicholson (1753-1815) and Anthony Carlisle (1768-1840) in England used it to separate chemical compounds. Andre Ampere (1775-1836) in France and Georg Ohm (1789-1854) in Germany added to the development of current electricity.

Northern Italy took part in military actions that involved the French emperor Napoleon I (1769-1821). Napoleon established the Cisalpine Republic in 1797, which included such cities as Bologna, Milan, and Pavia. Napoleon was essentially the ruler of northern Italy. He was fascinated by Volta's battery and encouraged the National Institute of France to ask him to lecture and give demonstrations in Paris. It took Volta 26 days to make the journey in September 1801. He stayed for four months. It was the beginning of a wave of rewards for the modest and talented university professor.

Volta received France's Legion of Honor and was made a count of the French Empire. Napoleon provided Volta with a lifetime pension of 4,000 lire per year and, in 1809, his annual salary increased to 24,000 lire. For comparison, at the time, a small house cost about 14,000 lire. Volta had the income of a wealthy man during the last 20 years of his life.

Volta was a pleasant person who made friends easily. Perhaps because of his noble heritage, Volta had the bearing of a gentleman. He did not marry until he was 49 and had a most happy life with the former Teresa Peregrini. She was an expert in agriculture and botany and spoke French and German. The couple had three children and were devoted parents. Volta's eldest son became mayor of Como and wrote many studies of his father's life. Volta retired to his family's country estate in 1819 and died at 82.

Volta was in his mid-50s when he invented the battery. After that, he came to a complete stop. He took no part in applying his discovery to the new fields it opened up. During the remaining 27 years of his life, he showed none of his earlier creativity. Before 1800, Volta had published 80 articles or books, many on significant topics. But after 1800, he produced only six unimportant works. Nobody knows why. The 1881 International Electrical Congress established the *volt* as the standard unit of electrical pressure.

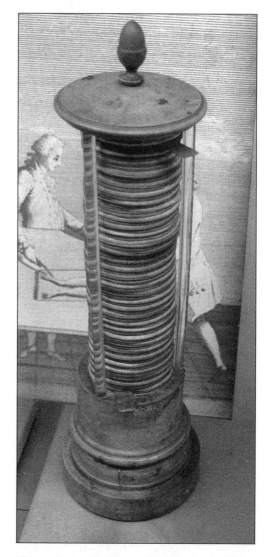

This voltaic pile has a label that describes it as a column. Volta obtained positive (+) electricity with a wire connected to the copper disk at the top and negative (-) electricity from the zinc disk on the bottom.

References

Allesandro Volta and the Electric Battery by Bern Dibner, Franklin Watts Publishers, 1964.

Electrical Engineers and Workers by P. W. Kingsford, Edward Arnold Publishers, 1969.

Electricity by Steve Parker, Dorling Kindersley Publishing for the Science Museum, 1992.

John Stevens

An eighteenth-century manufacturer who shipped goods in the eastern part of America could choose between using canals or horse-drawn freight wagons. Canals were preferred for heavy products, such as lumber, coal, or large machine tools, or for breakable ones, such as glass and pottery. The most successful canal was the 363-mile Erie Canal. An eight-year project completed in 1825, it went from Albany to Buffalo in New York State. The Erie Canal linked the Atlantic Ocean with the Great Lakes. But canal travel was slow and expensive. The Erie had 82 locks and horses pulled the boats. The entire trip took 10 days. Shipping rates were also high because canals were expensive to construct. Nonetheless, America had 4,500 miles of them by 1840.

Road transport by horse-drawn freight wagons also presented problems. Wagons could only carry relatively light loads and horses required changing every 15 to 20 miles. In evaluating transportation alternatives, some technologists thought about railroads. They held the promise of being cheaper to construct than canals while being faster than horse-drawn freight wagons. John Stevens was a pioneer in steam-powered boats and locomotives. He operated river ferries and in 1825, built the first locomotive made in America. It was a remarkable project for the 76-year-old man to undertake.

Stevens was born into a wealthy New York City family. His father was a ship owner, merchant, land owner, and politician. The younger Stevens was educated by tutors and graduated in 1768 from Columbia University with a law degree. He worked with his father on farming administration and political commissions. During the Revolutionary War, Stevens served as New Jersey's treasurer and collected money for the Continental

Born:
(n.d.) 1749, in New York, New York

Died:
March 6, 1838, in Hoboken, New Jersey

Army. He rose to the rank of colonel and for the rest of his life was known as Col. John Stevens.

In 1782, Stevens married the equally wealthy Rachel Cox. The couple lived in New York City at 7 Broadway Avenue, but Stevens wanted to establish a rural estate. In 1784, he purchased 789 acres of land across the Hudson River in New Jersey for £18,340. The Native Americans called the region Hopoghan Hackinge, which meant "land of the smoking pipe." The city of Hoboken now occupies the site. Travel between Stevens's new home in New Jersey and his offices in New York City required him to regularly cross the Hudson River by ferry. Other commuters did the same. The rates in 1799 covered 17 types of passenger and freight combinations. They included 6¢ per person, 20¢ for a person and horse, and 3¢ for a bag of flour. Some ferry boats had paddle wheels powered by horses that walked on an endless belt. Others used oars or sails. The service was often undependable and occasion-

This model shows how the locomotive's central gear engages the rack gear between the rails. The steam engine turned the central gear, which propelled the locomotive at 12 mph.

ally perilous in rough water. Steam engines were being made small enough for use for transportation purposes and Stevens wondered if one might be used to provide better ferry service. In 1788, he happened to see a steamboat built by John Fitch (1743-1798) operating on the Delaware River near Philadelphia. It was the first steamboat in America, possibly the first in the world.

Stevens began to correspond with Fitch's rival, James Rumsey (1743-1792), about improvements to his steam engine. From that time on, Stevens worked to advance the use of steam power for water and land transportation. He designed his own steam engine with an improved multi-tube vertical boiler. His boilers often had 12 to 20 tubes. This increased the heating area and produced more steam, which resulted in more power. Stevens wanted to legally protect his new invention. Through his contacts in government, he encouraged the passage of America's Patent Act in April 1790. He was among the first dozen applicants to receive a patent. Granted in August 1791, his was for a steam engine and boiler.

To develop steam-powered ferry service, Stevens went into partnership with Robert Livingston (1746-1813) and others in 1797. Livingston was a powerful New York politician who had acquired a 20-year monopoly on steam operations in New York State. He was also Stevens's brother-in-law. Livingston fancied himself an inventor and forced his ideas on others. He did not have Stevens's technical talent, but he did have the monopoly. When Livingston went to France as America's ambassa-

This single-cylinder steam engine powered the two propellers of Stevens's *Little Juliana*, which was not much larger than a row boat. The engine is the oldest surviving steam power plant built in America.

dor, Stevens took the opportunity to construct a truly revolutionary steamboat.

The 20-foot-long *Little Juliana* was named for one of Stevens's daughters. He and his wife had 11 children. Completed in 1804, the boat had a single-cylinder engine that turned two screw propellers. It was the world's first propeller-driven steamboat. Two of Stevens's sons operated it experimentally in New York Harbor. Its success spurred Stevens to build an even larger craft, a 100-foot-long side-wheeler named *Phoenix*. He planned to use it for freight and passenger service on the Hudson River. In the meantime, Livingston had teamed up with Robert Fulton (1765-1815) who then lived in England. Livingston used his political influence to remove Stevens's name from their original 1797 partnership agreement. While Stevens was quite angry, the law was clear. He could no longer operate his innovative steamboats in New York.

At first, Stevens decided to fight the new Livingston-Fulton monopoly, but he changed his mind. He sent the *Phoenix* to the Atlantic Ocean and to Chesapeake Bay for service on the Delaware River. That 1809 journey was the first time a steamboat went out to sea. It was a potentially dangerous voyage because most people thought the waves would lift the paddles out of the water, which could damage them and the engine. The boat's captain was the highly experienced Moses Rogers (1779-1821). Rogers would also go down in history as the captain of the first steamship to cross the Atlantic Ocean. He commanded the *Savannah* on its 1819 trip from Savannah to Liverpool. Rogers stopped often in protected coves, and the *Phoenix's*

13-day trip was uneventful. The *Phoenix* went on to great success as a riverboat operating between Trenton and Philadelphia.

Stevens decided to alter his focus and concentrate on steam-powered locomotives. The unnamed 4,575-pound locomotive he built in 1825 resembled a heavy-duty, four-wheeled cart, about 14 feet long by 4 feet wide. Its distinguishing characteristics included four 4-foot-9-inch-diameter wheels, a tall vertical boiler, a large wooden water barrel, and a gear sticking from its bottom. Its single-cylinder engine had a bore of 5 inches and a stroke of 12 inches. The multi-tube boiler developed a pressure of about 100 psi, which allowed the engine to move the locomotive at 12 mph. Wood fueled the locomotive.

Not knowing if the wheels would develop enough traction, Stevens used three tracks. The locomotive's iron-clad but unflanged wooden wheels rolled freely on flat rails. Four metal posts extended down from each corner of the locomotive to keep it on the tracks. Between the two main tracks lay a third track, actually a rack gear. The steam engine powered a gear in the middle of the small locomotive. It meshed with the rack gear and propelled the locomotive around a circular track on the Stevens estate. The locomotive did not pull any cars but could carry up to a dozen people. A concept vehicle, Stevens's locomotive was never used on a railroad line.

This replica locomotive was built in 1928 for a celebration at the Stevens Institute of Technology and then donated to Chicago's Museum of Science and Industry. The wooden cask was for water and passengers sat on the benches at the rear.

Many people dismissed it as a fantasy or a toy. Canals were viewed as the most efficient means of overland travel. But the Stevens locomotive helped inspire the railroad boom that swept America in the 1830s and guided the way for future technologists. New Jersey granted Stevens the first government charter to build a railroad line. It became the Camden and Amboy Railroad of 1830. A man who had many technical accomplishments, Stevens died in Hoboken at the age of 89.

The influence of the Stevens family continued after Col. John Stevens died. His son, Robert Livingston Stevens (1787-1856), invented the T-shaped railroad rail that has become the worldwide standard. Another son, John Cox Stevens (1785-1857) organized construction of an 1851 yacht named *America*. It defeated all English contenders to become the first winner of the trophy now known as the America's Cup. A third son, Edwin Augustus Stevens (1795-1868), provided funding and 55 acres of land for the 1870 establishment of the Stevens Institute of Technology in Hoboken. Millicent Fenwick (1910-1992) was Col. Stevens's great-great-granddaughter. She was a New Jersey congresswoman between 1975 and 1983 and the model for the Lacey Davenport character in Garry Trudeau's "Doonesbury" cartoon.

Replica of Stevens's multi-tube vertical boiler of about 1788. Early American boilers for land transportation were often vertical so that locomotives could be built short enough to make tight turns.

References

John Stevens—An American Record by Archibald Douglas Turnbull, Century Company Publishers, 1928.

"The First Family of Inventors" by Oliver E. Allen, in *American Heritage of Invention and Technology*, Fall 1987.

S.S. Savannah—The Elegant Steamship by Frank O. Brayard, Dover Publications, 1963.

Stevens Institute of Technology home-page. Available: www.stevens-tech.edu/about_stevens/history/

Count Rumford (Benjamin Thompson)

Born:
March 26, 1753,
in Woburn,
Massachusetts

Died:
August 21, 1814,
in Auteuil, France

Few eighteenth-century technical topics caused more confusion than heat, or *thermodynamics*. No one knew what heat was, what caused it, or how it flowed. Many people incorrectly thought substances had certain amounts of *caloric*, from the Latin *calor*, meaning "heat". But no one knew how to describe caloric. Scientists were generally unconcerned. Not until the appearance of internal combustion engines and central heating did thermodynamics become particularly important.

An American technologist was among the first to conduct experiments that reformed everyone's view of heat and thermodynamics. His other technical accomplishments included establishing the military field of ballistics, improving lighting, and inventing new ways of cooking. He organized a technical society in London, England, and built a huge city park in Munich, Germany. He established awards that were precursors to the Nobel Prizes. His accomplishments were legion. President Franklin D. Roosevelt rated him with Benjamin Franklin and Thomas Jefferson as "the greatest mind America has produced." But Benjamin Thompson was also a foreign agent for at least two countries. Better known by his German title, Count Rumford, he led a dangerous life of politics, intrigue, and technical accomplishment.

Rumford was his parents' only child. He was born into a poor farming family near Boston and his father died when he was an infant. His mother remarried a man who was equally poor. Rumford did not have a secure childhood. A small inheritance from his grandfather funded his public school education for two years. He apprenticed at 13 to a merchant in Salem and spent much of his free time reading technical books. He sometimes attended lectures at nearby Harvard

National Portrait Gallery, Smithsonian Institution

University. Rumford had faultless speech and the bearing of someone from a far more cultured background. Recognizing social status as a way to improve his condition, he decided to teach himself French.

A teaching position became available in Concord, New Hampshire. Rumford applied, got the job, and moved to Concord in 1772. He had a pleasant and polished personality, and he soon met and married a wealthy widow 14 years his senior. Sarah Rolfe had been married to an army colonel and her political connections extended to the governor's office. She used her influence to secure a colonial army position for her new husband. Rumford had almost no military experience. This sowed seeds of discontent among passed-over junior officers and may have resulted in Rumford's 1774 trial on charges of being unpatriotic. He was acquitted but emerged with a badly damaged reputation.

Rumford moved to Boston, leaving behind his wife and young daughter. The Revolutionary War was being fought at the time and Boston was a pivotal port city. Because Rumford was an American officer holding the rank of major, he could obtain sensitive military information. He lost no time in sharing it with British General Thomas Gage. Throughout his entire life, Rumford never neglected an opportunity for his own advancement. Gage sent four people to England in 1776 to report on the status of the war. One of them was Rumford.

The Colonial Secretary in London, Lord George Germain, had responsibility for conducting the war in America. He knew of Rumford's assistance to the British cause. Germain helped Rumford obtain a knighthood and the rank of lieutenant colonel in the British army. During this period, Rumford worked on standardizing gunpowder. He constructed a special cannon that had a bore of about one inch. With the cannon pointed straight up, Rumford poured in a controlled amount of gunpowder. He used weights over the muzzle to measure the explosive force. In one nine-day period, he ran 123 experiments. He was the first to fire a cannonball at a pendulum to determine its momentum effect. Rumford's tests were insightful and helped him gain membership to the Royal Society in 1781. The modern science of ballistics began with Rumford.

But Rumford also worked on other projects. He spoke several languages and often traveled around Europe. Under the appearance of scientific inquiry, he made observations of British ship deployments and sold the information to France. The British government found out. Only the intervention of Lord Germain kept Rumford out of prison, or worse. In a politically contrived move, he accepted a half-pay retirement in 1783 and moved to Munich, Germany.

Rumford's European contacts helped him get an appointment to reorganize the Bavarian state militia. The militia's clothing and food were poor, advancements rare, morale low, and professionalism almost totally lacking. To keep the militia busy, Rumford had them convert a large tract of swampy wasteland in Munich into a city park. It is now called the English Garden and has a large statue of Rumford. It was the world's first city park, and is still its largest.

In studying fabric for uniforms, Rumford conducted experiments on thermal conduc-

tivity of different materials and weaves. He discovered convection currents. No weaver would produce what he requested, so Rumford set up large workhouses. During a single day in 1790, he had 2,600 beggars arrested and sent to the workhouses. They made clothing and supplies for the militia. In return, the beggars received food, shelter, and education. Many went on to become self-supporting members of the community. Rumford's action was viewed by many as a successful social experiment and other countries later copied it.

The problems of operating workhouses involved Rumford with nutrition, heating, and lighting. He developed metal cooking stoves that we call *ranges*. The name comes from the placement of the firebox at one side. Moving a pot away from the heat provided a "range" of cooking temperatures. Rumford developed pots and pans to use with his new stove, including the double boiler. He invented the drip coffeemaker now found in countless American homes. He also designed tall, shallow fireplaces to reduce the loss of heated room air.

Rumford was the first to investigate the caloric value of food. He found that an inexpensive diet of heavy soup made from potatoes, barley, and other items was quite nutritious. Known as *Rumfordsuppe*, it still appears on German restaurant menus. To keep his large workhouses warm, Rumford invented the steam-heated radiator. To evaluate lighting, he invented a simple photom-

Although once in a rural setting, Rumford's home is now on a busy street in a Boston suburb less than a mile from Interstate 95. It has been open to the public since 1877.

These models show how Rumford's fireplace worked. The earlier design (left) has a large opening for the chimney. Rumford's design (right) has a restricted opening and was the first to use a smoke shelf to keep out cold air. When cold air came down the chimney, it hit the smoke shelf and turned back upward, thus improving the fireplace's draft.

eter that measured light intensity. Rumford coined the term *foot-candle*. The German government conferred a title on him in 1791. He chose the name Rumford, after the original town name of Concord, New Hampshire, where he had met his estranged wife.

Rumford regularly returned to investigating heat while boring cannon barrels. Others thought that metal had a fixed amount of heat potential, called a *caloric quantity*. Rumford thought otherwise. To demonstrate his theory, he had a dull boring bar rub against the inside of a rotating brass cannon that was cooled by a small water bath. The water continually boiled, even after the theoretical amount of caloric should have been used up. Rumford showed that heat was not a substance to be squeezed out of matter. It was a motion of particles, not unlike electron flow and electricity.

Rumford moved back to London in 1795, hoping to become a German diplomat to England. British politicians were flabbergasted because he had almost been arrested as a spy several years earlier. Rumford settled for being a distinguished scientist. He worked at establishing the Royal Institution in 1799 and helped finance the purchase of a large building for £4,850. With tall, impressive columns, it is not far from London's Buckingham Palace. Michael Faraday (1791-1867) lived there and was its most accomplished director.

In 1796, Rumford gave $10,000 to establish awards for technical achievement in America and England. Ironically, those two countries viewed him as somewhat traitorous. About 63 people have received Rumford Medals in America.

Rumford moved to Paris in 1804. His wife had died in America in 1792 and he married Marie Lavosier, widow of the renowned

Most early cooking was done in a fireplace. Rumford's cylindrical oven roasted meat and vegetables in a closed container. The small pipe is a vent for cooking vapors. The brass plate on the front says "Number 56, Rumford's Roaster, Made and Sold by Joseph Howe, Number 7 Harris Lane, Boston."

chemist Antoine Lavosier (1743-1794). Ironically, Antoine Lavosier was the person who had given *caloric* its name. Rumford and his second wife had a short quarrelsome marriage. She enjoyed socializing, while he liked quieter activities, such as growing roses. They separated in 1809. Rumford died five years later.

Because Rumford enjoyed intrigue and deceit, he had many close encounters with disaster. He paid a dear price for his lifestyle. He was wealthy, successful, and widely known. But Rumford had no friends and many enemies. Despite his tremendous accomplishments, he had earned great dislike in every country where he lived. Credited as one of America's greatest minds, Rumford gave much to the world but has been very nearly forgotten.

References

"Count Rumford: The Most Successful Yankee Abroad, Ever" by John Dornberg, in Smithsonian, December 1994.

American Science and Invention by Mitchell Wilson, Bonanza Books, 1960.

The House of the Royal Institution by A. D. R. Caroe, The Royal Institution, 1963.

Engines of Our Ingenuity, Number 4, "Rumford" by John Lienhard, National Public Radio, c. 1992.

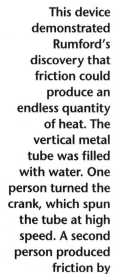

This device demonstrated Rumford's discovery that friction could produce an endless quantity of heat. The vertical metal tube was filled with water. One person turned the crank, which spun the tube at high speed. A second person produced friction by clamping the wooden paddles around the spinning tube. The water in the small tube quickly boiled.

Thomas Telford

The Science Museum/Science & Society Picture Library

Manmade elements that comprise the basic facilities for community growth are often called *infrastructure*. Such elements include school buildings, electrical power plants, railroad lines, harbors, and other installations. Few components of a country's infrastructure are more important than high-quality roads. America has about 2 million miles of paved roads, the most in the world. The Romans knew the importance of roads more than 20 centuries ago and were the first society to construct long, high-quality ones. They built 50,000 miles of roads and developed techniques still used today. The Appian Way connects Rome with Brindisi on Italy's southeast coast. Construction began in 312 B.C. and parts of the 350-mile-long road are still used by modern vehicles.

Although not the first to follow in Roman footsteps, no person in British history is more closely identified with road construction than Thomas Telford. He borrowed from the Romans' experience and added several new ideas. Telford knew that a good road had to consist of more than a dirt trail with stones on top. It had to have a solid foundation to support the high loading pressure of wagon wheels and horse hooves. A good road needed a solid surface to eliminate problems with mud. Telford designed and built many highways through the challenging terrain of his native Scotland. In a play on words with one of the seven wonders of the ancient world, some call him the "Colossus of Roads."

Telford was born in a small lowland village just a few miles north of the Scotland-England border. His father, a shepherd, died shortly after Telford's birth. The family's comfortable cottage had been provided as a condition of employment. aTelford and his mother had to leave and she went on to raise her only child in a single room provided by a benefactor. She earned some income by doing occasional farmwork. An uncle paid for elementary school and Telford earned extra money by herding animals. His youth was a poor but happy one.

A typical education for a young man during Telford's time involved working with a master to learn a craft. Telford apprenticed to a stone mason at 14 and learned his trade well. Each new bridge or farmhouse project added to his perspective on construction methods. He enjoyed reading in his spare time and a family friend shared her private library with him. The library had few technical resources but Telford enjoyed reading poetry. It became one of his lifelong interests and he wrote some poems. Mildly uncomfortable with his choice of a hobby, he often published his poetry anonymously.

Opportunities for full-time work and career advancement were limited in Scotland. Telford found a job in Edinburgh for a year

The young Telford first found employment in London by laying blocks for the Somerset House. Much larger than this picture suggests, the building is on a on a busy street called The Strand.

and then decided to travel south. He borrowed some clothes and rode a horse to London, about 400 miles away. At the age of 24, he found a job laying stone blocks for a large government building being constructed on the River Thames. Forty years later, as the first president of the Institution of Civil Engineers, he often proudly pointed out the very stones he laid. Called Somerset House, the huge building on a busy street still stands and is now used for document storage and other archival purposes.

The young Telford was a reliable worker who studied at night. Largely self-educated and confident of his abilities, he was made supervisor of public works in the west-central county of Shropshire in 1787. His first major project, in 1793, involved building a series of canals in that region. Many of the canals and aqueducts Telford built in England, Scotland, Wales, and Sweden are still used by recreational canal boats. His talents extended to constructing churches and prison buildings. Financiers soon found that he was excellent organizer and manager and that he could motivate people who worked for him. Telford had a reputation for fairness, honesty, and good judgment. Much of his technical success came from his knowledge of the problems of ordinary workers. He was so well liked that his subordinates often worked themselves to exhaustion.

Supervising diverse projects required so much traveling that Telford did not have a permanent residence until late in life. However, he always made time to visit his mother and his home in Scotland. His roots in his rural homeland ran deep and he provided a headstone for the grave of the father he never knew. Telford saw northern highlanders leaving to find employment in England, America, Australia, and other countries. Farmers could not easily get wool, cattle, and other products to market because the roads in that

mountainous region were mostly just dirt tracks. Telford often visited the highlands and wrote an 1802 report to the British Commission for Highland Roads. The commission accepted his analysis and he worked on the road-building project for the next 28 years.

Telford's roads were better than others because he followed methods pioneered by the Romans. He dug down a foot or more along the roadway and filled the base with a well-drained foundation of large stones. Each stone was placed by hand with its widest end pointing down. Smaller ones filled in the gaps. Depending on local conditions, Telford often followed with a layer of flat stones, then a crowned surface of broken stones that would shed water. This method gave rise to the word *highway*. The jagged edges of the rocks and stones tended to hold everything together. Also, the stone dust and water formed a mortar that bound the surface into a hard layer. In total, Telford's roads were about 20 inches thick at the center.

Fellow Scot John McAdam (1756-1836) developed a reputation for road construction at the same time. The major difference was that McAdam's roads did not have a stone foundation, which made them cheaper to construct. He dug a trench, as did Telford, but did not carefully position large stones. McAdam's name identifies a type of bituminous road covering. He never used such a material. He surfaced his roads in a manner similar to Telford.

Telford supervised about 1,100 miles of highland road construction. Although best remembered for his pioneering road work, he also constructed canals, suspension bridges, and docks. His most immense single project was the Caledonian Canal. It links the North Sea on Scotland's northeast coast with the Atlantic Ocean on the country's southwest coast. The canal opened in 1822 following 18 years of construction. It connects Scotland's major lakes, including Loch Ness, in a narrow waterway about 100 miles long. Telford built about 150 miles of shorter canals in the region south of Liverpool.

Telford also built several bridges, 40 of them between 1790 and 1796. Suspension bridges were in their infancy and were not generally well regarded by the technical community. They tended to sway with the wind and some even fell apart. Two of Telford's suspension bridges are particularly beautiful. One is his 1826 Menai Strait Bridge in north Wales. It has a single span of 579 feet and a total length of 1,710 feet. Another is well known partly because it is so photogenic. The 1827 Conway Bridge is not far from the Menai Strait Bridge.

Telford also constructed the St. Katherine Dock near the Tower Bridge in London. By Telford's day, shipping along the River Thames had increased to the point where loading and unloading at riverside docks proved impractical. Telford dredged a 10-acre lake adjacent to the river. A canal lock controlled its water level. He constructed several warehouses around the dockside. The project took more than two years to complete, was the fourth dock on the river, and the largest. It opened in 1828 and is now the site of upscale apartments and shopping areas.

Telford had a pleasing personality and could brighten up any gathering. He easily communicated with people from all walks of life. He had a great sense of humor, was quick to laugh at his own shortcomings, and made everyone feel comfortable in his presence. A bachelor, he spent his free time reading and writing poetry. He established several public libraries and was a benefactor to poets. Telford died at his London home at the age of 77.

Historians are amazed that Telford managed to successfully complete his many road, canal, suspension bridge, and dock projects. Subcontractors did much of the work following his precise specifications. Telford was the first to provide carefully written instructions. His subcontractors went on to form the core of Great Britain's modern construction com-

panies. Telford is one of few technologists interred in London's Westminster Abbey.

References

Thomas Telford by Rhoda M. Pearce, Shire Publications Ltd., 1972.

Engineers, Inventors, and Workers by P. W. Kingsford, Edward Arnold Publishers, 1964.

Telford's Britain by Derrick Beckett, David and Charles Publishers, 1987.

St. Katherine Dock in London was completed in 1828. It is now used as a harbor for pleasure craft and small commercial vessels. Apartments and shops are in the original warehouses ringing the small harbor.

The Menai Strait Bridge (above) in northern Wales connected Wales with the island of Anglesey. It was on the Holyhead Road that ended at a dock for boat travel to Ireland.

The lovely Conway Bridge (left) was named for nearby Conway Castle and its design blends with the castle's architecture.

Simeon North

Born:
July 13, 1765,
in Berlin,
Connecticut

Died:
August 25, 1852,
in Middletown,
Connecticut

Machine tool development may have been nineteenth-century America's most important technology. Machine tools allowed people to manufacture the products that moved the Industrial Revolution forward. Machine tools cut, drilled, ground, milled, and turned metal for use in countless items. Joseph Brown (1810-1876) used his machine tools in Rhode Island to make Willcox and Gibbs sewing machines. They were well received by the public and remained in production through the 1950s. Sewing machines were the first invention of the Industrial Revolution to find use in the home. Henry Maudslay (1771-1831) used his improved metal lathe in England to make precise screw threads. It was just a short step from there to making other power tools and accurate micrometers. Maudslay was the first to use machine tools for high-speed mass production. But firearms presented the greatest production challenge. They required close tolerance and precisely made parts, and their manufacture attracted the most talented metal workers.

Early pistols and rifles were handmade, and technologists searched for tools and techniques to allow mass production. The milling machine played a key role in manufacturing interchangeable parts. But with many claims and counterclaims during that period, the record of technical accomplishment has remained cloudy. The outspoken Eli Whitney (1765-1825) often received credit for inventing the milling machine and introducing interchangeable parts. He did neither. Only within the past few years has Simeon North been recognized as the 1816 inventor of the milling machine. He was also a key player in the manufacture of interchangeable parts.

North was born into a large farming fam-

Smithsonian Institution Photo No. 48698

ily in central Connecticut. He was a successful farmer for many years. Using a nearby abandoned textile mill, North branched out into metalworking. He made scythes for cutting grain and other agricultural implements. North was well known for the quality of his work. Several people asked him to make individual pistols, which were difficult items to produce. Only a highly skilled person could make a complete firearm. Such a technologist had to be talented enough to make the entire pistol: the lock, the stock, and the barrel.

Making the lock was the most difficult part of pistol construction. The lock held a piece of flint in a small vice-like clamp. Pulling the trigger caused the flint to strike a small iron plate, called a *frizzen*. The resulting spark ignited a charge of gunpowder in the barrel. North's customers were pleased with the pistols he made, which encouraged him to bid on a government contract. The army wanted 500 single-shot flintlock pistols, commonly called *horse pistols* because officers used them while riding horses. A surprised North won the contract in March 1799.

An innovative metalworker, North hired others to help him complete the order. Another for 1,500 came the following year and then one for 2,000 in 1802. Modern historians puzzle over why so few of these pistols have survived the years. Many saw service on naval ships and may have been lost at sea. The pistols may also have been scrapped for salvage when newer models became available. Collectors consider North's early pistols the most desirable of all American single-shot pistols.

Seeing the need for increased production, the government established two armories to manufacture and store firearms. One was in Springfield, Massachusetts, which opened in 1794. The other was in Harpers Ferry, Virginia (now West Virginia), which opened in 1798. The government also wanted to encourage independent arms makers by providing an environment to help them produce a salable product. It occasionally granted cash-advance contracts to help private manufacturers mechanize their pro-

Single shot, flintlock pistol of the type North manufactured. This 1780 model was made in Pennsylvania by Jacob Welshans and has a caliber of 0.62.

duction. North once received $30,000 in this manner. Production information was freely exchanged and both armories allowed visitors. The government wanted to have ample manufacturing infrastructure available should the need arise. It had the resources of the national treasury to support development of new machines and techniques.

By 1813, North had produced more than 10,000 pistols and employed 40 or 50 people at his Berlin plant. Because his work was invariably well done, he received a huge order for 20,000 pistols priced at $7 each. To

meet the production schedule, he opened a new $100,000 factory on Straddle Hill in Middletown. Many of his pistols had a 10-1/8-inch smoothbore barrel and a caliber between 0.64 and 0.67. *Caliber* is the diameter in inches of the spherical lead bullet.

At the time, all arms manufacturers were trying to make interchangeable parts and North's 1813 contract was the first to include that goal. It specified that "the component parts of pistols are to correspond so exactly that any limb or part of one pistol may be fitted to any other pistol of the 20,000." North had been working on interchangeable parts and may have had his milling machine operational at that time. It seems likely he was the motivating force behind the clause and it may have helped him get such a large order. But although he kept getting closer and closer, North never achieved true interchangeability of parts with the pistol.

The industry was working at transforming itself from craft to factory techniques. Hand weapons were delicate and complex mechanisms. All parts had to be precisely fitted to each other if the weapon was to work properly. Firearms were the world's most sophisticated product, but production methods were not accurate enough to make interchangeable parts. Most metal parts reached their finished shape by hand filing. Critical pistol components could not be uniformly made with such techniques. As long as parts were individually fitted together, repair in the field was almost impossible. North began to consider machine tools, which could make identical parts over and over.

During this period, North developed a crude milling machine, probably between 1808 and 1816. No original drawings exist and the record is incomplete, but the machine was described as making small flat cuts. Power came from a waterwheel that turned a line shaft along the ceiling. A leather belt connected the machine to the shaft. The milling machine had three different pulley diameters to control cutter speed. It featured

Single-shot flintlock pistol of the type North manufactured. This one was made in Pennsylvania in 1810 and has a caliber of 0.52.

The Springfield Armory in central Massachusetts was one of two government armories in the late 1700s. The other was at Harpers Ferry, Virginia. North often visited this armory to both give and receive information on firearms production.

A modern reproduction of North's milling machine of about 1816. The three wooden pulleys allowed the belt-driven machine to operate at three cutting speeds. It was built by United Technologies Corporation from drawings made by a former employee in about 1866.

a hand-cranked rack-and-pinion feed, which moved the workpiece under the cutting wheel. Cutting adjustments could not be made. Primitive by modern standards, the machine nonetheless performed a true milling function. It allowed flat or curved iron surfaces to be cut by simply passing a revolving steel cutter over a piece. It was a quick operation and resulted in mass-produced, uniformly sized parts.

North also had an enviable reputation as a manufacturer of high-quality rifles. His first rifle contract for 6,000 came in 1823. The rifles were to be delivered over a five-year period, but North completed the order in four years. He was the first to make a breech-loading rifle accurately enough so that it did not leak during firing. Earlier, flames often spurted between the chamber and barrel causing a loss of muzzle velocity. North's improvement appeared on 1843 rifles and was based on a patent by John Hall. Hall was a supervisor at the Harpers Ferry Armory. The rifles sold for about $17 each and North continued making them for the remainder of his long, productive life.

North kept well informed by regularly reading the New York City newspapers. He was a soft-spoken, hard-working, energetic, and honorable person who easily made friends. He was married twice, widowed once, and had a total of nine children. Four of his sons worked with him in the firearms business. His youngest son, also named Simeon, graduated from Yale University at the top of his class. He later became president of

Hamilton College in Clinton, New York. North died at age 87. For unknown reasons, all his private papers were destroyed shortly afterward.

North was an important contributor in reaching the goal of interchangeable parts. Modern historians point to the Springfield Armory in 1844. That was the year large-scale production began on the first musket made entirely from interchangeable parts. It was a collective effort among many individuals like North, Hall, and Roswell Lee, who was superintendent of the Springfield Armory for 18 years. North's skills and his milling machine were instrumental in the achievement. Merritt Roe Smith, the Cutten Professor of the History of Technology at the Massachusetts Institute of Technology, wrote: "If anyone deserves credit for conceiving and developing the milling machine in America, it is Simeon North."

References

"John H. Hall, Simeon North, and the Milling Machine" by Merritt Roe Smith, in Technology and Culture, January 1973.

The Collecting of Guns edited by James E. Servn, Stackpole Co. Publishers, 1964.

From the American System to Mass Production, 1800-1932 by David A. Hounshell, Johns Hopkins University Press, 1984.

Nuts and Bolts of the Past by David Freeman Hawke, Harper & Row Publishers, 1988.

"Swords into Plowshares: An Interview With Merritt Roe Smith" by Victor McElheny, *American Heritage of Invention and Technology,* Spring 1986.

Henry Maudslay

People sometimes say we are living in an Information Age, and there is evidence of its effect all around us. Information about natural disasters allows government groups to send necessary assistance in a timely manner. Information about a student's learning difficulties allows school systems to determine the cause quickly. Information is particularly important in manufacturing. Special sensors now often evaluate raw materials. Modern machine tools make continuous small adjustments to assure that critical parts are made correctly. Automated quality-control procedures at all stages of manufacture ensure that only products of the highest quality reach the customer.

Of all the technologies necessary to move the world through the Industrial Revolution, none was more important than production machine tools. The nineteenth century's power-driven metal lathes, boring machines, milling machines, grinders, and similar machine tools manufactured sewing machines, printing presses, steam engines, typewriters, and countless other products. At the time, manufacturers were just coming to grips with close tolerances and high precision. Power tools were being designed with quality mass production in mind. One person's pioneering efforts in precision machine tool work helped many others in the field. He was Henry Maudslay, often identified as the inventor of the slide rest lathe and the first to use machine tools for high-precision, mass-production manufacturing.

Maudslay's birth city is only a few miles east of London on the River Thames. His father had been injured while serving in the British Army as a machinist for 20 years. The government offered him a job as storekeeper at the large Woolwich Arsenal. The work did not pay well, but Maudslay's family was bet-

Born:
August 22, 1771,
in Woolwich,
England
Died:
February 14, 1831,
in Lambeth,
England

The Science Museum/Science & Society Picture Library

ter off than many others. Maudslay's father introduced him to tools and the basics of production. Maudslay himself went to work at the arsenal at the age of 12, three years after his father died. Starting out filling cartridges with gunpowder, he showed such mechanical aptitude that he quickly moved to the carpenter shop. His talent then brought him to the blacksmith shop where he received education and training almost equivalent to an apprenticeship.

A series of robberies in London and improvements in the technology increased the demand for well-made door locks. Self-made manufacturer Joseph Bramah (1749-1814) had patented a high-security lock in 1784 that required accurate machining. But Bramah could not make it a commercial success until he hired 18-year-old Maudslay as an apprentice locksmith. Bramah was a tireless worker who also invented the hydraulic press in 1813. He gave Maudslay excellent training. In return, Bramah received specialty machines that

This 1805 block shaper was one of 43 special machines used to make wooden pulley blocks for ropes on sailing ships. A plate on the front with "1906" indicates the machine's serial number.

Maudslay's most historically significant machine tool was his highly accurate 1797 screw-cutting lathe. The 1-inch-diameter lead screw has a narrow section square thread with a pitch of 1/4 inch. The lathe shows obvious evidence of heavy use.

Maudslay designed for lock manufacture. The three that still exist are each about the size of a small end table. One was a quick-release vise. Another was a metal saw for cutting precise slots. The third wound flat spiral springs. Maudslay designed and built the machines before he turned 23. The machines allowed Bramah to go into production with his complex lock. So certain was Bramah of its security that he offered £200 to anyone who could pick the lock. The reward went unclaimed until 1851.

Maudslay married Bramah's housekeeper, Sarah Tindale. Some time later, he asked for a raise to better support his wife and their children. Bramah foolishly refused his reasonable request. After eight years with Bramah, Maudslay quit in 1797 and opened his own machine shop. He took with him the designs for a new lathe he had been working on. It was his classic screw-cutting lathe. Precisely made screws were a vital component in machinery. They were not only used to fasten parts together, but for adjusting movable parts. The screw was also an essential component of the micrometer.

Maudslay's design exhibited characteristics that were soon standard on all lathes. Some of them came from earlier technologists who used delicate lathes to produce small decorative items like jewelry or for instruments like sextants. Maudslay was the first to combine many important characteristics in one heavy-duty production machine. They included stiff trian-

gular guideways, an adjustable tailstock, and an index dial to control the depth of cut. The cutting tool was attached to a threaded slide rest that moved from the rotation of a lead screw. Maudslay did not invent the slide rest lathe but he perfected it. Recognizing that sharp angles form stress concentration points, he rounded the internal edges of his cast-iron frames. He was the first person to make machine tools entirely from metal. Maudslay's lathes were large industrial tools that were more productive and accurate than others in common use.

Maudslay sometimes receives credit for inventing the powered slide rest. Although he certainly had the ability to create such an important accessory, slide rests had been in use for many years before Maudslay introduced his 1797 lathe. French instrument maker Antoine Thiout, for example, used one in 1741. The misconception came from James Nasmyth (1808-1890), an apprentice of Maudslay who all but worshipped him. (See pages 91-93.) After Maudslay's death, Nasmyth incorrectly credited Maudslay with inventing the slide rest. Since Nasmyth later rose to prominence in machine tool manufacture, his word carried much credibility.

By 1805, Maudslay used his machine tools to make affordable bench micrometers to 0.0001 inches, an almost unheard of accuracy. He and his business partner Joshua Field (1786-1863) employed 200 people at their London factory. They made machine tools and measuring equipment for manufacturing companies and for scientific use. Maudslay won an award of £1,000 for personally making a 5-foot-long screw for calibrating instruments at the Royal Observatory in Greenwich.

The screw was 2 inches in diameter, had 50 threads per inch, and was a masterpiece of screw cutting. The long 12-inch nut for supporting the instruments had 600 internal threads.

Marc Isambard Brunel (1769-1849), father of Isambard Kingdom Brunel (1806-1859), asked Maudslay for assistance in developing some new machinery in 1800. The British navy needed 100,000 new wooden pulley blocks each year and production was barely keeping up with demand. The expensive handmade elm blocks with durable lignum vitae pulleys controlled the ropes on sailing vessels. Ropes were used with the sails, for hauling heavy items, and for moving guns into and out of position. The average-size ship had about 1,000 pulley blocks and they wore out quickly.

Brunel developed a process for automatically making pulley blocks and Maudslay designed the machinery. The project took six years to finish. Located in the southern seacoast city of Portsmouth, the factory where the pulley blocks were made featured the world's first fully mechanized, large-scale, mass-production operation. Ten unskilled workers controlled 43 machines and produced up to 160,000 pulley blocks per year. This was a remarkable achievement. The British government saved about £17,000 per year from a capital outlay of only £54,000. Industrialists

from all over the world regularly visited the factory. Powered by a single 30 hp steam engine, the belt-driven machines were so well made that several were in continuous use for 145 years.

Machine tools of Maudslay's era operated from a single powerful steam engine. A leather belt around the flywheel turned a shaft running along the ceiling. Each machine tool had its own belt connected to the shaft. To make it easier for smaller workshops and factories to use machine tools, Maudslay patented a compact steam engine in 1807. It was called a table engine because it had iron legs. Maudslay made the cast iron frame in pieces, accurately fitted and fastened with screws. Although the engine and frame were strong and stiff, they were not as bulky as others and required no elaborate foundation. Built for moderate power in the 2 to 30 hp range, table engines remained in production for 25 years after Maudslay's death. By then, the business had been transferred to his two sons, Thomas and Joseph, and his partner, Joshua Field.

Maudslay's accurate screw threads showed an appreciation for precision that influenced a generation of British technologists. Dozens of the most gifted young people were attracted to his company and Maudslay willingly provided their initial training. All helped Britain become the nineteenth-century leader in manufacture. Three of Maudslay's brightest stars were Joseph Clement (1779-1844), Nasmyth, and Joseph Whitworth (1803-1887). Clement made machine tools and many critical parts for Charles Babbage's (1791-1871) ill-fated 1834 mechanical computer. Nasmyth invented the steam hammer and Whitworth developed standardized screw threads. Many people around the world worked on improving machine tools. But nobody accomplished more than Maudslay. His passion for accuracy resulted in tools that have been called the cornerstone of the Industrial Revolution.

References

"The Maudslay Touch: Henry Maudslay— Product of the Past and Maker of the Future" by F. T. Evans, *Newcomen Society Transactions*, Vol. 66, 1994/95.

Tools for the Job—A History of Machine Tools to 1950 by L. T. C. Rolt, Her Majesty's Stationary Office, 1986.

Called a table engine because of its general appearance, this 5 hp, 30 rpm, 1815 steam engine pumped water for a hospital. Maudslay's table engines were successful because they were relatively lightweight and occupied little floor space.

Eli Terry

Born:
April 13, 1772, in
South Windsor,
Connecticut

Died:
February 26, 1852,
in Terryville,
Connecticut

Almost all measurements made before the year 1900 had some degree of substance associated with them. Most people had an intuitive appreciation of how long a *mile* was. They could lift a 5 *kilogram* bag of potatoes and get a feel for its weight. People could drink a *pint* of milk from cows grazed in a 10-*acre* field and understand those quantities. But the measurement of time was more mysterious because it had no substance. And an hour was not particularly important to people involved in farming, the most predominant occupation in early America. They were more concerned with the dates on calendars, which helped them schedule planting and harvesting. People generally accepted an hour as one-twelfth the time between sunrise and sunset. The number 12 was a useful figure to ancient people since it allowed grouping things evenly in twos, threes, fours, and sixes. But using such an imprecise measure, called a *variable hour*, meant the length of an hour changed as winter days shortened from summer days.

People who lived in cities thought they knew the correct time because the town clock had bells that rang on the hour. But some town clocks used a variable hour and others used an *average hour*. An average hour is the one we use today. Two situations caused clockmaking to become a particularly important occupation during the nineteenth century: factories and train travel. The Industrial Revolution resulted in the construction of many factories and working hours had to be consistent throughout the year. Scheduled train travel started in 1831 between Charleston and Aiken, South Carolina, with different trains using the same tracks. In the days before telegraph communication was universally available,

Smithsonian Institution Photo No. 18091

rigid timetables and accurately kept time were essential for safety.

Most early clocks were bulky, complex, and expensive museum-piece designs. Some were called *astronomical clocks*. Eli Terry brought time measurement to common people in 1800. He started the world's first clock factory to produce accurate, affordable wall clocks. America's contribution to timekeeping was Terry's mass-produced clock.

Terry was born into a farming family a few miles north of Hartford, Connecticut. The eldest of 10 children, he had very little formal schooling before becoming a clockmaker's apprentice at the age of 14. He worked for several master clockmakers over the next six years. Among them was Daniel Burnap. He was one of the foremost clockmakers in the country, but his average production rate was only four clocks per year. After completing his apprenticeship in 1793, Terry settled near the small western-Connecticut town of Plymouth. He opened a business to make and repair clocks, engrave metal parts, and sell eyeglasses. It was a small shop and Terry was only 21 years of age.

As was the common practice, Terry made a clock only when a customer ordered one. He typically worked on one or two at a time. He wanted to expand the clock market but saw a slim market among country people in his area. He chose to concentrate on making clocks with wooden movements. Using locally available cherry wood, he could sell clocks for half the price of those with brass parts. Although metal clocks were more accurate, Terry viewed tight accuracy as a secondary consideration for rural customers. His shop had two primitive machine tools. One was turned by hand to cut many of the gears. The other was a treadle-operated lathe. Terry made his clocks without firm purchase orders but easily sold all he produced.

Because of his success, Terry decided to open a clock factory in 1800. He took over an unused textile mill and designed equipment to mass produce wooden movements. Nothing is known about the machinery involved except that it operated from rotating shafts running along the ceiling. The shafts received their power from a waterwheel in a channel of the Pequabuck River. This was the world's first clock factory. With three apprentices using automated tools, Terry could make 20 clocks at a time, instead of only 1 or 2. He manufactured 200 hang-up clocks in 1802, a large number for such a small facility. Unlike others, he did not personally sell his products. He consigned them to peddlers to sell on his behalf. The clocks were priced at $30 with the case included, and the price later dropped to $15. A clock was a status symbol to most people. A modestly well-off family that could afford to purchase a clock would feel they had moved up a notch in the world. By the early 1800s, wooden clocks accounted for most of American clock production.

Terry's clock factory used an entirely new method of production, which was ridiculed by his neighbors and other clockmakers. Most people viewed clockmaking as an art form and felt Terry was damaging its image by making clocks as if they were ordinary products. Terry was unperturbed, though, maintained high standards of manufacture, and easily sold all the clocks he made.

Retailers were quick to notice the consumer demand. Edward and Levi Porter operated a business in Waterbury that purchased clock movements made by others. In 1807, they asked Terry if he could manufacture 4,000 movements in three years for $4 each. It was a huge request, but Terry said he could do it. The Porters provided the working capital and Terry spent the first year improving his machinery. With the help of 12 workers, he made 1,000 clock movements in the second year.

To keep production costs down and improve his clock's accuracy, Terry used as few components as possible in his wooden mechanisms. Weights operated the escapement in this early design. The bell struck on the half hour and counted on the hour.

One of those workers was Seth Thomas (1785-1859), a future world-renown clockmaker. Terry completed his contract in the third year. He sold his factory in 1810 for $6,000 to Thomas and Silas Hoadley, another fine clockmaker.

Terry retired for several years and spent time planning his next project. Then, he opened another factory to work with his sons and brother Samuel. He wanted to try out some new ideas. His early clocks had been tall and formal looking. In 1814, he brought out a 30-hour shelf clock that had greatly simplified construction. Resembling a plain rectangular box, it could be manufactured by unskilled workers and required less assembly work. It was the most radical development in clockmaking during the early nineteenth century. It cost as little as $7.50.

But the clock that shook the industry to its roots was Terry's patented 1822 "pillar and scroll" shelf clock. This clock became the standard of the industry for the following 15 years. The beautiful, accurate, and inexpensive clock cost as little as $15. At 31 inches high and 17 inches wide, its compact size made it easy for peddlers to carry. With its attractive design, the clock almost sold itself. Its familiar wooden construction materials made it simple to repair. Annual production peaked at 12,000 and it was the main reason Terry accumulated a personal fortune estimated at $100,000. However, the clock was so popular and the patent law so weak at the time, that the movement and case design were openly pirated by many dishonest competitors.

Most Terry clocks had a small painting beneath the face. It sometimes showed a factory because the clocks were used to regulate manufacturing during the Industrial Revolution. But it more typically depicted farm life or a rural village. The pendulum was visible and mimicked nature's cycle in an agricultural world.

Terry was married to the former Eunice Warner for 44 years and then to the former Harriet Pond Peck for the last 12 years of

his life. He had 11 children. He was an early user of the 1790 U.S. Patent Law and took out his first patent in 1797. He held a total of 10. Terry made not only wooden movement clocks, but also fine brass clocks he sold to watchmakers as regulators. He built several tower clocks. He charged $200 for the 1826 town-hall clock he built for New Haven. Terry spent his entire adult life in an area of western Connecticut that came to be called Terryville. He died of natural causes at the age of 79.

Terry played the role of teacher to the clockmaker trade, just as Henry Maudslay (1771-1831) had done for machine tool manufacture in Great Britain. He instructed his apprentices in both product design and production, and none learned better than Seth Thomas. Although they were once involved in a patent suit, each gave the other subcontracting work and licenses to make their clocks. Each also had a town named after him. Thomaston, Connecticut, is about four miles from Terryville. There were eventually 275 firms related to clockmaking in the region.

References

Keeping Watch–A History of American Time by Michael F. O'Malley, Smithsonian Institution Press, 1996.

"Clocks for the Masses" by Brooke Hindle and Steven Lubar, *Tools and Technology*, American Precision Museum Association, Fall 1996.

Eli Terry Pillar & Scroll Shelf Clocks by Lockwood Barr, National Association of Watch and Clock Collectors, December 1952.

The American Clock, 1725-1865 by Edward Battison, New York Graphic Society Limited, 1973.

Terry's beautiful pillar and scroll clock was the most popular clock of its time. The clock was inexpensive and reasonably accurate. Its three top finials were turned out of brass.

Andre Marie Ampere

The Science Museum/Science & Society Picture Library

Born:
January 22, 1775,
in Lyon, France

Died:
June 10, 1836, in
Marseilles, France

People who work with new ideas in technology, often fall into one of two categories. Those who work with pencil and paper are *theoretical* investigators. Those who work with hardware are *experimental* investigators. As a practical matter, most scientists, engineers, and technologists perform both types of investigation. When inventors sketch an idea, for example, they act as theoreticians. When they construct their prototypes, they act as experimentalists. But many technologists have tended to focus on one approach. Josiah Willard Gibbs (1839-1903) was America's only native-born theoretical scientist with a worldwide reputation. He did most of his work in physical chemistry with pencil and paper. By contrast, Ferdinand Zeppelin (1838-1917) of Germany was an experimentalist. No one before him had ever constructed a successful rigid-frame airship. His first was built mostly by trial and error, and was an experimental activity.

The field of electricity has provided plenty of theoreticians as well as plenty of experimentalists. Italy's Allesandro Volta (1745-1827) made the first battery and Germany's Georg Ohm (1789-1854) constructed the first electrical circuits. Both were primarily experimentalists. The unit of electrical pressure, the *volt*, is named for Volta, and the unit of resistance, the *ohm*, is named for Ohm. The *amp*, or *ampere*, is a unit of electrical flow named for the theoretician Andre Ampere. Ampere's best work came through his use of advanced mathematics to analyze the relationship between electricity and magnetism. He established the field known as *electrodynamics*, which studies the effects caused by electricity flowing in a wire.

Ampere was born in south-central France, the second of three children in a prosperous business family that dealt with textiles. His father taught him at home and Ampere quickly developed a huge appetite for printed materials. He read the entire French encyclopedia and showed an aptitude for mathematics. Most of the city library's books on mathematics were in Latin. So at the age of 12, Ampere taught himself Latin so he could read the books. He had an uncommon talent for languages and a lifelong interest in the possibility of a universal language.

Ampere was a teenager during the French Revolution (1789-1799) and probably understood little about it. A revolt against the monarchy of Louis XVI (1754-1793) and Marie Antoinette (1755-1793), it started in the north of France. Ampere's father supported the monarchy and was involved with the city of Lyon's attempt to maintain the old government. Revolutionaries captured the city, convicted Ampere's father of treason, and executed him in 1793. The younger Ampere was devastated and did not speak for a year. It was the first of an number of personal tragedies Ampere would experience.

Ampere met a young lady, Jule Carron,

These three unlabeled Leyden jars probably date from around 1800. They stored electric charge and were precursors to modern capacitors. They were metal-foiled glass jars first made in 1745 at the University of Leyden in the Netherlands.

who brought him out of his depression. The couple married in 1799. They had one son. Ampere at first taught mathematics at a high school in Lyon. Then, he found a better-paying job at a similar school in Bourg, about 40 miles away. The distance was too great for him to return home every evening. He spent the week in a small apartment and went home on weekends. The forced separation from his family gave him time to work on publications about the statistics of gambling. The theory of probability got its start with the mathematics associated with games of chance, like dice and cards. Ampere expanded his work into advanced topics like probability calculus and began to gain a reputation in mathematics.

After several years of illness, Ampere's wife died in 1803 of an unknown ailment. Ampere again endured the loss of a loved one. He began to look around for a job that would take him away from Lyon and its unhappy memories. Based partly on the quality of his published works, Ampere found a position as a mathematics instructor at the Polytechnic [Technical] School of Paris, about 250 miles north of Lyon.

Unhappy in a large and unfamiliar city, he eagerly sought companionship with the Potot family, which befriended him. He married Jeanne Potot in 1806. But Ampere's new father-in-law swindled him out of some money and his marriage proved unhappy. He and his wife had a daughter before they dissolved their marriage. From that time, Ampere raised both his children in Paris with the help of his mother and an aunt.

Ampere's professional life followed a few twists and turns over the next several years. In 1808, he became inspector general for the newly formed university system. He became a member of the mathematics department of the Imperial Institute in 1814. Ampere's job required frequent travel and he chose to name his scientific findings after the places where he first thought of them. Ampere had a "theory of Avignon," a "demonstration of Genoble," and a "theorem of Montpellier." He was a member of many professional organizations. The French Academy of Sciences was among the most important. He was named assistant professor of astronomy at the University of Paris in 1820. That same year, he heard about a remarkable electricity and magnetism observation made by Hans Oersted (1777-1851) in Denmark.

Current electricity was a new idea and technologists often argued over any connection it might have with magnetism. Many people thought there was no relation between the two. Oersted was professor of physics at the Polytechnic Institute in Copenhagen. He conducted an experiment in which he placed a magnetic compass needle directly below a wire. When electricity flowed through the wire, the needle deflected slightly. This easily repeated experiment suggested a direct link between electricity and magnetism. Oersted's minor observation put Ampere on the path to worldwide recognition.

Within two weeks of hearing of Oersted's discovery, Ampere was working out the math. He began a series of six weekly reports to the Academy of Sciences. He showed that electrical flow in a circle acts like a bar magnet and that an iron bar placed inside the coil becomes magnetized. Ampere called this a *solenoid*, from the Greek word *solen*, which means "tube." The word is still used for electromagnetic relays.

Ampere proved mathematically that the force between two current-carrying wires is a function of the electrical flow, wire lengths, and distance between the wires. Electricity was too new for technologists to be certain of the flow direction. But Ampere showed that the direction of flow between two nearby wires influenced whether they were attracted or repelled. He called this Ampere's rule. Flows in the same direction attracted two parallel wires and Ampere built a simple apparatus to demonstrate the effect. He discussed magnetic forces around wires carrying electricity. All of this showed how clearly Ampere understood current electricity. His was the first work on electrodynamics.

Ampere was the first person to apply advanced mathematics to electricity and magnetism. In 1823, he stated a theory that magnetism came from the action of tiny electrical charges in the wire. In spite of Ampere's reputation, his colleagues were skeptical. Ampere was trying to describe an effect that was similar to saying that electricity is a flow of electrons. The electron was not discovered until 1897 by Joseph Thomson (1856-1940) in England.

Ampere's successful professional life contrasted with his personal life. His son, Jean-Jacques, had an involvement with a wealthy widow twice his age. He became one of many in her entourage. Ampere's daughter, Albine, married an alcoholic army officer and had a difficult life. Though he lived in Paris, Ampere died in Marseilles at the age of 61 while visiting a university. The 1881 International Electrical Congress established the *amp* as the standard unit of electrical flow rate.

Electricity is one technical area whose development was shared by people from many countries. Volta was Italian, Ohm was German, and Ampere was French. The unit of inductance, the *henry*, was named for the American Joseph Henry (1797-1878). The unit of magnetic flux, the *tesla*, was named for Nikola Tesla (1856-1943) a Croatian-

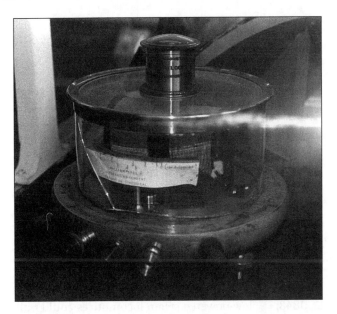

This early, unlabeled, moving coil milliammeter probably dates from about 1850. The manufacturer was Baird and Tatlock.

American. The unit of capacitance, the *farad*, was named for Michael Faraday (1791-1867) of England. The unit of energy, the *joule*, was named for James Joule (1818-1889), also from England.

The unit of power, the *watt*, came from James Watt (1736-1819) of Scotland. The unit of electrical charge, the *coulomb*, was named for Charles Coulomb (1736-1806) of France. The unit of frequency, the *hertz*, was named for Heinrich Hertz (1857- 1894) of Germany. (See pages 25-27 and 166-168.) No field of human endeavor has been more international than electricity.

References

Andre-Marie Ampere by James R. Hoffman, Blackwell Publishers, 1995.

Pioneers of Electrical Communication by Rollo Appleyard, Macmillan Publishers, 1930.

Electricity by Steve Parker, Dorling Kindersley Publishing for the Science Museum, 1992.

George Stephenson

Born:
June 9, 1781, in
Wylam, England

Died:
August 12, 1848,
in Chesterfield,
England

Railways, or railroads, are such a common type of transportation that it is hard to imagine that early technologists were unsure of their potential. They had two specific concerns. First, they did not know if low-powered steam locomotives could pull a heavy load up a hill. Second, they thought the locomotive's metal wheels might not properly grip metal rails, especially if the surfaces were wet or the load large. With no experience to rely on, these were reasonable concerns.

To accommodate them, railway companies sometimes used stationary winding engines. A large engine in a building at the top of a hill rotated a drum. A rope attached to railway cars at the bottom of the hill wound on the drum and pulled the cars up the hill. But foresighted individuals like George Stephenson thought that steam-powered locomotives were more practical than winding engines. Several early locomotives appeared promising, but it was Stephenson's 1829 *Rocket* that launched the railway age.

Stephenson was born near Newcastle in northeastern England, an important coal mining area. Stephenson was surrounded by all the elements of that industry. He was barely out of childhood when he joined his father working at a local mine. He received no formal education, was illiterate until age 18, and never really read well as an adult. He spoke with a strong local accent that was hard for people outside the region to understand. In 1802, he married Frances Henderson, a house servant of the mine owner. Their only child to survive childhood, Robert, was born a year later.

Fascinated by steam engines, Stephenson spent his free time disassembling and rebuilding them. He developed a reputation

The Science Museum / Science & Society Picture Library

for outstanding knowledge of machinery. Others had been working on locomotive construction and Stephenson brought up the subject with the owners of the mine where he worked. They agreed to finance some experiments and Stephenson built his first locomotive in 1814. He named it *Blucher* for Gebhard Blucher, a German military leader who helped the British in their victory against France at the Battle of Waterloo. The locomotive hauled a 30-ton load up a slight incline at 4 mph. Stephenson took out a patent in 1815 titled "Construction of Locomotive Engines."

This success came to the attention of Edward Pease, a coal mine owner in Darlington. The town was about 25 miles from Stockton where ships received coal for delivery to various European ports. Low-capacity, horse-drawn wagons could not support an expanding local market. Stephenson was hired to help lay out and construct equipment for the 1825 Stockton and Darlington Railway (SDR), the world's first public railway. Possessing great strength and endurance, Stephenson personally selected and surveyed the entire route.

The distance between the two rails used in early coal mines was 4 feet 8-1/2 inches. That dimension was used for the SDR and for almost all future railway construction. It is still the worldwide standard. The SDR used a combination of horse- and steam-operated equipment. There were some locomotives, but stationary steam winding engines pulled loaded wagons up hills. The line mainly carried coal, but people occasionally rode in open carriages.

Stephenson was assisted by his son, who went on to become a famous bridge builder and railway designer in his own right. He laid out the first railway into London, the 1838 London-to-Birmingham line. The elder Stephenson was aware of his educational shortcomings and made sure his son received a proper education. Robert Stephenson (1803-1859) attended Bruce Academy in Newcastle and Edinburgh University. He was also apprenticed at a coal mine to receive basic steam engine and railway instruction. The father-and-son team was a good match and they worked well together.

Textiles were Britain's most important factory product and large quantities of raw cotton from America entered Liverpool's harbor. It went by barge on the Bridgewater Canal to spinning and weaving factories in Manchester, 30 miles away. But canal shipping was slow, unreliable, and expensive. Factory owners asked Stephenson in 1824 to lay out the Liverpool and Manchester Railway (LMR) for them. With complete confidence in his decision, he recommended using only steam-powered locomotives—and no winding engines . That was a technical gamble and Stephenson worked to keep the railbed as level as possible. Not knowing how metal wheels would grip metal rails under heavy loads, he was concerned about the gradient (slope). The route Stephenson laid out required the construction of 63 bridges and the removal of large quantities of dirt from miles-long trenches. He sent the line through a large, unstable bog, which had to be partly filled in. With only human and animal labor, the unbelievably difficult project employed thousands

of workers and took five years to complete. But it achieved Stephenson's prime objectives. The railbed was solid and almost perfectly flat with a gradient that never exceeded 1:880.

To select a proper locomotive design, LMR officials held a competition in October 1829. They offered a prize of £500 and a contract for a series of locomotives. The trials were held on a two-mile length of new railway at Rainhill, just east of Liverpool. Four locomotives competed. One, named *Novelty*, was submitted by John Ericsson (1803-1889), the designer of the Civil War ironclad *Monitor*. Stephenson named his locomotive *Rocket*, in part because of the sparks that left its smokestack. An individual favoring winding engines had once

Full-size replica *Rocket* at Great Britain's National Railway Museum in York. The elegant train was constructed in 1980 to commemorate the 150th anniversary of the Liverpool and Manchester Railway.

Original *Rocket* locomotive at London's Science Museum. Beautifully accented with yellow and white paint in 1829, it was heavily modified in the mid-1800s. The locomotive is now displayed in a largely unrestored form.

suggested that passengers would be safer riding a rocket than being pulled by a locomotive. The Stephensons took up the challenge, and for that reason, too, decided on the name *Rocket*. The locomotive was constructed at the family's factory in Newcastle under the direct supervision of Robert Stephenson. It was the eighteenth steam locomotive the Stephensons built.

The beautiful 20 hp, 4.5 ton bright yellow locomotive with its white chimney was the first to exhibit modern characteristics. It had a 25-tube boiler to increase the heat-

The 1830 Manchester passenger railway station (left) for the Liverpool and Manchester Railway is the oldest in the world. The 1830 warehouse on the right was large enough to store 10,000 bales of cotton. The railway tracks are now not used, but they connect to the main line where a commuter train is barely visible.

ing area. Steam exhausted into the firebox to improve its draft. Preheated water came from a container that surrounded the boiler. Two angled cylinders, one on each side, rotated the wheels instead of the axle. With a top speed of 29 mph and experiencing no breakdowns, *Rocket* easily won the competition. It was a pleasant victory because Stephenson's opponents quickly saw the advantages of his locomotive and became loyal supporters.

The experimental *Rocket* was not meant for everyday traffic. The original steeply inclined cylinders made the locomotive ride unsteadily, so their position was changed. A smokebox and front bumper were also added, and the chimney was redesigned. Other changes were made to piping, valves, and supports. A coal company in Carlisle bought *Rocket* in 1836 and the locomotive passed to London's Science Museum in 1862. It is now on public display in largely unrestored form.

The LMR opened in 1830 with eight locomotives similar to the *Rocket* and was an immediate success. Passenger travel was more popular than anyone had anticipated and within one year, more than 2,000 passengers per day rode LMR trains. A one-way second-class ticket between Manchester and Liverpool cost 5 shillings and 6 pence. Nine daily trains ran in both directions. Passenger service was so successful that it unexpectedly earned more income than the transportation of cotton, the original purpose for the line. The 1830 station in Manchester became overworked and another opened nearby in 1844. As a result, the original station survived largely unaltered. Now preserved as part of an industrial museum, it is the oldest purposely built railway station in the world.

The LMR greatly enhanced Stephenson's reputation and his services were in constant demand. There was hardly any railway project in which he was not consulted. He played an active role in many professional societies and was a founding member of the British Institution of Mechanical Engineers. Stephenson was chief engineer on five other major railway lines until his retirement in 1843. He respectfully declined a knighthood and offers of honorary governmental positions. Stephenson turned his business over to his son and took on the role of country gentlemen for the last few years of his active life. He particularly enjoyed experimenting with agricultural crops.

George Stephenson did not invent the steam locomotive. But between 1814 and 1825, he was alone in developing it. Although his first efforts were aimed at improving the steam locomotive, from the late 1820s, he was more concerned with selecting railway routes and building the necessary infrastructure. It was his vision that transformed an industrial conveyor system into a national transport network.

References

Stephenson's Britain by Derrick Beckett, David and Charles Publishers, 1984.

George and Robert Stephenson by Michael Robbins, Her Majesty's Stationary Office, 1980.

Robert Hoe, Richard Hoe, and Robert Hoe III

Between 1898 and 1927, about 60,000 Stanley Steamer automobiles were manufactured in Newton, Massachusetts. The company was started by twins Francis Stanley (1849-1918) and Freelan Stanley (1849-1940). It was America's first successful automobile company. The Stanleys production rate was adequate for the early twentieth century, though it would be quite inadequate today. Approximately 10 million motor vehicles are currently made in America every year. As the population increases and lifestyles change, consumers often want more or better products. Newspapers and other printed materials were not immune from those demands.

Johann Gutenberg (1397-1468) made his printing press in the 1450s by adapting a grape press used in wine making. Turning a large screw with a broomstick-sized handle forced two horizontal plates together. Between the plates lay a sheet of paper and a clamped form holding inked type. The pressure printed an image on the paper. Gutenberg's press had a printing rate of about 30 pages per hour. Centuries later, Benjamin Franklin used a similar screw-type press to print his *Pennsylvania Gazette* in 1730. In England, Charles Stanhope (1753-1816) made the first all-metal flat-bed press in 1800. It was stiffer and made sharper impressions than the wooden variety. But all hand-operated, flat-bed presses had low printing rates. Robert Hoe's introduction of his rotary cylinder press in 1830 made low printing rates a thing of the past. The high-speed cylinder press became the workhorse of all newspapers during the 1800s. It allowed newspapers to expand their readership and to provide more pages of information per issue.

Robert Hoe was the first member of a

National Portrait Gallery, Smithsonian Institution

Robert Hoe III (1839-1909)

printing family that greatly affected American life. He was born into a prosperous farming family in central England. His father apprenticed him to a carpenter when he was about 15. Robert heard about advancement opportunities in America and he emigrated in 1803, while still a teenager. Hoe made his living as a carpenter and soon married Rachel Smith. He changed professions to help his wife's brothers, Peter and Matthew Smith, manufacture screw-type printing presses. The type that Gutenberg and Franklin used, they sold for about $400.

The Smith brothers both died in 1823 and Hoe inherited the business. He went looking for a new printing product to manufacture. He heard of the flat-bed cylinder press invented in Germany by Friedrich Koenig (1774-1833). In 1814, *The Times* of London was the first newspaper produced on the new steam-powered press. A clamped

Robert Hoe:

Born:
October 29, 1784, in Hoes, England

Died:
January 4, 1833, in New York, New York

Richard March Hoe:

Born:
September 12, 1812, in New York, New York

Died:
June 7, 1886, in Florence, Italy

Robert Hoe III:

Born:
March 10, 1839, in New York, New York

Died:
September 22, 1909, in London, England

This Hoe newspaper press was built around 1900 from designs developed in the 1870s. It printed and folded 48,000 four-page newspapers every hour.

This is a drawing from Richard Hoe's patent for a four-impression cylinder rotary press. The cylinder at the bottom center carries the type. Workers stand on scaffolds and feed individual sheets of paper along the four ramps labeled "M."

frame held the inked type on the flat bed of the printing press. A steam engine rotated a cylinder that rolled sheets of paper against the type. An operator hand fed each sheet onto the cylinder at a printing rate of 1,200 sheets per hour. The previous rate with high-speed hand-operated flat-bed presses was 200 sheets per hour. But the Koenig press cost $4,000 without the steam engine, a high price. Hoe sent one of his best workers to London to see the press in action.

Steam power was not widely available in America. Based on his worker's reports, Hoe made some changes to Koenig's design. He offered his own hand-cranked cylinder press for sale. The first machine was purchased in 1830 by the *Temperance Recorder* of Albany, New York. The printing rate was about 400 pages per hour. Hoe was starting to introduce steam-powered presses when his health began to fail. He transferred the business to his son Richard, just before he died prematurely at age 48.

Richard Hoe, his family's oldest son, was educated in the public schools of New York City. His father brought him into the factory at 15 and gave him various jobs to help him learn the business. He was the most technically proficient member of the family and expanded on his father's work. Richard was the first person to successfully attach printing type to the cylinder. The cylinder rotated continuously, which greatly

increased production rate. His patented hand-cranked single-cylinder flat-bed press printed 2,000 pages per hour. He followed it with a double-cylinder press in 1837. All hand-operated presses required strong men to turn the cranks. They had to be frequently relieved. The increase in business encouraged Richard Hoe to construct more buildings for manufacturing his presses. His company soon had four acres under roof.

Richard Hoe made a huge leap forward with his 1846 patented steam-engine-powered press. He eliminated the flat bed entirely. Four impression cylinders contacted one cylinder carrying type. At each impression cylinder, a worker fed individual sheets of paper into the press. One rotation of the type cylinder printed four sheets. Its production rate was 10,000 pages per hour. The workers had to be fast because at top speed, the cylinders drew paper at the rate of about one sheet per second from each person. The first new press was installed in 1847 by the Philadelphia's *Public Ledger* newspaper. Later Hoe presses had as many as 10 impression cylinders and sold for $25,000.

Also called a type-revolving press, Hoe's cylinder printing press overshadowed all others. It soon replaced those used by newspapers throughout America, Europe, and Australia. With higher production rates, newspapers became more profitable. The R. M. Hoe Company was the leading manufacturer in the world. In the late 1860s, Hoe opened a branch factory in London that soon employed 600 persons. William Bullock (1813-1867) found a way to print from a continuous roll of paper in 1865. He called his innovation a *web feed,* after bolts of woven fabric, which were called webs. The method proved particularly useful in printing newspapers. The Hoe family quickly arranged to use Bullock's invention with their presses.

The company was an enlightened organization and had an early employee benefits program. Workers received a free lunch, partly subsidized medical care, a cooperative company store where they could purchase food at reduced prices, and free night school classes. Richard Hoe married twice and had three children. He watched over every detail of his factory until he suffered from overwork. He went overseas to rest and recover. He died at the age of 75 while visiting Italy. His nephew Robert Hoe III, grandson of the man who started the company, was the next family member to take over.

Robert Hoe III did not have the technical insight of his predecessors. However, he had the ability to select people who could carry out the improvements he thought most important. Hoe directed his energies toward meeting customer's requests for improved production rates. In 1891, his company constructed a cylinder press of 16,000 parts for the *New York Herald*. Using a continuous roll of paper, the web press printed, cut, folded, and counted 72,000 eight-page newspapers per hour. Such presses cost between $40,000 and $80,000.

One of the founders of New York's Metropolitan Museum of Art, Robert Hoe III appears at the start of this chapter in a 1891 drypoint engraving. He married Olivia James in 1863 and they had five children. Robert Hoe III wrote and published *A Short History of the Printing Press* in 1902. An avid collector of rare books, he had over 20,000 when he died unexpectedly at the age of 70.

Production printing has seen more changes than most other technologies. Flat-

bed printing gave way to cylinder printing using hand-set type. Hoe presses were supreme throughout the world for about 25 years after 1847. Cylinder printing easily merged with the Linotype method that used melted type to form a page. It was invented by German-American Ottmar Mergenthaler (1854-1899). In July 1886, a portion of the *New York Tribune* was set with an experimental Linotype typesetter.

Hoe's presses were very important to the spread of news throughout an expanding nineteenth-century America. In a free society, literate and informed citizens depended on inexpensive newspapers. Between 1830 and 1860, the number of American newspapers increased from 863 to 3,725. Although there are fewer modern newspapers, each of the top four print more than one million copies a day. Their computerized techniques are far removed from Gutenberg's original printing press, but they provide more flexibility. All printing methods had the same objective, to get information to people on a printed page. Each accomplished that objective as best it could during its time in history.

References

The Story of Printing by Irving B. Simon, Harvey House Publishers, 1965.

American Journalism by Frank Luther Mott, Macmillan Publishers, 1962.

A History of American Manufactures from 1608 to 1860 by Leander Bishop, Samson Low, Son and Co. Publishers, 1868.

The Printers by Leonard Everett Fisher, Franklin Watts Publishers, 1965.

The first Hoe cylinder press was used to print a short-lived specialty publication in Albany, New York, called the *Temperance Recorder*.

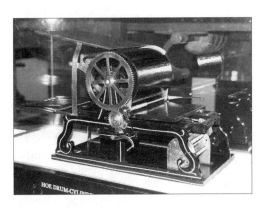

This is a model of Robert Hoe's early hand-cranked rotary press. The inked type is locked into the flat bed. The paper is fed from the plate at the top and carried along the outside of the cylinder. Cranking a handle rotates the cylinder and moves the type under the paper, to form an impression.

Georg Simon Ohm

Born:
March 16, 1787,
in Erlangen,
Germany

Died:
July 7, 1854,
in Munich,
Germany

It would have been impossible for early nineteenth-century technologists to predict the future importance of electricity. They saw it only in the form of bolts of lightning and harmless shocks that resulted from scuffing shoes over a rug. The word *electricity* comes from the Latin *electrum*, which means "amber." Amber is fossilized tree sap similar to modern plastic. Rubbing it with fabric produced static electricity that attracted small scraps of paper and lifted animal fur. In the early nineteenth century, most people considered electricity to be more of a curiosity than a potentially useful form of power.

It became an important emerging technology after the invention of the storage battery in 1799. Alessandro Volta's (1745-1827) battery was a small stack of dissimilar metals and cardboard. Volta was not the only person investigating electricity. Others around the world were adding their bits and pieces. Michael Faraday (1791-1867) built the first electric motor in England in 1821. Andre Ampere (1775-1836) mathematically analyzed electricity in France in 1827. Joseph Henry (1797-1878) worked with electromagnetism in America in 1830. Another international player gave us a simple electrical formula that many people remember all their lives. Georg Ohm experimented with electrical currents and included his results in an 1827 German book. We now call his formula Ohm's law: voltage equals current times resistance. It is the most widely used relationship governing the behavior of electrical circuits. In formula form, it is $E = I*R$. E stands for electromotive force, I for intensity, and R for resistance.

Georg Ohm was born in south-central Germany, not far from Nuremberg. (His first name is pronounced "gay-org.") Ohm's par-

The Science Museum/Science & Society Picture Library

ents had seven children, though only three survived to adulthood: Georg, younger brother Martin, and sister Elisabeth. They were raised by their father after their mother died in 1797. Georg and Martin enjoyed technology and supported each other's activities. Martin became a respected mathematician at the Military College in Berlin. Their father was a master locksmith who liked science. He encouraged his children to read and taught the brothers the use of precision tools. He showed them how to fabricate the high-quality parts needed to construct a reliable lock. That skill would become the foundation of Ohm's future success.

Ohm attended the local high school, hoping to learn about the technical concepts his father introduced. But his education stressed classical instruction such as art, literature, and Latin. He grew up at a time in German history when educators felt that speculation was the only way to answer important technical questions. They had not yet accepted the idea that answers should come through experimentation and data evaluation. After high school, Ohm attended the university in his hometown, but he could not discipline himself. After he wasted time on social pursuits for a year and a half, his father decided that he was misusing the family's limited resources. Ohm's father insisted that he leave school. Ohm moved to rural Switzerland and started earning a living. He was a mathematics teacher and tutor for several years before returning to the University of Erlangen. This time he was better prepared for his studies and earned a Ph.D. in physics in 1811. He specialized in the subjects of mechanics and light.

Germany was then involved in the 1806-1814 Napoleonic Wars with France, which depressed civilian employment opportunities. Erlangen had a population of just 8,000 but had to shelter more than 33,000 soldiers. The possibility of a civil war was on everybody's mind. It was a difficult time for any young person looking for a job. Ohm wanted to teach at a university, but he decided to stay at home and work as a tutor for a short time. He finally found a full-time position teaching mathematics and physics at a high school in Bamberg. He stayed there for three years, working on his first book during his free time. It was a geometry book with a 13-word title that came out in 1817.

Although not outstanding, Ohm's textbook proved adequate. More important, its publication helped him obtain a similar position in a better setting. He taught at a religious high school in Cologne. The school had enthusiastic students and Ohm could use its laboratory. His first investigations were into the new concept of electromagnetism. It was 1825 before Ohm realized that research, followed by publication, might be the key to obtaining a university teaching position. Technical research and publication was a new practice just beginning to take place in Germany.

Focusing on how resistance influenced electrical current flow, Ohm was particularly interested in the conductivity of metals and their behavior in electrical circuits. But the science and practice of electrical measurement simply did not exist. Electrical technology was not well established and

Ohm had to make his own meters and resistors. His resistors were of copper wire that he had learned to make while studying under his father. They were of remarkably consistent quality. Ohm laboriously wrapped the bare wire with insulating silk thread. Some wires were as thin as 0.025 inches in diameter and some as long as 75 feet. Each length was wrapped around a nail or wooden peg. Ohm's early crude amme-

The resistance value of Ohm's resistors varied with length and diameter. He wrapped them with insulating fabric, usually silk, and then twisted them around a nail or wooden peg. This resistor was made and used by Ohm in January 1826.

ter was a compass needle hanging on a thread. His zinc-copper battery was approximately the shape of a cube, six inches on a side. Brine-soaked cardboard separated the metal plates.

Ohm conducted tests for several years and was the first to suggest an analogy between electrical flow and liquid flow in a pipe. His initial conclusions were disappointing because he could not repeat them. Ohm determined that the high internal resistance of the battery was influencing his readings. He was the first to make such an observation. When he felt confident enough, he published his conclusions. His 1827 book was titled *The Galvanic Circuit Mathematically Treated* and included what we now call Ohm's law. Ohm wrote, "The magnitude of the current (amperes) in a galvanic circuit is directly proportional to the sum of all tensions (volts) and inversely to the total reduced length (ohms) of the circuit." He used the term *total reduced length* for resistance because his resistors were wires of varying lengths.

Ohm's conclusions drew little attention from the technical community. Others could not duplicate his results. In part, they lacked the skills needed to make accurate instruments and test specimens. They were also held back by the early-nineteenth-century German belief that experimentation and mathematics were irrelevant to understanding natural phenomena. Influenced by German philosopher Georg Hegel (1770-1831), many thought Ohm's formula too simplistic. They treated him with indifference and even hostility. Ohm was crushed. He felt he had done everything necessary to advance the field of electrical technology, but others misread his results.

Although Bavaria is now a state of Germany, it was a separate kingdom in the mid-1800s. Bavarian King Ludwig I had far-sighted technical advisors. He gave Ohm a professorship at the Polytechnic School of Nuremberg in 1833. This was the first positive acknowledgment Ohm received concerning his electrical discoveries. He was also appointed to the position of State Inspector of Scientific Education where he could oversee necessary educational reforms. The British used his book to guide many of their electrical experiments. They were so impressed with Ohm that in 1841 they awarded him the coveted Copley Medal, a precursor to the Nobel Prize. The photograph at the beginning of this chapter shows Ohm with the medal around his neck. German scientists saw their past errors and lobbied to offer Ohm a professorship at the University of Munich. He accepted this high honor in 1849 and became department head in 1852. Ohm died in 1854 while working on an optics experiment. He had understandably turned his back on electrical research.

Ohm was of average height, sturdy, and physically strong. He often wore a long, dark-blue coat with large pockets for his electrical equipment. He was an almost tireless worker who found comfort in his professional work. He never married. Ohm was an excellent teacher who approached problems with students as if he did not yet know the answer. He had a pleasant voice and an excellent sense of humor. He never let his professional disappointments affect his relationship with students or supportive colleagues. Although Ohm was displeased that powerful leaders worked against him, he never gave up on technology. And he achieved world recognition in his own lifetime. Few people have been as courageous as Ohm. Fewer still have both a law of nature and a unit named after them. In 1881, the International Electrical Congress in Paris, established the *ohm* as the basic unit of electrical resistance.

References

Pioneers of Electrical Communication by Rollo Appleyard, Books for Libraries Press, 1968.

The Communications Miracle by John Bray, Plenum Press, 1995.

Famous Names in Engineering by James Carvill, Butterworth and Co Publishers, 1981.

Louis Jacques Mande Daguerre

It would be hard to imagine modern life without pictures. Colorful static images from newspapers, magazines, billboards, personal photographs, and the Internet almost demand our attention. Dynamic moving images in motion pictures, on commercial television, and in instructional videotapes appear everywhere. Imagery informs, entertains, and inspires.

Nineteenth-century technologists knew the power of the picture and looked for easier ways to capture life on paper, glass, or metal. The earliest technique was the *camera obscura*. The name comes from two Latin words. *Camera* means "chamber" and *obscurus* meant "dark." It was a box with a lens at one end and a 45° mirror at the other. Aimed at a building or rural scene, the mirror focused the reflected image onto a piece of ground glass on top. A person could then place a piece of thin paper onto the glass and sketch or paint the subject, using the reflected image as a guide.

Another image-recording technique was named after Etienne Silhouette (1709-1767), a French minister of finance. Although it is uncertain whether Silhouette originated the idea of forming an outlined profile of a person, his name describes such an image. Another Frenchman, Joseph Niepce (1765-1833), produced the first permanent photograph in 1822. But his technique required exposures of up to eight hours. Louis Daguerre introduced a more practical method in the 1830s. Requiring exposures of only a minute or less, it was fast enough to photograph people. His daguerreotypes caused a worldwide sensation.

Daguerre was born about nine miles northwest of Paris and grew up during a difficult period in his country's history. The French Revolution (1789-1799) affected the

Born:
November 18, 1787, in Cormeilles, France

Died:
July 10, 1851, in Petit-Bry-sur-Marne, France

country's social programs and Daguerre received little formal education. His father, a minor government official, apprenticed Daguerre to a draftsman in his early teens. Daguerre had an aptitude for sketching and wanted to study painting. His father relented in 1804 and Daguerre went to Paris. He was fortunate in locating an apprenticeship with the chief stage designer at the Paris Opera House. He stayed at the opera house for three years and then went to work for Pierre Prevost, a painter well known for his panoramas.

Part of Daguerre's responsibility was to help paint large cycloramas that had become Prevost's specialty. In an era before photography and motion pictures, the viewer sat in the center of a large cylindrical painting of a single expansive subject. The paintings required great attention to accuracy of scale and perspective. Daguerre's experience with drafting served him well and he stayed at the Prevost Stu-

This 1839
daguerreotype
camera has a
peep hole to help
the photographer
compose the
image.

dios for nine years. He married Louise Smith in 1810 and chose to strike out on his own in 1816.

Daguerre made a living by painting scenery for many of the best-known theaters in Paris. The prestigious Academie Royale de Musique contracted with him on several occasions. He was one of its chief designers for two years. Daguerre perfected a clever scene-painting technique that allowed the background to appear to change, such as from summer to autumn. One of his best-remembered efforts was a mountain slope. Carefully turning lights on and off gave the set the appearance of a landslide occurring in the background. This was the most exciting optical illusion before motion pictures.

Such dramatic results began with Daguerre painting a scene onto a large piece of semi-translucent cotton fabric. A carefully applied thin coat of paint was almost transparent. To show summer changing to autumn, Daguerre would complete two paintings and place one in front of the other. Sometimes he had three in a row. A turpentine wash on the fabric made it even more transparent. By illuminating various sections of the painting from different angles and through different filters, Daguerre obtained a variety of exciting motion effects. The huge fabric sheets were as wide and high as an opera stage. Daguerre's work

Daguerreotype
cameras were
little more than
lighttight boxes.
The back of this
one slid to allow
focusing on a
ground glass
before the
sensitized plate
was put in place.

achieved an astonishing degree of three-dimensional realism and he received many forms of recognition. But none of his early work remains. Because of the combustible nature of the materials, a theater fire in 1839 destroyed everything.

In Daguerre's era, a person's financial condition was all important. He had no social status on which to build and felt

obliged to use his talents and whatever else he could muster. Driven by a desire for wealth and publicity, Daguerre saw photography as a way to reach that end. He learned of Niepce's work in 1826 and read what he could on the topic. He knew that long exposure time was a limiting factor. Daguerre had no photographic experience and arbitrarily thought he could shorten exposure time to take pictures of people. He wrote a letter to Niepce suggesting a partnership. Daguerre misrepresented himself by saying he had some photographic successes when he actually had none. Niepce had received little financial reward from his innovation. He felt that the younger Daguerre was an energetic and optimistic person who could improve his process. The two signed an agreement in 1829.

Although Daguerre worked diligently on the project, he did not make any notable progress until after Niepce died in 1833. Sometime around 1837 and entirely by accident, Daguerre made two critical discoveries. He found that highly polished iodized plates could be used as a surface to capture images and that mercury vapor would develop the image. The mercury technique came about because Daguerre unknowingly broke a mercury-filled thermometer in a cabinet where he stored his plates. Today we would say the vapor fogged the plates. The effect gave Daguerre a useful clue.

Daguerreotype photography was the first method by which people could have their pictures taken. The exposure time was around one minute or less in bright sunlight. Lengthy by modern standards, it explains why so many early photographs show people with almost no expression. They had to sit perfectly still for a long time.

There were different ways to generate a daguerreotype. Here is one:

- Polish a silver-plated piece of copper.
- In a lighttight box, coat it with iodine vapor for 30 seconds.
- Using a camera, expose the plate in bright sunlight for 30 seconds.
- In a lighttight box, develop with vapors from heated mercury for 1 to 3 minutes.
- Fix the image in hyposulfite and water for 2 to 3 minutes.
- Protect the image by pouring on a gilding solution of 1 gram gold-chloride and 1/2 liter of water.

Since there was no negative, daguerreotypes could not be duplicated. Each photograph resulted in only one print. They were typically made in six standard sizes. The largest was 6-1/2 by 8-1/2 inches, and the smallest was 1-5/8 by 2-1/8 inches. The gilding solution gave some measure of protection to the highly polished image. The gold coating and the silver plating give daguerreotypes their metallic appearance.

Like Daguerre, telegraph inventor Samuel Morse (1791-1872) started life as an artist, primarily as a portrait painter. Before becoming wealthy with his 1840 invention, Morse met Daguerre while traveling in Europe. He learned how to take daguerreotypes and opened one of America's first photographic studios in New York City in 1838.

Soon after he made it practical, Daguerre turned over his photographic rights to the French government. They had offered a lifetime pension of 6,000 francs per year, equivalent to about $1,200 and a very comfortable amount. Daguerre retired to a country estate in 1840 and let others improve his process. He returned to painting for personal enjoyment and died at the age of 63.

The daguerreotype dominated photography for more than 10 years. But it was a blind alley that was put to rest by William Talbot's (1800-1877) invention of negative-positive photography. Daguerreotypes were fragile and could neither be duplicated nor enlarged. But deadend designs have often helped point inventors to a better direction. Elias Howe's 1846 sewing machine was also a blind alley. Its cumbersome horizontal needle could make about a dozen stitches before the fabric had to be repositioned. Thomas Edison's 1877 phonograph used cylinders instead of more practical flat records. The Wright Brothers 1903 airplane had rear-mounted pusher propellers with the flight control surfaces in front.

Beautiful daguerreotypes have had unusual staying power. Museum displays make them look as if they were exposed only yesterday. Photographer Matthew Brady (1823-1896) had studios in New York City and Washington, D.C. During the Civil War, he took countless poignant, troubling, and historic images. Many were daguerreotypes.

Daguerreotypes were commonly used for portraits during the mid-1800s. They were kept behind glass and often in velvet cases like this one because they scratched easily.

References

Daguerre by Beaumont Newhall, Winter House Publishers, 1971.
The History of Photography by Helmut Gernsheim, Thames and Hudson Publishers, 1969.
American Handbook of the Daguerreotype by S.D. Humphrey, Arno Press, 1973.

PUBLISHED SEMI-MONTHLY, AT THREE DOLLARS PER ANNUM, IN ADVANCE.

THE
DAGUERREIAN JOURNAL :
Devoted to the Daguerreian and Photogenic Art.

Also, embracing the Sciences, Arts, and Literature.

NOVEMBER 1, 1850.

NEW-YORK:
S. D. HUMPHREY, EDITOR AND PUBLISHER,
NO. 235 BROADWAY.

SINGLE NUMBERS TWENTY-FIVE CENTS   SUBJECT TO NEWSPAPER POSTAGE.

Title page of *The Daguerreian Journal*, printed in New York City. It was America's first photographic journal.

Michael Faraday

Born:
September 22,
1791, in
Newington Butts,
England

Died:
August 25, 1867,
in Hampton
Court, England

Electricity was the most mysterious technical force in the nineteenth-century world. It was invisible, poorly understood, and potentially harmful. The only kind of readily available electricity was static electricity. Incompetent individuals who used rotating equipment to produce it, often generated high voltages that shocked people. In the eyes of many, electricity was no more than a laboratory curiosity. But Benjamin Franklin (1706-1790) knew better. He was once asked of what good was electricity. The philosophical Franklin answered with a question of his own, "Of what use is a newborn baby?"

Alessandro Volta's (1745-1827) invention of the storage battery in 1799 went a long way toward convincing people that electricity might have future applications. His battery was the first practical source for current electricity. But most people remained unsure until a poorly educated, quiet technologist uncovered its major secrets. No one did more to advance the field of electricity than Michael Faraday. His two greatest accomplishments were inventing the electric motor in 1821 and the transformer in 1831.

Faraday's family had moved from northeastern England to a London suburb shortly before his birth. He was the third of four children. His father was a blacksmith and suffered from poor health. He could not work regularly and the family had only the bare necessities of life. Faraday later said his education consisted of "little more than the rudiments of reading, writing, and arithmetic." He was apprenticed to bookbinder George Riebau when he was 13.

Quick to learn the profession, Faraday developed considerable manual dexterity. Riebau was sensitive to Faraday's lack of education and allowed the young man to

The Science Museum / Science & Society Picture Library

read everything that passed through the shop. He enjoyed scientific books and the entries on electricity in the *Encyclopedia Britannica*. But one book in particular convinced him to work in the field of technology. It was *Conversations in Chemistry*, written in 1806 by Jane Marcet (1769-1858). Marcet was one of the very few authors at the time who wrote books aimed at young people. Her other diverse subjects were economics, philosophy, and Africa.

The City Philosophical Society consisted of a group of young men who met regularly to discuss scientific matters. Faraday joined in 1810 and his interest in technology quickly grew. One of Riebau's customers gave him tickets in 1812 to attend four public lectures by Britain's greatest chemist, Sir Humphry Davy (1778-1829). Faraday took careful notes at the lectures and wrote them out in his fine handwriting. He bound them in a book and presented them to Davy as a way of seeking employment. As it turned out, Davy needed an assistant and was impressed with the book. Faraday went to work for Davy as a chemical assistant at London's Royal Institution. The position provided an environment that allowed Faraday to develop his full potential and he stayed there the rest of his life.

Davy took an 18-month European tour beginning in 1813, taking Faraday with him. Faraday met such important people as Volta, Andre Ampere (1775-1836), Count Rumford (1753-1814), and others in the field of chemistry. Those contacts led to others and helped Faraday throughout his life to better define his research path. He started his career in the field of chemistry and his first important achievement was to liquefy chlorine in 1823. He discovered benzene in 1825 and was the first to describe the compounds of carbon. Faraday introduced such words as *electrode, anode, cathode, ion,* and *electrolysis.*

Experiments with chemicals led Faraday to electricity and magnetism. Hans Christian Oersted (1777-1851) in Denmark had shown in 1820 that electricity and magnetism are related. He placed a magnetized compass needle directly under a wire. When electricity passed through the wire, the needle moved slightly. Oersted's discovery was like a shot heard around the world. Technologists everywhere began investigating the field of electricity. It was the first time people took the subject seriously since Franklin had shown that lightning was electricity in 1752. Many were confused by electricity and magnetism, but not Faraday. An insightful experimenter, he made a device in 1821 that included a battery, a magnet, a dish, a small amount of mercury, and a wire hanging from a pivot. Once started in motion, the wire continued rotating around the magnet. Though it did not look like its modern counterparts, this was the first electric motor. It showed people that electricity could be used as a source of mechanical power.

The technical community was puzzled by Faraday's achievements. He was born poor, had minimal education, and did not understand advanced mathematics. Faraday

succeeded because he was the best experimental scientist the world had ever produced. To this day, he continues to receive that lofty praise. His laboratory skills were unmatched and in 1825, he became head of the Royal Institution. Faraday was primarily responsible for the institution's future success. The Royal Institution had been founded in 1799 by the American-born Count Rumford. It is a large building for scientific research only a few blocks from Buckingham Palace. Faraday had a small laboratory in the basement and an apartment on the third floor that he shared with his wife. He had married the former Sarah Barnard in 1821.

Much of Faraday's work dealt with chemistry, but he always returned to electricity and magnetism. Faraday had been trying to use electromagnetism to induce a flow of electricity and finally succeeded in 1831. He had a soft iron bar bent into a ring six inches in diameter. He wound the left side with many turns of wire, separating each

This replica of Faraday's first electric motor is at the Royal Institution. A cylindrical magnet is set in a beaker of mercury and a length of thin wire hangs from a pivot. A battery provides electrical power. Faraday started the wire swinging around the magnet and it kept rotating on its own.

Faraday induced a flow of electricity by pushing a cylindrical magnet through a coil of wire. This is the principle used by dc generators (dynamos).

This is one of Faraday's original iron-ring transformers. Each half is separately wound with wire. Connecting one set of wires to a battery induces electrical flow in the other set, which is read with a galvanometer.

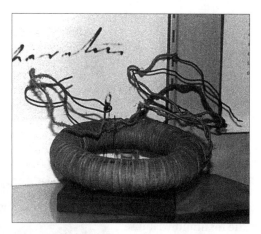

layer with string or fabric. The right side was similarly wrapped. Faraday connected the wire terminals from one side of the ring to a galvanometer. When the wire terminals from the other side of the ring were connected to a battery, the galvanometer needle swung toward full scale. Disconnecting the battery caused the galvanometer to swing in the other direction. He discovered that a current change in one coil produced a current flow in the other. It was the first transformer and many consider it Faraday's greatest contribution to technology. He called the effect *electromagnetic induction*, a term still in use.

Possessing great personal energy and expansive interests, Faraday had many different successes. He made an electrical generator in 1831. It was essentially a copper disc that rotated between the poles of a horseshoe magnet. Connections went from the edge and the center of the disc to a galvanometer. Faraday made his generator at about the same time Joseph Henry (1797-1878) made one in America. Faraday also produced electrical current by pushing a magnet into a coil of wire connected to his galvanometer. He developed the idea of magnetic fields and lines of force. He designed a device for measuring electric charge and called it a *voltameter*. Faraday discovered the two laws of electrochemistry and the effect of magnetic forces on light beams. He worked on alloy steels and optical-quality glass. He taught chemistry for more than 20 years at the Royal Military Academy in east London. The list goes on and on. To compensate for his weakness in mathematics, Faraday developed a sense of how things should occur. He organized his

thoughts in graphic terms. His application of magnetic lines of force is still used today.

Faraday and his wife had no children, but several nieces lived with them and their apartment was always full. They frequently attended the theater. Faraday suffered a physical breakdown in 1841 that may have been due to mercury poisoning. He fully recovered after three years of rest and continued working with electricity's effect on optics. He retired to a house provided by Queen Victoria in 1858. Victoria's husband, Prince Albert, had suggested the offer of housing. Prince Albert was the driving force behind the first world's fair in London in 1851. When Faraday died in 1867, his funeral was as simple and private as his life had been.

Faraday was gentle, calm, and deeply religious. He had little interest in money or personal recognition. He turned down many awards and honors, including the presidency of the prestigious Royal Society and a British knighthood. While others were feverishly working to be the first at some aspect of electricity, Faraday created Christmas Lectures for young people. Beginning in 1826 at the Royal Institution, the artful Faraday dazzled a young audience with technical experiments and colorful stories. He did much to popularize science and technology during his era. The Christmas Lectures continue to this day and are now televised. Faraday is the only person honored by the naming of two units of electrical measurement. The *farad* is a unit of capacitance and the *faraday* is a unit of charge.

References

Michael Faraday of the Royal Institution by Ronald King, Royal Institution Press, 1973.

Faraday Rediscovered edited by Frank Gooding, American Institute of Physics, 1989.

Michael Faraday by L. Pearce Williams, Basic Books Publishers, 1965.

The Engines of our Ingenuity, "Michael Faraday," Number 741, by John Lienhard, University of Houston, circa 1995, broadcast on National Public Radio.

Charles Babbage

The Science Museum/Science & Society Picture Library

Born:
December 26,
1791, in
Teignmouth,
England

Died:
October 18,
1871, in
London, England

When technologists of the mid-twenty-first century look back at twentieth-century inventions, it is impossible to predict which will emerge as the single most important. But, without doubt, the digital computer will be in the running. These electronic workhorses have a very wide range of uses. They control factory machine tool operations, monitor hospital patients, and, under the command of pilots, fly commercial airplanes. They function on a more personal level by helping individuals write letters and connect to the world through the global Internet system. They also entertain people with video games. The digital computer seems to have endless potential.

But the digital computer was not a twentieth-century invention. It was born in England, before the reign of Queen Victoria. All modern computer pioneers credit Charles Babbage as being the first in their field. He designed all the essential parts of a digital computer in 1834. Working in an era before the use of electricity, he planned to operate his gear-driven device with a steam engine.

Babbage came from a wealthy banking family and never had to work for his living. Both he and his sister, Mary Anne, received excellent primary education at boarding schools. Their parents' only surviving children, they shared a close lifelong relationship. Self-educated in mathematics from an early age, Charles Babbage knew more than his college instructors at Cambridge University. After graduating in 1814, he married Georgiana Whitmore. In 1815, the couple moved to London, where they lived for the rest of their lives.

Babbage's father had helped finance business ventures that included transportation by ships. Navigators of the era used printed navigation tables to determine safe sea routes. The tables were both calculated and typeset by hand, which produced some errors. It was estimated the tables may have contained thousands of undiscovered errors. When, on occasion, a ship did not return from a voyage, it was common to assume that incorrect navigation tables might have caused it to hit rocks and sink. Babbage grew fascinated with the possibility of making a machine that could calculate and print navigation tables, which would eliminate errors in calculation.

In the early 1820s, Babbage started working on what he called a *difference engine.* This was actually a calculator that worked by adding and subtracting numbers over and over again. A *difference* is the amount by which one number varies from another. The word *engine* simply meant a device or machine. Babbage spent most of the next 10 years designing his difference engine. His plans called for 25,000 carefully made parts. The brass machine would be 6 feet tall, 6 feet long, and almost 3 feet deep.

Thousands of gears, levers, cams, and linkages had to be made to exceedingly close tolerances. Babbage had his own machine tools and used them regularly. He even had a forge. But there was no way he could have made the necessary high-quality parts for his difference engine. Such a large and demanding project required the services of talented machinists. The best available worked for Joseph Clement (1779-1844) in Newington.

Babbage contracted with Clement, using his own money to finance the work. He soon saw that it would cost far more than he had expected. Babbage approached the directors of the Civil Contingencies Fund of the British government for a grant to continue the project. At the time, government funding of technical research was a new idea, and many politicians disapproved of Babbage's request. Other opinions prevailed, however, and the money became available in 1823.

Clement's machine shop worked on Babbage's calculator for four years. In later years, Joseph Whitworth (1803-1887), the person who standardized threaded fasteners, would proudly relate that he personally made many of the critical parts. Unfortunately, Babbage's difficult personality upset the gifted machinists with whom he worked. It has been speculated that the unexpected deaths of his wife, father, and two children in 1827 had a devastating and permanent effect on Babbage's temperment. For whatever reason, his calculator was never completed. He had spent about £6,000 of his own money and £17,000 of government money on the project.

Babbage's difference engine was built in 1832 and is the earliest known automatic calculator. It was the first mechanism that included mathematical logic. It operated by turning the crank at the top.

To show some progress to British officials, he had a one-seventh size demonstration model assembled in 1832. It was 72 by 59 by 61 centimeters, and it had 2,000 parts. Powered by a hand crank, it operated faultlessly and was the earliest automatic calculator. For the first time, mathematical logic was a basic part of a mechanism. The beautifully made brass assemblage of gears, levers, cams, and rods has survived the years and is on display at London's Science Museum. It remains the finest example of nineteenth-century precision machining and is the most celebrated icon from the earliest history of computing.

Babbage's difference engine processed numbers the only way it could, by adding or subtracting them in a particular sequence. After designing it, Babbage thought about a more general purpose machine—one that would allow the user to determine all the mathematical tasks. He called this an *analytical engine*. Today it would be called a *digital computer*. The five processing sections Babbage developed in 1834 are the same as those used by modern computers:

- input device—punched cards,
- arithmetic unit—the calculating section,
- programmable unit—the part that determines the calculating sequences,
- memory—the gear positions, and
- output mechanism—printer, plotter, or card puncher.

The idea for the punched cards came from Joseph Jacquard (1752-1834), who used them to control weaving patterns in his French textile looms. Babbage used one style of card to specify arithmetical operations like adding, subtracting, multiplying, or dividing. Larger cards had data and controlled where the answers were placed in the *store*, Babbage's word for memory. He developed conditional branching, looping, and subroutines. All are important operations in modern computer programs. Although Babbage worked on it for over 30 years, he published little technical detail about his computer.

A teenager interested in Babbage's work attended an 1833 presentation he made in Turin, Italy. Ada Lovelace (1815-1852) was a mathematically proficient and dynamic young woman. (See pages 106-108.)The daughter of esteemed British poet Lord (George Gordon) Byron, she assumed public relations duties on Babbage's behalf. Babbage kept poor records and most people were unaware of his progress. He asked Lovelace to translate into English his Turin lectures, which had been transcribed and printed in French. She added her own insights regarding Babbage's invention, and the final document was three times the size of the original. It is only from that publication that the world found out the details of Babbage's computer. Lovelace and Babbage were close friends until Lovelace died at the young age of 35.

Babbage made many precise cardboard templates of the estimated 200,000 parts for his analytical engine. Its construction turned out to be vastly more challenging than that of his difference engine. Babbage's personality again got in the way of his invention, and the analytical engine was never built. A crude iron demonstration model was under construction at the time of his death. There is little doubt that Babbage could have constructed a legitimate computer, if only he had had the financial resources. With a bit more government funding, he could have changed the course of history. He was twice visited by the American Joseph Henry (1797-1878), the first director of the Smithsonian Institution. Henry wrote that "[Babbage] more, perhaps, than any [person] who has ever lived, narrowed the chasm separating science and practical mechanics."

Babbage and his wife had eight children. Four survived infancy. He did not remarry after his wife's death and devoted all his energies thereafter to his calculator and computer. To commemorate the 200th anniversary of his birth (1991), the Science Museum in London commissioned the construction of a Babbage-designed difference engine using nineteenth-century materials and machining methods. The materials used were cast iron, steel, and bronze. Modern manufacturing techniques were used, but care was taken to make parts no more precisely than Babbage could have. The difference engine cost a half million dollars to build, has 4,000 parts, weighs three tons, and is accurate to 30 decimal places.

Babbage saw too far beyond his time and his efforts were never seriously appreciated. He enjoys more widespread esteem today than he did during his lifetime. His computer work was an isolated episode in technology that did not resume development until the late 1930s. Because Babbage had such a difficult personality and worked in an unappreciated area of technology, many ordinary citizens belittled him. Museums rejected his hardware following his death. Most of his many small models of gear trains and cam arrangements were broken up for scrap. Babbage's grandchildren used some as toys until they were damaged beyond repair and discarded. Today, museum curators would kill for mere fragments of those models that their predecessors refused over a century ago. Charles Babbage was the ultimate "computer pioneer."

Babbage never completed the "analytical engine," his name for a digital computer. This one was built by the Science Museum in London in 1991 to commemorate the 200th anniversary of Babbage's birth. It works just as he had designed and is accurate to 30 decimal places. The mechanical computer operates by turning the crank on the side.

References

Charles Babbage and His Calculating Engines by Doron Swade, Science Museum Publication, 1991.

Portraits in Silicon by Robert Slater, MIT Press, 1989.

The Making of the Micro by Christopher Evans Van Nostrand Publishers, 1981.

William Henry Fox Talbot

Born:
February 11, 1800,
in Melbury,
England

Died:
September 17,
1877, in Lacock
Abbey, England

Songwriter and singer Jim Croce (1943-1973) recorded a lovely song just before he died in an airplane crash. Titled "Time in a Bottle," it included the phrase: "If I could save time in a bottle, the first thing that I'd like to do is save every day." Although it's not possible to literally accomplish what Croce's popular song suggested, people can figuratively get close. Photography lets us stop time by recording life as we see it. The technology has been applied by the technical world in a variety of ways. Satellite photography helps road builders find the best route in undeveloped areas. Companies employ technical photographers to show customers and suppliers the details of a project's progress. The quality of a building's insulation can be evaluated by infrared photography.

Like almost all technologies, photography did not spring fully formed from the mind of a single person. It was an evolutionary process marked by milestones. In 1727, German scientist Johann Schulze discovered that silver salts turn dark when exposed to light. Joseph Niepce (1765-1833) photographed the first crude image in France in 1822. His countryman Louis Daguerre (1787-1851) developed the first practical method for producing permanent images in the 1830s. (See pages 61-63.) Daguerreotypes were particularly popular for family portraits. But they were a one-time image, like an oil painting is. People could not make additional copies from a daguerreotype. W. H. Fox Talbot established photography's most important milestone when he invented the method now used all over the world. Talbot was the first person to produce a positive image from a negative.

Talbot was born near Dorchester in

The Science Museum/Science & Society Picture Library

southern England. His father was a British army officer who often lived beyond his means. He died when Talbot was an infant. His mother was a member of an aristocratic family, but they had fallen on hard times. They owned a large estate but were forced to rent it to a wealthy member of Parliament to pay their bills. Although not poor in the traditional sense, Talbot and his mother often moved from the home of one friend to another. His mother married admiral Charles Feilding when Talbot was four and he settled down to a more stable childhood. He grew up with two half-sisters, Caroline and Horatia. Talbot had a close lifelong relationship with them and his stepfather.

Talbott attended boarding schools, where his grades always put him near the top of his class. His methodical approach to problem solving impressed his teachers. He entered Trinity College at Cambridge University and graduated in 1821 after studying classical languages and mathematics. His family was now financially secure, and Talbot did not have to work for a living. He was in line to inherit the ancestral home at Lacock Abbey near Chippenham, though it was still under lease. Talbot spent his time traveling while he studied his favorite subjects of mathematics, astronomy, optics, and photography. He published more than 50 scientific papers and eventually took out 12 patents.

Talbott's friends included Charles Wheatstone (1802-1875), who invented the first practical telegraph and Charles Babbage (1791-1871), who worked on an early computer. (See pages 67-69 and 76-78.) Another friend, astronomer John Hershel (1792-1871), later suggested the word *photography*. It comes from two Greek words meaning "light" and "painting." The Royal Astronomical Society selected Talbot for membership in 1822 as did the prestigious Royal Society in 1831. Very few people were accepted into the Royal Society at the relatively young age of 32. But Talbot's insightful work with math caught the notice of the members. He had written extensively on the subject for several years.

Lacock Abbey finally became available in 1827. Talbot moved in with his mother, stepfather, and half-sisters. He remained there for the rest of his life. He married Constance Mundy in 1832 and they eventually had four children. Talbot became interested in photography during their honeymoon in Italy. He used the camera obscura technique to make drawings of Lake Como. A lens focused the scene onto a sheet of paper and Talbot sketched the image poorly with pen and ink. Realizing that he lacked artistic ability, Talbot wondered if the image could imprint itself onto a special paper. When he worked out a method, his wife became almost as interested in the subject as he was. She later became the world's first woman photographer.

As scientists and technologists had done for centuries, Talbot built on the work of others. He studied their experiments to produce his own light-sensitive materials. In a completely darkened room, he brushed ordinary paper with a solution of silver and iodine compounds. He placed the dried paper, with delicate ferns and flower petals on top of it, under glass. Exposing these

items to sunlight produced dark reversed images that are now called *photograms*. No camera was involved and Talbot produced only a silhouette. He took that paper negative, laid another sheet of sensitized paper on top of it, and took it outdoors into the sunlight. The sandwiched combination took more exposure time because light had to penetrate the paper negative. He used 100 grains of silver nitrate dissolved in six ounces of water as a developer and saltwater as a fixer. Talbot's longtime assistant, Nicholaas Henneman helped him in his work.

The next step involved using the procedure with a camera to record shades of gray. Talbott ordered the construction of a light-tight box. The box had a lens in front and a sliding back for focusing. In August 1835, Talbot placed a sheet of light-sensitive paper inside the camera. He aimed the lens at a latticed window in Lacock Abbey and opened the shutter for several minutes. He processed the paper negative and then made a print from it with sunlight. It was the first negative-to-positive photograph ever made and was surprisingly sharp. On one print, Talbott wrote, "When first made, the squares of glass [in the window], about 200 in number, could be counted with the help of a [magnifying] lens."

Talbot originally called his process *photogenic drawing*, but changed its name to *calotype* from the Greek *kalos*, which means "beautiful." He improved the process and took out a patent in 1841. His developer was a combination of silver nitrate and gal-

An early camera used by Talbot, 1840. It had a simple lens with an inspection hole in the upper right to check that the image was centrally located on the sensitized paper and correctly focused.

lium, which increased the sensitivity of the paper. Talbot fixed the image with sodium thiosulfate, the same chemical used today. Although calotypes were grainier than daguerreotypes, they could be used to make multiple copies and larger prints. Talbot knew that natural light might not always suffice for exposing sensitized paper and was the first to experiment with flash photography. He fastened a piece of newspaper to a spinning disk in 1851. Then he discharged a spark to expose his photographic emulsion. A portion of the paper was readable in the final print.

Talbot himself took hundreds of photographs, often with a camera of the type shown in the photograph of him at the beginning of this chapter. His favorite produced a 4-1/16-inch square negative. A book

Copy of the earliest paper photograph in existence, made in 1835. Celluloid was not used as a carrier for light-sensitive photographic emulsion until after 1870.

he wrote and published in 1844 titled *The Pencil of Nature* was the first book illustrated with photographs. Talbot established a photographic studio in Reading, but calotypes had only limited commercial success. After 1851, more detailed images were possible with wet solutions on glass plates. But the calotype process was instrumental in Talbot's receiving the 1842 Rumford Medal, a precursor to the Nobel Prize, from the Royal Society.

A man of diverse interests, Talbot was elected to the British parliament in 1832. He held patents for internal combustion engines and the chemical plating of metal. He had an appreciation for nature and many of his photographs involved natural settings. He also enjoyed traveling and reading about history. Talbot was fascinated when the Rosetta Stone was used to decipher Egyptian hieroglyphics in the 1820s. Possessed of a natural flair for languages, he translated written works from Assyrian and other ancient languages. He wrote five books on the subject and had 62 articles published. Talbot died at his home at the age of 77.

Talbot not only developed photographic technology, he used it brilliantly. His images had a modern character to them. They were not stiffly posed indoor portraits like daguerreotypes became. His pictures were often interesting candid shots of people in typical outdoor settings. Talbot tried to include natural elements in his pictures. He wrote poetically and philosophically about the potential of photography. Few early photographs are more compelling than those he took in the mid-nineteenth century.

References

Fox Talbot by John Hannavy, Shire Publications, 1976.

Fox Talbot and the Invention of Photography by Gail Buckland, David R. Goodine Publisher, 1980.

The History of Photography by Helmut Gernsheim, Thames and Hudson Publishers, 1955.

Catharine Esther Beecher

Born:
September 6, 1800, in East Hampton, Long Island, New York

Died:
May 12, 1878, in Elmira, New York

The word *pioneer* often brings to mind an eighteenth-century American farmer or hunter. These people were among the first to help tame a new country. The word also can be used in a more general way, as in describing an early piece of hardware. The aluminum *Pioneer Zephyr* was a beautiful, streamlined diesel-electric train that did much to rekindle America's interest in rail travel during the Great Depression. It made an unprecedented 1,015-mile nonstop run from Denver to Chicago in 1934.

It is not unusual for the word "pioneer" to also describe a person's professional activities. Ellen Swallow Richards (1842-1911) was the first American woman to earn a technical degree. It was an 1873 B.S. in chemistry from the Massachusetts Institute of Technology. Richards was also the first woman member of an engineering society. She became a member of the American Institute of Mining Engineers in 1879. Both were major pioneering professional accomplishments.

Another woman made a significant impact on the technical world in the early 1800s. Catharine Beecher was probably America's first female technologist. Her technical accomplishments rest on a series of innovative house designs she developed in the 1840s. She saw the house as a workplace for many women. Existing houses were often poorly designed, which resulted in people wasting time or energy during the completion of various tasks. Beecher designed efficient, moderately priced houses for America's expanding middle class. Architect, designer, drafter, and efficiency analyst, Beecher was a pioneer in establishing a place for women in the field of technology.

The eldest of eight children, Beecher was born near the tip of Long Island. Her father was Lyman Beecher, a minister who gained

Rendering by Tim Harmon

fame in the early 1800s for his work with moral reform movements. Her mother, the former Roxana Foote, assumed responsibility for Beecher's early education. The family moved to Litchfield, Connecticut, when Beecher was 10 years old. There, her famous sister Harriet Beecher Stowe was born (1811-1896). Stowe wrote the controversial book *Uncle Tom's Cabin* in 1852. When introduced to Stowe, President Abraham Lincoln supposedly said, "So you're the little woman who wrote the book that started this great war," referring to the Civil War.

Beecher's mother died in 1816. As the eldest, Beecher felt a responsibility to help raise the seven younger children. She was a particular favorite of her father but readily accepted the elegant Harriet Porter, whom her father married in 1818. The large family continued to be primarily raised by Beecher. In this setting, Beecher received her initial exposure to the home as a workplace.

Beecher briefly attended a finishing

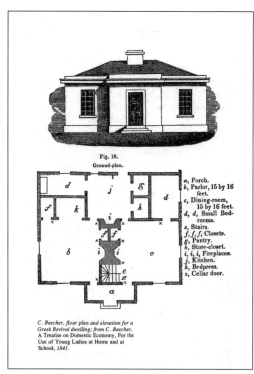

Fig. 18.
Ground-plan.

a, Porch.
b, Parlor, 15 by 16 feet.
c, Dining-room, 15 by 16 feet.
d, d, Small Bedrooms.
e, Stairs.
f, f, f, Closets.
g, Pantry.
h, Store-closet.
i, i, i, Fireplaces.
j, Kitchen.
k, Bedpress.
z, Cellar door.

C. Beecher, floor plan and elevation for a Greek Revival dwelling; from C. Beecher, A Treatise on Domestic Economy, For the Use of Young Ladies at Home and at School, 1841.

From A Treatise on Domestic Economy, by Catharine Beecher, 1841

One of Beecher's floor plans for a relatively small two-bedroom house with about 1,200 square feet of floor area. Note the central location of the kitchen ("j") and the small bedrooms ("d"). Both were characteristics of Beecher's designs.

school in Litchfield and received part of her education at home. But much of what she learned came through self-study. She taught herself mathematics, drawing, music, Latin, and philosophy. Her youth was happy and she had a zest for life that stayed with her all her years. Beecher's upbringing in a small New England town was very pleasant and almost idyllic. It laid the foundation for her future campaigns to improve household conditions for American women.

Like all young people, Beecher wanted to be self-sufficient. In 1821, she took a job in New London as a teacher. Teaching was one of the few professions open to women in the early nineteenth century. Beecher met Alexander Fisher, a young mathematics professor from Yale University. They became engaged in 1822, but he died during a sea voyage the next year. His ship, the Albion, was lost in a storm off the Irish coast. Beecher was devastated and never fully recovered from the tragedy. She never married.

Over the next few years, she worked at several schools and often helped her father in his religious work. Beecher lived in Cincinnati, Boston, Hartford, and other cities. Her defining moment in technology seems to have occurred while she was teaching in Hartford in the 1820s. An architect provided drawings of a new facade for her school. Beecher was impressed with his effort and she went on to formulate some ideas of her own. She slowly developed revolutionary floor plans for apartments, churches, schools, and houses. She particularly enjoyed working on house plans.

Beecher visualized an expanding middle class as the Industrial Revolution took hold in America. Her designs appealed to people interested in purchasing moderately sized,

single-family dwellings in urban areas. The housing market had typically focused on apartments and larger luxury houses. Much of Beecher's work emphasized improved kitchen designs. Kitchens had often been located outside or in a separate building. The lack of adequate urban fire departments made people very concerned about house fires.

As is often the case today, nineteenth-century kitchens frequently served as a focal point of activity. So Beecher located her kitchens near the center of the house. They typically had doors into both the living room and the dining room. Her kitchens had single-surface work spaces and ventilating equipment. Other rooms were more compact and often included built-in closets, cabinets, and shelves. Beecher recommended efficient use of space, such as through hassocks that stored knitting materials and removable screens that hid extra beds. People accepted Beecher's designs because the exteriors of her houses looked similar to those of other houses. From the outside, they did not appear as revolutionary as they truly were.

Just as Karl Benz (1844-1929) was not alone in working with motor cars, Beecher was not the only person developing designs for houses. (See pages 148-150.) But like Benz, she was in the first wave of her field and had many innovative ideas based on personal experience. Male architects did not typically share that advantage. With each house design, Beecher provided efficient methods to complete household tasks. Her ideas were similar to those of Frederick Taylor (1856-1915), who worked with factory efficiency. But Beecher preceded Taylor by about 50 years. Her writings also addressed such modern topics as building model homes for inspection, use of various heating fuels, and air-tight stoves. For all her influence, it was not until 1977 that her work as an architect had its first major museum showing.

Beecher publicized her ideas through books, pamphlets, and journal articles. Much of her income came from her skill as an author. She wrote six major books, one of which went into 15 editions. Another, written with her sister Harriet Beecher Stowe, The American Woman's Home, is available as a reprint. Ever an education champion, Beecher collaborated with William McGuffey

Although the original Beecher house no longer exists, a sign in Litchfield, Connecticut, commemorates the family's homesite as: "Homesite of Lyman Beecher / Birthplace of Henry Ward Beecher, Harriet Beecher Stowe."

(1800-1873) in writing his *Eclectic Fourth Reader* (1837) for grade schools. Beecher advocated for higher education for women and helped establish colleges in the western reaches of the country. She involved herself with organizing women's colleges in Illinois, Iowa, and Wisconsin. A woman of great charm and sparkling wit, she was also a physical fitness advocate. Beecher considered physical education a vital part of any educational program and included it with musical accompaniment. That was a most modern instructional method. Beecher moved to Elmira, New York, in 1877 to live with a half-brother. She died shortly afterward at the age of 77.

The nineteenth and early twentieth centuries were challenging times for women with technical skills. Technology was dominated by men. Many women found it less troublesome to enter the field by assisting or continuing the work of their husbands. Ellen Swallow Richards is one example, as is Lillian Gilbreth (1878-1972), who worked with factory efficiency. Other women forged ahead on their own, without husbands. They in-

cluded Alice Hamilton (1869-1970), who established the field of industrial medicine. Astronomer Maria Mitchell (1818-1889) was America's first famous woman scientist. (See pages 115-117.) Like Hamilton and Mitchell, Beecher had no husband to assist her in her chosen profession.

Beecher's work came during a time of great change for women in all aspects of life. She was liked by some groups and disliked by others. The advancement of women's rights did not interest her, for example. Some early feminists could not accept the identification of women with the home, as Beecher did. Others viewed her architectural designs as the height of efficiency. All that aside, in the field of technology, Catharine Beecher was a first. She was the most influential domestic designer of the nineteenth century. This talented and indomitable woman deserves recognition for helping to open new professional paths for women. Beecher was a genuine pioneer.

References

Women in American Architecture: A Historic and Contemporary Perspective, edited by Susana Torre, Watson-Guptill Publications, 1977.

Pioneers of Women's Education in the United States, edited by Willystine Goodsell, McGraw-Hill Publishers, 1931.

The American Woman's Home, by Catharine E. Beecher and Harriet Beecher Stowe. Reprint by Arno Press, 1971 (originally published in 1869).

The drawing below of an efficient kitchen, is part of a display at the Eli Whitney Museum in Hamden, Connecticut. It shows a perspective of the type of kitchen that Catharine Beecher designed.

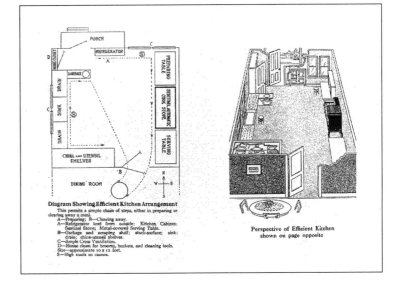

Diagram Showing Efficient Kitchen Arrangement
This permits a simple chain of steps, either in preparing or clearing away a meal.
A—Preparing; B—Clearing away.
A—Refrigerator iced from outside; Kitchen Cabinets; Sentinel Stove; Metal-covered Serving Table.
B—Garbage and scraping shelf; stack-surface; sink; drain; china-utensil shelves.
C—Ample Cross Ventilation.
D—House closet for brooms, buckets, and cleaning tools.
Size—approximate 10 x 12 feet.
S—High stools on casters.

Perspective of Efficient Kitchen
shown on page opposite

Charles Wheatstone

Born:
February 6, 1802, in Gloucester, England

Died:
October 19, 1875, in Paris, France

It is common for an inventor to stay with one technology and become an expert in that particular field. But some develop several different technical accomplishments. Joseph Swan (1828-1914) of England invented an incandescent lamp, photographic paper, and synthetic fibers. Gustave Eiffel (1832-1923) of France designed and supervised the construction of the Eiffel Tower. He also worked on aeronautics and constructing the Panama Canal. American-born Benjamin Thompson (1753-1814) established the field of ballistics and improved heating and cooking equipment. He also built a large city park in Munich, Germany, named the English Garden. (See pages 34-36, 130-132, 136-138.)

More unusual was Austrian-American Michael Pupin (1858-1935). He invented telephone and x-ray equipment and won a 1924 Pulitzer Prize for his autobiography. Charles Wheatstone also had diverse accomplishments. Trained as a musician, Wheatstone invented the small bellows-operated concertina and the stereoscope. He devised a secret code used to rescue a future American president. Wheatstone had a working telegraph system five years before Samuel Morse (1791-1872). But he is best known for an electrical circuit that accurately measures resistances.

Wheatstone was born about 100 miles northwest of London. His father was in the retail music trade and moved the family to London when his son was four. The elder Wheatstone taught music, and he made and sold instruments. The home environment introduced the young Wheatstone to music, sound transmission, and wave propagation. He received his primary education at a private school and was apprenticed in 1816 to an uncle in the music business. He had no formal technical instruction.

The Science Museum / Science & Society Picture Library

After his uncle's death in 1823, Wheatstone and his brother William took over the business. They specialized in making flutes and other instruments that used moving air. Wheatstone invented and patented the concertina in 1829. His company had others produce the parts, which they assembled and sold as completed concertinas. Wheatstone was an experimenter at heart. He built a small pipe organ to analyze air columns. He wanted to understand the properties of tone in terms of vibration. He spent much of his time trying to send sound through solids. He built a small harp he called an "enchanted lyre." He suspended the harp from a ceiling with a wire the "thickness of a goose quill." A piano on an upper floor was attached to the wire and caused the harp to make music when the piano was played. Wheatstone worked at sending music and speech over considerable distances. But he concluded that electrical methods were more promising than mechanical methods. He began investigating the field of electricity but did not neglect his other varied interests.

Wheatstone measured the speed of electrons in a conductor and tried to slow down a spark. He examined light given off by burning metals, a practice we now call *spectrum analysis*. He invented a small telescope-like polar clock that used the sun and a prism to indicate the time. He invented the stereoscope in 1832. Resembling a pair of binoculars, it allowed a person to view two images at the same time in a way that made them appear three dimensional. Wheatstone occasionally wrote technical papers and made presentations. In 1834, he was appointed professor of experimental philosophy at King's College in London. At first a part-time position, it developed into a full-time one. Wheatstone married Emma West in 1847 and they had five children.

Because Wheatstone was so shy that he rarely appeared in front of a class, he never did much teaching. As a youngster, he qualified for a French award. But he forfeited the prize because he was too bashful to recite a speech in front of an audience. In his adult years, he spent most of his time with laboratory work.

Stringing several miles of wire in the college corridors, Wheatstone experimented with different telegraph methods. He teamed up with William Cooke (1806-1879) to invent a practical system. Michael Faraday (1791-1867) introduced the two men. Cooke was a professor of anatomy at Durham University and had no formal technical education. His travels made him aware of the value of instantaneous, long-distance communication. Since he was having little success trying to achieve it, Faraday advised him to contact Wheatstone. The two pooled their ideas and together invented an indicating telegraph in 1837. Instead of using dots and dashes, their telegraph used five needles to point out specific letters on a diamond-shaped lattice. Cooke and Wheatstone received a joint patent in 1837. Theirs was the first practical telegraph.

The indicating telegraph used six wires, one for each needle, and a return wire. Sending a letter, such as "F," caused two needles to deflect slightly. They would point in the direction of two lines. The letter "F" lay at their intersection. Wheatstone and Cooke established the Electric Telegraph Company. Their system linked the port city of Liverpool with the manufacturing city of Manchester in 1839. This was the world's first public telegraph service. Operators required little training and could send about 22 words per minute.

Railways showed particular interest in rapid communication. George Stephenson (1781-1848) built the first major line in 1830 between Liverpool and Manchester. Early railway lines used only a single set of tracks and it was necessary to communicate about late departures or damaged equipment. British railways adopted the Wheatstone-Cooke system extensively. By 1852, more than 4,000 miles of telegraph lines were in use throughout the country. But Wheatstone and Cooke had a personal conflict and separated in 1845.

Wheatstone went on to develop another type of indicating telegraph that he called the *ABC*. Using Morse code took great skill and Wheatstone felt the future of telegraphy lay in unskilled operators. It was a reasonable assumption based on the methods used by factories. In Wheatstone's ABC system, the

Wheatstone ABC telegraph of 1860. Received letters were read on the top dial. To send a letter, the operator pressed the proper button and turned the crank.

Wheatstone and Cooke five-needle indicating telegraph of 1837. For each letter received, two needles deflected a small amount and pointed at a common letter. The top letter, for example is "A." An operator read that letter when both the far left and far right arrows pointed toward "A."

Wheatstone telegraph transmitter of 1870. Paper tape was prepunched according to standard Morse code. This clockwork mechanism then automatically sent the message at a high transmission rate.

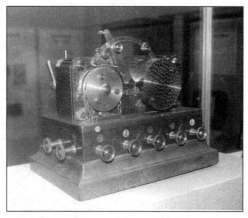

telegraph operator pressed a button corresponding to the desired letter. Cranking a handle sent electrical impulses. A dial rotated at the destination and indicated the transmitted letter. Wheatstone intended the system for use on private business lines, similar to a modern telephone. Like Wheatstone's joint design with Cooke, the ABC telegraph lost out to the simpler, single-wire Morse equipment.

In 1860, Wheatstone invented a printing telegraph. It was based on the Morse code, which had gained universal popularity by that time. Operators punched Morse code dots and dashes onto narrow paper tape. The tape then rapidly ran through a clock drive mechanism and sent a high-speed signal along the wire. Thin metal fingers sensed holes in the paper and opened or closed relay contacts as indicated.

Wheatstone is most commonly identified with the Wheatstone Bridge for measuring resistance, a circuit he did not develop and one for which he never claimed credit. It was actually devised in 1833 by his close friend Samuel Hunter Christie (1784-1865). Wheatstone's name was associated with the

bridge circuit because he often used it in his experiments. He called it a "differential resistance measurer" and said in 1843, "To Mr. Christie must, therefore, be attributed the first idea of this useful and accurate method of measuring resistances."

A tireless worker, Wheatstone also had other notable accomplishments, including invention of the variable resistor in 1843. He improved dc generators by making one that delivered a more constant current. In 1844, he was the first to experiment with underwater cables. He invented the chronoscope, an instrument to measure small time intervals. Wheatstone discovered in 1867 that the earth's magnetic field was strong enough to start a generator. Previously, batteries were necessary. Queen Victoria knighted Wheatstone in 1868 and he received at least 34 awards of distinction. Wheatstone died while on a business trip to Paris in 1875.

Of all Wheatstone's accomplishments, perhaps the most colorful was his development of a secret code, or *cipher*. Cryptography was a hobby of his and he came up with a simple and reasonably secure cipher in 1854. It used a five-inch by five-inch grid of 25 letters of the alphabet. The letters "i" and "j" were assumed to be interchangeable. The cipher was named after Wheatstone's friend and look-alike Lyon Playfair (1818-1898). (The men looked so similar that even Wheatstone's wife once confused the two.) Almost a century later, Lt. John F. Kennedy's torpedo boat was destroyed in World War II combat in the South Pacific. Australian coast watcher Lt. Arthur Evans received a Playfair-enciphered radio message in August 1943. It read, "PT boat 109 lost in action in Blackett Strait two miles SW Meresu Cove. Crew of 12. Request any information." Kennedy and his crew were rescued five days later.

References

Pioneers of Electrical Communication by Rollo Appleyard, Books for Libraries Press, 1930.

The Codebreakers by David Kahn, Macmillan Publishers, 1967.

Electric Telegraphs, William Cooke and Charles Wheatstone, British Patent 7390, issued 1847.

Making of the Modern World edited by Neil Cossons, Science Museum Publications, 1992.

A simplified example of the Playfair cipher developed by Wheatstone in 1854:

Code word is "machine." Letters "i" and "j" are interchangeable.

```
M A C H I
N E B D F
G K L O P
Q R S T U
V W X Y Z
```

The message to be sent is "Meet Alex at museum"
1. Break message into pairs of letters: ME ET AL EX AT MU SE UM
2. Each letter pair defines a rectangle.
3. The enciphered text uses the letters at the other corners of the rectangle. For example, ET becomes DR.
4. Plain text: ME ET AL EX AT MU SE UM
   Cipher text: AN DR CK BW HR IQ RB QI
5. Send enciphered message: AN DR CK BW HR IQ RB QI

John Ericsson

National Portrait Gallery, Smithsonian Institution

Born:
July 31, 1803, in Langbanshyttan, Sweden

Died:
March 8, 1889, in New York, New York

No invention of the Industrial Revolution was as truly revolutionary as the steam engine. It was the first form of power that did not directly rely on natural forces, such as wind or water, or on muscle power. The first steam engines had one large vertical cylinder and were as big as a garage. Technically minded people were fascinated by their operation and visualized making steam engines small enough for use in transportation. The first steamship to make a transatlantic voyage was the *Savannah*. The American-built vessel sailed from Savannah, Georgia, to Liverpool, England, in 1819.

Nineteenth-century steam engines operated without modern safety features and explosions were not uncommon. To address the problem, a Swedish-born American designed an external combustion engine that used hot air in a manner similar to that in which steam was used in a steam engine. John Ericsson built many such engines. They were used to operate small water pumps, printing presses, and lighthouse foghorns. Ericsson also built a huge hot-air engine that powered a 260-foot ship from New York to Washington in 1853. But he may be best remembered as the designer of the steam-powered *Monitor,* used by the Union forces during America's Civil War.

Ericsson was the youngest of three children. His older brother was destined to become supervisor of the Swedish railway network. His father was a college-educated mining engineer who provided his family with solid financial security. Ericsson's mother was highly intelligent, and she made sure her children had all available educational advantages. Ericsson's schooling came from occasional tutors and home instruction. He enjoyed sketching the hard-

ware used in coal mines and showed an aptitude for technology. His father allowed him to work with mining equipment, and at 13 the younger Ericsson developed a minor improvement to a water pump. This so impressed a visiting Swedish admiral that he offered Ericsson a position as a cadet in the Corps of Engineers. Accepting the offer, Ericsson worked on government canals and received advanced instruction in mathematics, drafting, and the English language. He showed particular skill on the drawing board, and became an army topographic officer in 1820. It was during this period in his late teens that Ericsson developed an interest in the hot-air, or *caloric*, engine.

Caloric was a term for what we now know as thermal or heat energy. It comes from the Latin word *calor*, which means *heat*. The Stirling engine is the best known external combustion engine that uses heated air as a driving force. It was named for its Scottish inventor, Robert Stirling (1790-1878),

Ericsson provided this model with one of his patent applications for a two-cylinder hot-air engine. The cylinders operated at different temperatures and were of different diameters. He called them high pressure and low pressure cylinders.

who received a patent in 1816. Publicity surrounding Stirling's invention may have brought it to Ericsson's attention. But he was unable to do more than read about Stirling's engine and develop his own imaginative designs.

Wanting to improve his technical abilities, Ericsson moved to London in 1826. London was at the time the world's engineering center. Ericsson worked for several different companies that specialized in steam engines. His technical performance was so good that he received 14 British patents for inventions such as steam engine boilers, centrifugal blowers for forced ventilation, and a steam locomotive. His patents helped establish the Braithwaite and Ericsson Company. With a hard-won technical reputation in hand, Ericsson put much of his time into designing equipment for water transportation. He correctly felt that paddle wheels wasted engine power and developed the first successful blade-type propeller in 1836. It was used on the canal boat *Novelty,* which ran between Manchester and London, and represented the first use of a propeller in commercial service. Ericsson did not invent the ship propeller, but he did more to improve and publicize it than any other person. His most dramatic experiment was an 1845 tug-of-war between two comparable British warships with 200 hp steam engines. With both ships connected by iron cables at their sterns and under full power, the propeller-driven *Rattler* pulled the side-wheeler *Alecto* at a speed of 2.8 knots.

An 1869 Ericsson caloric engine is on display at London's Science Museum. It developed about 1 horsepower and drove machinery in a workshop.

During his 13 years in Britain, the British navy showed no interest in propellers, and in 1839 Ericsson moved to America. He expected to stay only a short time but remained in the United States for the rest of his life, 50 more years. He became an American citizen in 1848. He had married Amelia Byam in 1836, and she moved to America with him. But Ericsson was completely immersed in his work and Amelia felt isolated. Soon after their arrival in America, she returned to England and they never saw each other again. Still, they regularly exchanged letters and Ericsson supported Amelia until she died in 1867.

Ericsson continued refining the propeller and by 1844, 25 vessels on the Great Lakes used them. Some operated with twin propellers. Although he was continually busy, Ericsson always found time for his pet project, the hot-air engine.

Hot-air engines were most suitable for low-power applications, and Ericsson's Massachusetts Caloric Engine Company sold more than 4,000. They were primarily used to pump water for homes. Through the modern field of thermodynamics, we now know that hot-air engines are an inefficient power source. Ericsson was unaware of this because the laws of thermodynamics had yet to be discovered. Like electric motors, hot-air engines were useful in small sizes but could not be readily scaled up for large power applications. Ericsson did not know that at the time he built a huge four-cylinder engine for his 1853 demonstration.

Easily obtaining financial backing, he had the 2,200-ton, $300,000 *Ericsson* built in New York City. It was a gleaming white ship with impressive gold trim. The passengers during the January 1853 voyage included friends, investors, politicians, scientists, and businesspeople. The ship's engine had been widely promoted as safe, quiet, clean, and economical. A total of 2,400 hp was expected, but the engine developed only about half that. Its long six-foot stroke necessitated an uncommonly slow 9 rpm rotational speed. Many of the passengers stood on the large vertical pistons as they moved up and down at a leisurely pace. Side-mounted paddle wheels 32 feet in diameter drove the ship through the water at 6 to 8 knots. Ocean-going steamships typically traveled at 12 knots or more. The *Ericsson's* fuel consumption was double the expected amount but still less than that of comparable steam engines. However, before any conclusive tests could be conducted, the ship sank in a storm just outside New York harbor in 1854. Later refloated, the *Ericsson* was fitted with less expensive steam engines.

An iron frigate Ericsson built for the navy in 1844 was the first warship ever powered by a propeller. The 1,000-ton *Princeton* was also the first to have its engine and boiler below the waterline, protected from enemy fire. Ericsson regularly worked with the navy, and during the Civil War Union officials requested that he construct an iron-clad vessel. He accepted the assignment and received a budget of $275,000 and a time limit of 100 days. This was an incredibly short schedule, but Ericsson was undisturbed. A physically powerful man who often excercised 2 hours a day, he typically worked 12 to 14 hours per day all his life.

The boat Ericsson designed displaced almost 1,000 tons, used two 160 hp steam engines, and had a top speed of 9 knots. Ericsson built it on the East River in New York City. He chose the name *Monitor* because he said it would prove a "severe monitor," or warning, to the Confederacy. The 172-foot boat was launched in January 1862. It met the Confederate *Merrimac* off Norfolk, Virginia, in March. Their battle was the first between steam-powered iron clads.

Ericsson was a brilliant and persistent inventor. His ship the *Ericsson* was his only failure. Many great inventors have at least one such experience. And Ericsson cannot be totally blamed for not understanding the new and unclear field of thermodynamics. Like successful people in all walks of life, Ericsson recovered from this experience and continued his work and research. He still had over 35 years of technical accomplishment ahead of him.

References

Yankee From Sweden by Ruth White, Henry Holt Publishers, 1960.

"The Big Engine That Couldn't" by Michael Lamm, *American Heritage of Invention and Technology,* Winter 1993.

"The Monitor is Mine" by Oliver E. Allen, *American Heritage of Invention and Technology,* Winter 1996.

Internal Fire by C. Lyle Cummins, Society of Automotive Engineers Press, 1989.

Joseph Whitworth

Born:
December 21,
1803, in
Stockport,
England

Died:
January 22, 1887,
in Monte Carlo,
Monaco

Many people consider *quality* one of the most important characteristics of a product. They may have trouble defining it in specific terms because the requirements of different items can be so wide ranging. A bicycle, for example, is far less sophisticated than an automobile, although both provide transportation. A quality bicycle to one person might be one that shifts easily and stops well. For another person, it might refer to a fast bicycle. To attain a desired level of quality, modern manufacturers follow certain standards. The General Secretariat for the International Organization for Standardization (ISO) is in Geneva, Switzerland. ISO was established in 1947 and later expanded to provide common standards for the original 12 countries of the European Union. But other countries use ISO standards, too, and America is an ISO participant. The Department of Defense and NASA require their contractors to use certain ISO standards.

ISO has about 180 technical committees (TC) continually upgrading different fields. TC 1 deals with "screw threads," TC 39 with "machine tools," and TC 176 with "quality management and quality assurance." Those three are all particularly important to manufacturers because machine tools produce the high-quality products people demand. Even in the nineteenth century, technologists realized the importance of standards and product quality. One person was a driving force for standardization and accurate measurement. Joseph Whitworth made his living building machine tools. He was the first person to establish standard screw threads in 1841. He also built a micrometer in 1855 that was accurate to one millionth of an inch.

Whitworth was born near Manchester, in the Midlands region of England, which once produced more than half the world's manu-

The Science Museum / Science & Society Picture Library

factured goods. His father was a schoolmaster who educated him until he reached the age of 12. Whitworth spent the next two years at an academy in Leeds. Then, he apprenticed to learn business with an uncle at his textile mill. But Whitworth was more interested in machinery and often left his clerk's desk to roam the factory. He learned the operation and construction of every machine before going to work at 22 for Henry Maudslay (1771- 1831) in London. (See pages 43-45.) Maudslay's machine tool factory was the best place to learn the trade. He was the first to design machine tools for precision mass production and he had a great effect on Whitworth.

Whitworth was one of Maudslay's most insightful young employees. He exhibited an exceptional aptitude for machine tool work and Maudslay made sure that his talent was not wasted. Whitworth personally worked on tools for turning, planing, milling, and drilling. It wasn't long before he made his first great achievement: producing a truly flat surface.

Planing and grinding were the accepted methods for making flat surfaces. Whitworth chose to hand scrape two different surfaces. He periodically laid one surface on top of the other with a coloring agent between them to highlight uneven surfaces. By introducing a third scraped surface for comparison, Whitworth eliminated the possibility of one surface being concave and the other convex. He was the first to use a method that was to become standard procedure. Flat surfaces are very important because they greatly reduced friction in machine tools.

After several years, Whitworth went to a shop operated by Joseph Clement (1779-1844). Clement had also worked for Maudslay. He had a contract to make Charles Babbage's (1791- 1871) complex and highly accurate calculator. Whitworth made many of its precise and oddly shaped parts. He left in 1833 to open a business in Manchester, which was becoming an important manufacturing city. He rented a section of a textile mill where steam power was available for power tools. His business steadily grew, eventually becoming a world model for machine tool precision. By 1844, he employed 200 people and by 1862, 700.

Manufacturers typically built lathes, planers, milling machines, and other machine tools to satisfy a customer's specific requirement. Whitworth designed general-purpose machines that could be used in a variety of industries. He worked on the entire range of production machine tools. He developed automatic mechanisms to improve accuracy and speed production. One of Whitworth's most innovative designs concerned metal lathes. In 1835, he was the first to drive the cutting tool both horizontally and back and forth. He built a planer in 1842 that cut in both directions. Whitworth was soon acknowledged as the most progressive machine tool builder of his time. He held patents for machine tools, textile machinery, and road sweeping equipment. The secret of his success was his careful attention to every detail of a machine's design. The Crystal Palace Exposition in London in 1851 was the first

world's fair. Whitworth's 23 exhibits totaled far more than any other company's.

Whitworth knew that a tool was only as good as the parts from which it was made. Early nineteenth-century specifications were often imprecise. A "full 1/32 inch" and a "scant 1/16 inch" are examples of confusing requirements once mentioned by Whitworth. Aiming for better measuring equipment, his production bench micrometer was accurate to a ten-thousandth of an inch. He made a special demonstration model in 1855 that was accurate to a millionth of an inch. That micrometer did not make absolute measurements–it w as a comparator. It measured the difference between a standard size and a manufactured size. Ever mindful of his customers' problems, Whitworth also invented the first go-no-go plug and ring gages. They were accurate and easier to use than micrometers.

Successes with his micrometer led Whitworth to establish standard thread sizes. In his day, each factory and machine shop cut threads to its own specifications. Each factory selected fastener diameters, pitch, and threads per inch. Such lack of consistent standards led to chaos when machines required repair or modification. Whitworth collected screws and bolts from different machine shops and developed reasonable average sizes. He recommended a fixed thread angle of 55 de-

This die has designations stamped into it in various locations. It says "J. Whitworth and Co," and "1-5/8 inch," and "5th," and "Manchester."

This 1842 Whitworth planer was belt driven through the wide flat pulley at the front right. The planer required only 0.40 hp to cut at 10 feet per minute. The ropes were part of the reversing mechanism and did not affect tool accuracy.

grees, a standard pitch for a given diameter, and a standard number of threads per inch. He began the project in 1835 and formally recommended its adoption in 1841. So-called Whitworth threads were in general use by 1860 and commonly found on machine tools, cars, locomotives, and other items until the early 1950s.

Whitworth turned his attention toward making improved weapons for his country's use in the 1853-1856 Crimean War. Whitworth concentrated on rifles and large rifled guns. He developed a special steel by compressing the molten metal as it solidified. With that steel and his precise manufacturing methods, Whitworth produced weapons that fired more accurately and with greater range. But political considerations kept them from going into significant production. His company merged with a competitor in 1897, after Whitworth's death, and formed the Armstrong-Whitworth Company for making armaments.

A man of few words, Whitworth said that his happiest day was when he became a journeyman at 18. He married Fanny Akers when he went to work for Maudslay at 22. She died after 45 years of marriage. He then married Mary Orrell. He had no children. Whitworth's working life showed him the value of an education. A self-made man from humble roots, he was knighted by Queen Victoria, made a member of the prestigious Royal Society, and three times elected to the presidency of the Institution of Mechanical Engineers. He received honorary degrees from schools such as Oxford University. Following his retirement, he purchased an estate near Matlock and spent time with landscape gardening and farming. He never tired of showing off the grounds, trotting horses, short-horn cattle, and farm buildings. Whitworth often spent winters in warmer climates and died at 84 in Monaco. His will provided financial sup-

Whitworth's one-millionth-inch comparative micrometer. The micrometer is such a precious technical icon that a Science Museum curator wore cotton gloves to remove it from its cabinet for this photograph

port for 30 college scholarships at Manchester area schools. They are still awarded and still called Whitworth Scholarships. Whitworth also included in his will £500,000 for four technical schools, libraries, hospitals, and other charities.

Whitworth's accomplishments were the best that nineteenth-century British technology produced. His company prospered, he became wealthy, and he shared his good fortune with others. But Whitworth is less well known than people who accomplished far less. That may be because all his knowledge went into his machines, not onto a printed page. People today might tend to value old books and paintings more than old factory equipment. But Hugh Pearman, a journalist for *The Times* (of London) made a pointed comment in 1996. He wrote that "Whitworth's 12-foot planer has had more effect on all our lives than Charles Dickens's [1854] book *Hard Times*."

Joseph Whitworth, later in life

References

Sir Joseph Whitworth—The World's Best Mechanician by Norman Atkinson, Sutton Publishers, 1996.

Engineers, Inventors, and Workers by P. W. Kingsford, Edward Arnold Publishers, 1964.

Tools for the Job by L. T. C. Rolt, Her Majesty's Stationery Office, 1986.

ISO—Compatible Technology Worldwide, International Organization for Standardization, Geneva, Switzerland, 1993.

Isambard Kingdom Brunel

The Science Museum/Science & Society Picture Library

Born:
April 9, 1806, in
Portsmouth,
England

Died:
September 15,
1859, in London,
England

The thin line separating nineteenth-century British technology from engineering was so narrow that technical innovators often crossed back and forth. An *engineer* of the period was someone responsible for building roads, railways, bridges, and ships. A *technologist* could have been similarly described had the word been in common use. To this day, the word *technologist* in England has a more global interpretation than it does in America.

One Briton succeeded remarkably well in many aspects of Victorian transportation technology. He built bridges and tunnels that continue to carry road and rail traffic today. He laid out railway lines and stations still used by modern high-speed trains. He built the world's first all-metal ship, now on public display. In a trading nation like Great Britain, improvements to transportation benefited the entire country. Isambard Kingdom Brunel worked on many large projects at the same time. Flamboyant and outspoken, he was a hero of the Industrial Revolution and Britain's most highly regarded Victorian-era engineer.

Brunel's father, Marc Isambard Brunel (1769-1849), was a successful construction engineer and inventor who left his native France in 1793. He settled in New York City for several years but moved to England to marry Sophia Kingdom in 1799. The elder Brunel was successful at projects such as canal and building construction, as well as designing factory machinery. Brunel's family was financially comfortable and he was first educated by private tutors in England. He then received technical instruction in Paris and entered his father's business at 19. Brunel served his apprenticeship by helping build the first tunnel under the River Thames in London.

Not far from the Tower Bridge, the tunnel used a metal liner designed by the elder Brunel. Serious problems with flooding arose as the hand-dug tunnel progressed. Showing great personal courage, the young Brunel placed himself in danger to save the life of a co-worker in 1828. The event established Brunel's reputation as an honorable and trustworthy person. He spent several months recovering from injuries and did not return to the difficult project. Marc Brunel finished the tunnel in 1843. It is now used by the London subway system.

While recovering from his injuries, the younger Brunel designed a suspension bridge to span the Avon River gorge near Bristol. He won the competition, in spite of competing with the most respected bridge designers of the period. Work did not start until 1835 and did not reach completion until five years after Brunel's death. The Clifton Bridge continues to carry road traffic and many consider it the most beautiful of the

The unique Royal Albert Bridge in Plymouth completed the railway connection along England's south coast. With two great spans of 455 feet, it did much to enhance Brunel's reputation with the general public.

1892 all broad-gauge locomotives had been scrapped and all rails relaid.

Partly because of his father's reputation, Brunel was often asked to design and supervise tunnel and bridge construction projects. Most were built as part of his railway projects. One of his most striking was the Royal Albert Bridge in Plymouth. He started it in 1847. Crossing the wide Tamar River, the bridge had to be high enough to allow sailing ships to pass underneath. It is 2,200 feet long and uses tubular wrought-iron arches to support the railbed. The single rail line still carries traffic, although the crossing speed is restricted to 30 mph for safety.

One of Brunel's most colorful failures was his attempt to construct an atmospheric railway in 1848. A 15-inch-diameter iron tube ran between the tracks near Exeter. Pistons under the cars fit into the slotted tube, which was kept at a vacuum by nearby stationary engines operating air pumps. Because atmospheric pressure pushed the cars along, no locomotive was needed. It was a clever idea, but air leakage, low speeds, and high construction costs doomed the project.

Of all Brunel's varied accomplishments, the design and construction of three steamships dominate our memory of him. They

The 1843 *Great Britain* was the first all-metal steamship. On public display in Bristol, it is in an almost constant state of restoration. Here it is shown without its 42-foot bowsprit at the front.

many bridges Brunel designed. Four million vehicles crossed it in 1973, the year before a new bridge opened nearby.

Brunel was just 26 when the national Railway Committee appointed him to supervise construction of the Great Western Railway (GWR) between London and the port city of Bristol. Brunel walked and surveyed the entire length of its route. Unsure of whether locomotives could pull heavy loads up a grade, he made it as flat as possible. The route required removing large quantities of dirt in long trenches. Finished in 1841, the railway was so flat that some people called it "Brunel's billiard table." Its gradient does not exceed 1/1,320 for 70 of its 110 miles. Brunel also designed stations at each of the railway's ends. Both the Bristol Railway Station and London's Paddington Railway Station are still in use.

At the time, the standard distance between railway rails was 4 feet 8-1/2 inches. That unusual dimension evolved from the dimensions of horse-drawn carts used in early coal mines. Brunel preferred a broader gauge, 7 feet. He wanted the wider spacing so the cars could travel faster with less power loss at the wheels.

But the narrower gauge was popular because it allowed sharper turns. Narrower roadbeds, bridges, and tunnels were also easier to construct. The broad gauge was incompatible with other rail lines and by

were the *Great Western* (1837), the *Great Britain* (1843), and the *Great Eastern* (1859). The 236-foot, 2,300-ton *Great Western* connected with the GWR. That allowed passengers and goods to travel easily between London and New York City. The side-wheeler helped show that long-distance steamships could operate profitably. Previously, ship builders had thought that steamships could not carry enough coal for an ocean crossing. The Great Britain nearly became the first to make a transatlantic crossing on steam power alone. The *Sirius* entered New York City harbor a few hours ahead of the *Great Western* on April 23, 1838. The *Great Western* was broken up in 1856.

The innovative and commercially successful *Great Britain* was probably Brunel's greatest achievement. It was the first all-metal, screw-propeller steamship. It carried up to 252 passengers from Liverpool to New York City and Australia. The ship had a displacement of 3,600 tons, a length of 322 feet, and width of 50 feet. Massive for the time, its size required Brunel to modify the Bristol harbor entrance so the ship could be floated to the ocean. Like all transition steamships of that era, the Great Britain was fitted with both sails and an engine. The huge four-cylinder steam engine developed 1,600 hp and used steam at a pressure of only 5 psi. The ship made a total of 47 round-trip ocean crossings, including two to San Francisco. Following a long career, it was beached in Port Stanley on the Falkland Islands in 1886. The deteriorating ship was used for storage until it sailed to England in 1970 on a huge pontoon boat. Restored to its original appearance, it is now on public display in Bristol.

The *Great Eastern* was five times larger than any other ship of its day and a terrific gamble. A 680-foot side-wheeler with a displacement of 27,000 tons, it was built for long-distance trade with Asia. Its double-hulled design with watertight compartments is still a standard feature in shipbuilding. The ship was unprofitable, but it became an important part of communication history. Cyrus Field (1819-1892) used the *Great Eastern* to lay the first transatlantic telegraph cable in 1866. Expensive to maintain, the ship was broken up in 1889.

Brunel had so many projects going at

once that he had little time for a normal life. He married Mary Horsley in 1836 and had three children. They established a large combination home and office in London. Brunel enjoyed painting and entertaining his children with magic tricks. The photograph at the beginning if this chapter shows Brunel in front of the enormous hand-forged launching chains used for the *Great Eastern*. Problems with finances and suppliers dogged him throughout the ship's five years of construction. Physically exhausted, Brunel died of chronic kidney failure just 10 days after its launch in London.

Brunel continues to be remembered and well regarded in Great Britain. The country has many statues and plaques commemorating his accomplishments. His bridges and tunnels still carry traffic. Londoners traveling by train to Wales and western England leave from busy Paddington Railway Station, which Brunel completed in 1854. It has a breathtaking central span with a triple arch that emphasizes its wrought-iron and glass roof. In late 1996, some of Brunel's drawings, letters, and other personal items were auctioned by Christie's of London. They estimated sales would total £100,000. The items actually sold for £144,112.

Only a few items remain from Brunel's 1848 atmospheric railway. This section of pipe was part of a 25-mile-long iron tube whose interior was kept at a vacuum. Atmospheric pressure pushed an interior piston, which was connected to a railway car.

References

What's Left of Brunel by Jonathan Falconer, Dial House Publishers, 1995.

Isambard Kingdom Brunel by John Adams and Paul Elkin, Jarrold Colour Publications Ltd., 1988.

Isambard Kingdom Brunel by Richard Tames, Shire Publications Ltd., 1972.

John Augustus Roebling

Born:
June 12, 1806,
in Muhlhausen,
Germany

Died:
July 22, 1869,
in New York,
New York

Graceful and incredibly strong, suspension bridges are among the most beautiful structures that people have built. They often span great distances or deep gorges where other designs would be impractical. The first ones built used fiber ropes for support and a floor of wood. These early bridges had a problem that still concerns modern technologists: They were not stiff and, consequently, tended to sway with the wind or when people and animals crossed them.

The first modern metal suspension bridge was a little-documented 70-foot span built across Jacob's Creek in western Pennsylvania in 1796. It was designed by James Finley who used wrought iron chains for support. Other early suspension bridges also used metal links, similar to modern bicycle chains, but these bridges only worked for short spans and light-duty foot traffic.

People were uncertain about the strength of suspension bridges until 1855 when John Augustus Roebling built a successful one for trains crossing the Niagara River. The first to use spun-metal cables, Roebling is best known as the designer of the beautiful Brooklyn Bridge. That bridge was the world's first steel-wire suspension bridge. Today, more than half of the world's suspension bridges are in America.

The youngest and brightest of five children, Roebling was born in central Germany. His family lived above his father's tobacco shop. Roebling attended the local elementary school, where he so impressed his teachers that they advanced him to high school at an early age. His closest friend had an interest in engineering, which influenced Roebling to follow a similar career path. Bright, serious for his age, well

National Portrait Gallery, Smithsonian Institution

read, and with an artist's ability for sketching, he was clearly his mother's favorite. She saw in him the potential for future greatness and encouraged him in his studies. She did everything she could to save enough money to send him to college. In the structured German social classes of the nineteenth century, it would have been unusual for a shop owner's son to afford the expense of attending the prestigious Royal Polytechnic Institute in Berlin. But Roebling's mother's perseverance paid off and he left home at 17 to study civil engineering in his nation's capital.

Roebling was fascinated with suspension bridges and wanted to work in bridge building after graduation. But opportunities in Germany were limited, and he worked on road construction projects instead. He found the work boring. After several years, he began to consider alternatives. He and his brother Karl decided to pool their savings with others to establish a new farming community in Pennsylvania. His mother was crushed. She had expected something different from her son but accepted his decision.

The 11-week ocean voyage from Bremen ended in Philadelphia in August 1831. It wasn't long before the party of about 40 people arrived by canal boat to the 7,000-acre undeveloped tract they had purchased near Pittsburgh, which was almost the frontier region of America at that time. The group named their new town Saxonburg, after a district in Germany. Roebling spoke fluent English and served as unofficial leader.

Everyone built log houses and began new lives as farmers. Roebling married Johanna Herting in 1836. Their first son was born the next year. They named him Washington in honor of their new country, and Roebling became an American citizen.

Roebling was not a good farmer, so when he heard about a canal construction project in Harrisburg in 1837, he applied for a position. He was hired by the state of Pennsylvania. The Allegheny Portage Railroad provided an important link between the eastern and western sections of the Pennsylvania Canal. Part of the canal system included steep railway ramps used to move produce over the Allegheny Mountains. Canal boats were hauled up and down the ramps in special wheeled dollies. Expensive three-inch-diameter natural-fiber rope controlled their movement, but the rope quickly wore out from abrasion. Roebling came up with the idea of substituting woven wire. He developed complex machinery to make the woven wire, conducted experiments, and eventually took out a patent titled "Wire Rope Mach." He opened a factory in 1841 in Saxonburg to manufacture

the first wire rope made in America.

By the mid-1840s, Roebling's reputation for engineerng success had grown, and he became an independent contractor. The first bridge he designed and built was actually an aqueduct that carried water for the Pennsylvania Canal Company. He made the lowest bid for a contract to replace an existing, but unsafe, structure over the Allegheny River. When he completed the aqueduct in 1845, it had seven spans of 162 feet each and a floor suspended from two seven-inch-diameter wrought-iron wire cables. It was unlike anything else in America, and the publicity Roebling gained from the project led to other contracts.

Roebling's most dramatic achievement was a railroad bridge across Niagara Falls. The New York Central Railroad wanted to connect its line with Canada's Great Western Railway. A light-duty bridge across the falls already existed for foot travel and small carriages. Roebling was invited to supervise the design and construction of a railway bridge. The idea of using suspension bridges for heavy-duty use had been condemned by other engineers because the bridges shook and swayed, and several had

Roebling's patented wire-rope machine rolled along a 3,500-foot track. It twisted iron or steel strands as it rolled. The Brooklyn Bridge has 5,296 steel wires in each of its major suspension cables.

collapsed. Roebling attributed the failures to poor design. The span of his bridge at Niagara Falls was 825 feet. He stiffened it with trusses. Two floors suspended from

Its skyline competing with the hustle of modern America, the 1866 John A. Roebling Bridge spans the Ohio River at Cincinnati. A plaque on the bridge states that it was the prototype for the Brooklyn Bridge.

four 10-inch-diameter cables, each cable composed of 3,640 individual iron wires. Rail traffic used the upper floor and other traffic used the lower floor. Connecting the two floors with struts and diagonal tension rods produced a continuous hollow girder. Stiff enough to support the action of a rolling load, it was the first successful railway suspension bridge in the world. The bridge stood until 1897, when it was taken down. Although it never showed signs of weakness, train loads by that time had increased to three times what the bridge was designed to handle.

To be closer to transportation centers, Roebling moved his wire rope factory and home to Trenton, New Jersey, in 1848. His large house easily accommodated his family of nine children, as well as his sister's family of seven children. Besides bridge-building activities, he found time to write many professional articles and a technical book. For recreation, he played the flute and the piano.

Anticipating the Brooklyn Bridge project, which he had recommended in 1857, Roebling constructed a prototype at Cincinnati. The first bridge to span the Ohio River, it was completed in 1866. It is officially called the John A. Roebling Bridge, making it one of very few bridges named for its designer. A brass plaque near a cable anchor describes it as the "Prototype for the Brooklyn Bridge."

New York City officials appointed Roebling chief engineer in 1867 for a bridge that would connect the boroughs of Brooklyn and Manhattan. As Roebling envisioned it, the bridge would have a span of almost 1,600 feet. Each of the four 16-inch-diameter supporting cables would weigh nearly 1,000 tons and have 5,296 galvanized and oil-coated steel wires. Heavy wire ropes hanging from the cables would support the six-lane roadway. Other radiating wires would add rigidity.

Construction work began in the early summer of 1869. While Roebling was taking transit readings in July, a boat entered the Fulton Ferry slip. It accidentally rammed the pilings on which Roebling was standing, and the impact crushed his right foot. Although he received immediate medical attention, tetanus set in and Roebling died a few weeks later. However, the Brooklyn Bridge was built almost exactly as he had designed it. The $15 million project for the largest bridge in the world at that time was completed by Roebling's oldest son, Washington, and Washington's wife, Emily. It opened to the public in 1883.

John Roebling's mother, Friederike, died shortly after John had arrived in America. It was her faith in his ability to succeed that put him on the path to technical greatness and gave us one of the world's most beautiful bridges. The National Park Service designated the Brooklyn Bridge a National Historic Landmark in 1964.

References

Engineers of Dreams, Great Bridge Builders and the Spanning of America by Henry Petroski, Alfred A. Knopf Publishers, 1995.

The Builders of the Bridge by D. B. Steinman, Harcourt, Brace and Co. Publishers, 1950.

The Smithsonian Book of Invention, Smithsonian Exposition Books, 1978.

Works of Man by Ronald W. Clark, Viking Penguin, 1985.

James Hall Nasmyth

The Science Museum/Science & Society Picture Library

Born:
August 19, 1808,
in Edinburgh,
Scotland

Died:
May 7, 1890, in
London, England

When thinking of high-quality products, people often consider items with which they come in personal contact. Examples might include electronic equipment, home appliances, clothing, and automobiles. But quality is not exclusively reserved for consumer products. There are many items for which it has critical importance: structural steel in bridges, large plastic drainage pipes, airplanes, ships, and others. Such things are so large, or have such large parts, that shaping and testing them for quality presents a challenge. George Ferris (1859-1896) developed the profession of materials testing. (He is better known for building the first Ferris wheel in 1893.) Bridge builder James Buchanan Eads (1820-1887) often returned structural steel when it did not meet his strength requirements. He built the 1874 Eads Bridge over the Mississippi River at St. Louis.

Before the appearance of powerful tools, large metal pieces were almost impossible to shape. Raw iron could be made in huge ingots, but there was no easy way to forge it into useful shapes. The first powerful machine for forging large metal pieces was the steam hammer invented by James Nasmyth in 1842. Although Nasmyth built a variety of machine tools, his 20-foot-tall steam hammer overshadowed his other accomplishments.

In Nasmyth's time, Scotland was a pioneer in education at all levels and he was fortunate to be born in that country's capital. His parents were artistic painters who founded the Scottish School of Landscape Painting. Nasmyth inherited their artistic talents. His father was a successful portrait and landscape painter, but he also had skills in technology and spent many free hours in his well-equipped workshop. Nasmyth was the last son of 10 children and he received special educational considerations. His parents ex-posed him to art, drawing, and mechanics. Nasmyth's ability at mechanical sketching contributed to his technical successes. He also became an excellent amateur oil painter.

Through his parents, Nasmyth met steam engine builder James Watt (1735-1819) and author Walter Scott (1771-1832), who both lived near Edinburgh. (See pages 25-27.) He was educated privately and at age 13 entered the Edinburgh School of Arts, a unique art and technical college. Nasmyth and a friend spent their spare time with his father's machine tools and often visited the iron foundry owned by his friend's father. Nasmyth wrote in his autobiography, "I look back to those days, especially to the Saturday afternoons spent in the. . . foundry as the most important part of my education." He never completed a formal education program because he enjoyed machine shop work much more. Nasmyth's skills were so well developed that at the age of 18 he made several working models of steam engines.

From *James Nasmyth—Engineer—An Autobiography* edited by Samuel Smiles, John Murray Publishers, 1889

Nasmyth painted this image of his steam hammer. The painting shows how many people were still required to handle the metal being forged. It is unclear who was operating the steam valves, but 14 people appear to be involved with controlling the workpiece.

Nasmyth hoped to obtain an apprenticeship in London under Henry Maudslay (1771-1831). (See pages 43-45.) Maudslay had a worldwide reputation and received many such requests. He turned them all down. Nasmyth's father had met Maudslay, and father and son traveled by boat in 1829 to Maudslay's factory. They took a small model steam engine the younger Nasmyth had built. Maudslay was impressed with the engine but even more impressed with Nasmyth's attitude and demeanor. Maudslay quickly hired Nasmyth as his personal assistant. Maudslay trained many of Britain's emerging technologists and Nasmyth always referred to him in the most glowing terms.

Maudslay died in 1831 and Nasmyth remained at Maudslay's factory for another year. The factory owners allowed him to have castings of a production lathe, which Nasmyth took home. He finished the castings using his father's old treadle lathe and then made a planer and boring machine. He and his older brother George found financial backing and opened a small machine shop in Manchester in 1834. The city was a terminal for the new Liverpool and Manchester Railway, which became more popular than anyone imagined it would. All the city's manufacturing businesses prospered. Nasmyth's grew so quickly that he moved to a six-acre site only two years later. At the age of 28, he built the factory where he would spend the rest of his working life. He married Anne Hartop four years later.

Nasmyth had a reputation for making elegant machine tools and excellent steam

engines. Following Maudslay's recommendation, he was among the first to use standardized fasteners. Nasmyth also manufactured standardized machine tools instead of expensive custom-built ones. Fasteners and components among similar Nasmyth-built machine tools would interchange. Nasmyth also used the fewest possible parts and designed his products for ease of repair. Maintenance and adjustments were straightforward and required little disassembly.

In 1838, Isambard Kingdom Brunel (1806-1859) set about designing a large steam-powered paddle ship. (See pages 85-87.) The ship would require a strong, high-quality forged-iron power shaft that measured 30 inches in diameter. No way to fabricate such a massive piece of metal then existed and Nasmyth began to consider options. In 1839, he sketched his idea for the first forge to be driven by a steam-powered piston. His drawing was so precise and well crafted that Nasmyth's combination of artistic and technical skill is obvious. He saw the steam hammer as the best way to forge large, high-quality iron pieces.

Brunel later abandoned the paddle wheel in favor of a propeller for his *Great Britain* steam ship, so there was no longer an immediate need for the steam hammer. The country was in an economic depression in 1840 and processing a patent would have

A small Nasmyth 1857 planer. The table moved back and forth by chains passing over a pulley wheel (hidden from view). Power came from a rotating metal shaft near the ceiling.

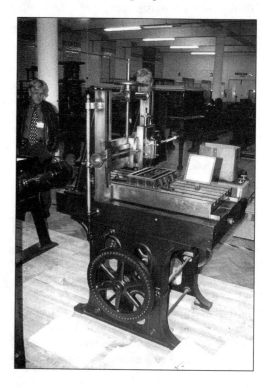

cost £500. Nasmyth did not build the steam hammer and did not immediately patent it. But he showed his drawings to many people, as he did with most of his inventions. A French machine tool builder in Creuzot used Nasmyth's idea without his permission. When Nasmyth found out, he quickly patented his idea in 1842 and built his own steam hammer. No one ever contested the patent.

Nasmyth built his first steam hammer in 1843. It was about 20 feet high and generally shaped like an inverted letter Y. Steam pressure of 20 to 50 psi under the piston lifted the 3,000-pound iron hammer. Releasing the steam pressure allowed the hammer to drop freely and strike the hot workpiece positioned on the anvil. The hammer had a four-foot range of fall. The single-acting piston had a diameter of about 18 inches. Easily controlled, the steam hammer was extremely powerful, but it could be adjusted to crack an egg in a wine glass without breaking the glass. Nasmyth occasionally made such demonstrations.

Although designed for shaping iron, the steam hammer was adapted to drive wooden piles for dock and bridge foundations. An 8,000-pound hammer delivering 80 blows per minute took 5 minutes to drive a 70-foot-long 18-inch square pile in place. Such an operation had previously taken 12 hours. Nasmyth's steam hammer received a gold medal at London's 1851 Crystal Palace Exhibition. The exhibition was the first world's fair. Nasmyth was proud of his invention and used it as the subject of one of his paintings.

Nasmyth's steam hammer was so unique and dynamic that it eclipsed his other technical accomplishments. He also invented the coiled-wire flexible shaft, and he improved the design of milling, planing, and shaping machines. He used his expansive mind to develop steam safety valves, new iron welding techniques, and mine ventilators. Nasmyth invented a hydraulic punch that slowly and quietly punched a hole through five inches of iron. Michael Faraday

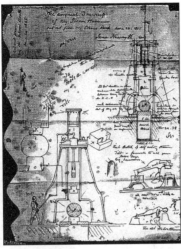

From *James Nasmyth —Engineer —An Autobiography* edited by Samuel Smiles, John Murray Publishers, 1889

(1791-1867) exhibited it during a lecture at the Royal Institution. (See pages 64-66.)

Because of the many steam hammers he sold, Nasmyth's business proved very successful. Nasmyth retired at 48 to a farm he named Hammerfield in southeast England. He spent most of his last years there occupied with astronomical work. He co-wrote a book in 1874 titled *The Moon Considered as a Planet, a World, and a Satellite*. Nasmyth had a lifelong interest in astronomy and made several reflecting telescopes. He drew an accurate map of the moon's surface that received a medal at the 1851 world's fair. When Nasmyth died in 1890, his estate was valued at £243,805.

Nasmyth's family name came from a nobleman who asked a blacksmith ancestor to become a soldier. After several battles where Nasmyth's relative exhibited great courage, the nobleman said he was, "Nay smith but a great warrior." Nasmyth was among the last of the learn-by-doing technologists. He disliked the practice of parents purchasing ready-made models of ships and steam engines for their children. He defined *technology* as the application of common sense to materials. Nasmyth saw it as an experience with tools that transferred information from the hands to the head. He wrote, "The eyes and the fingers . . . are the two principal inlets to sound practical instruction."

References

"Industrial Revolutionary: James Nasmyth" by Henry Petroski in *Mechanical Engineering,* February 1990.

James Nasmyth—Engineer—An Autobiography edited by Samuel Smiles, John Murray Publishers, 1889.

1839 drawing by Nasmyth of his steam hammer. With fragments of machinery, words, and calculations, it shows how clearly he visualized the steam hammer right from the beginning.

Nasmyth's steam hammer was a huge device capable of forging large pieces of iron. The heavy block of iron next to the nameplate was the hammer. Steam pressure applied to the piston at the top pulled up the hammer. It dropped onto the workpiece from its own weight.

Isaac Merritt Singer

Born:
October 27, 1811,
in Pittstown, New
York

Died:
July 23, 1875, in
Torquay, England

The Industrial Revolution began around 1760. The movement focused attention on improving manufacturing methods. The textile industry was the first to experience its changes. Early factories made yarn, later ones made thread, and then came fabric. Techniques for weaving fabric by hand on a loom were well known. It wasn't long before technologists started working on inventions that could be used to sew fabric into useful shapes like shirts, skirts, and trousers. They first tried to build machines that mimicked hand motions. But no one succeeded at making a practical one.

Another approach was to develop a new stitch and use a shuttle, as in weaving. A shuttle is a device that makes frequent back and forth movements. The idea was to use a threaded needle to force thread through layers of fabric. When the needle reversed direction, the thread formed a loop. A small metal shuttle sent another length of thread through the loop. As the needle pulled out of the fabric, the thread tightened, forming a locked stitch. It sounded simple enough, but the mechanical ingenuity required to accomplish that process was anything but simple. Many inventors developed mediocre methods, and Elias Howe (1819-1867) held the basic sewing machine patent, issued in 1846. But Isaac Merritt Singer is more closely associated with its invention. Singer had technical expertise and knew how to promote a product. He went after the consumer market, not the clothing manufacturers Howe pursued. The sewing machine was the first invention of the Industrial Revolution to find use in the home.

The youngest of eight children, Singer was born in eastern New York but mostly raised near Oswego. His father was a millwright who provided his children with an

National Portrait Gallery, Smithsonian Institution /
Gift of the Singer Company

appreciation for machines. That early introduction headed the younger Singer toward work in technology, although he occasionally strayed into the acting profession. His mother left the family when Singer was quite young and he had no recollection of her. Singer received only a small amount of formal schooling and left home at 12. He briefly tried a machine shop apprenticeship, then decided to wander for a while. Between the ages of 19 and 28, he moved from state to state, making a good living because of his abilities with machinery. He worked for a time as an actor and particularly enjoyed that profession.

However, he wanted to be an inventor and spent his free time drawing mechanical devices. At 28 and living in Illinois, Singer patented a rock-drilling machine. He sold the rights for $2,000. Unaccustomed to dealing with such an impressive amount of money, he soon spent it and then went back to acting. Singer used the stage name

"Isaac Merritt" and he organized a touring company called the Merritt Players. It was unsuccessful and on one occasion, the group became stranded in Ohio. Singer took work at a company that manufactured wooden type for printers. There, he saw the need for an automatic print-carving machine, so the industrious Singer invented and patented one. He took his invention to the Orson C. Phelps Machine Shop in Boston because he heard that the company might have an interest in manufacturing it. Instead, he found work as a machinist.

Phelps manufactured Blodgett and Lerow sewing machines under license. The machines generally resembled Howe's, with a horizontal curved needle that made just a few stitches at a time. Fabric had to be continually repositioned for each series of stitches. Blodgett's machine used a shuttle with a circular motion that took the twist out of the thread with each stitch. Another motion was required to put the twist back in, which made for a complicated assembly. It was an awkward arrangement that easily went out of adjustment. When Singer first saw Blodgett's machine in 1851, he criticized the clumsy shuttle design. Like some other people, he also thought a vertical needle would be easier to control.

Within one day, Singer sketched his ideas and took them to Phelps for review. His design incorporated the linear shuttle and vertical needle, but the major improvement was his feed mechanism. Singer's drawing showed a feed wheel with short pins that projected from a slot near the needle. The wheel pulled the fabric along with each stitch. The Howe machine, and the many others based on its method, could make only about a dozen stitches before the material had to be manually repositioned. Phelps was impressed and agreed to donate his shop and some employee time to make a prototype. A bookseller friend of Singer's from his type-carving days, George Zieber, agreed to put up $40 for supplies. After 10 18-hour days, the machine was ready. It perfectly sewed five stitches before breaking the thread. After a few minor adjustments, it worked continuously. Singer, Phelps, and Zieber were suddenly in the sewing machine business. They pooled their resources to form the I. M. Singer Company in New York City.

However, they soon discovered that finding a market was difficult. Previous inventors of imperfect machines had disappointed textile factory owners, who were often unwilling to even watch a demonstration. But Singer's acting ability combined salesmanship with his technical knowledge, and his company slowly prospered. No one had considered the consumer market because sewing machines—at $300 each—were very expensive for the time. Sewing machines were the most complex item manufactured before the Civil War.

To attract individual purchasers, Singer built large, pleasant salesrooms. With huge windows, fancy chandeliers, and expensive furniture, his New York salesroom resembled the interior of a palace. He hired skilled and knowledgeable women to demonstrate the machines to potential customers. His machines used a foot treadle for power instead of the less-convenient hand crank used by his competitors. Also, he was the first to use installments as a payment method. The combination of consumer marketing, pleasant sales environment, and an easy payment plan proved more successful than Singer expected. When sales increased, production costs decreased to the point where a typical machine cost $50 to $100. As its use increased, the sewing machine had an effect on fashion. Clothes became much more complicated and elaborate.

Singer was the first to spend $1 million a year for advertising to maintain sales. All

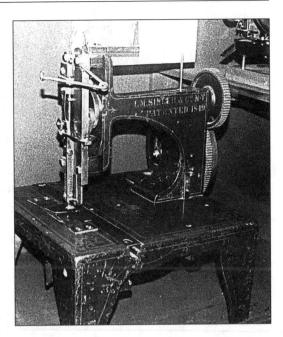

Singer's patent model with its vertical, eye-pointed needle is on display at the Smithsonian Institution's National Museum of American History. Because it was a full-sized working model, it was often demonstrated and shows evidence of being well used.

that publicity brought him to the attention of Elias Howe, who had almost gone broke trying to market his 1846 patented sewing machine. Howe spent several years in England searching for sales and was unaware of the booming American market. He was so poor, he sold everything he owned and still had to borrow money to return his family to America. Practically every sewing machine being sold infringed on Howe's patent in some way, and he settled with other manufacturers. But Singer refused,

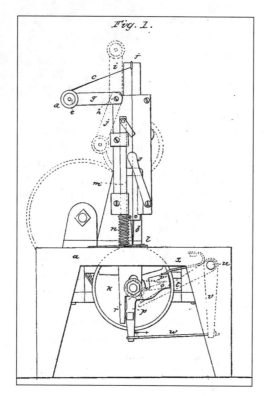

Fig. 1.

The unique characteristic of Singer's sewing machine patent was the fabric feed method. His was the first to automatically move the fabric forward with each stitch. He also made major modifications to existing shuttle techniques, shown attached to the circular feeding wheel marked 'k'.

and a court battle pitted the stronger Singer against the weaker Howe. Reason prevailed, and Singer was forced in 1854 to pay Howe a flat fee of $15,000 and up to $25 per machine from then on. To avoid similar situations in the future, the sewing machine industry established a patent pool. This approach was intended to keep competitors from dissipating resources through court suits. The automobile industry established a similar arrangement in 1911.

In 1864, the year he retired, Singer's company produced 1,000 machines per week. His wife's father came from Paris, France, and Singer decided to move there. After a few years, he then moved to Torquay on the south coast of England in 1870, not far from the birthplace of steam-engine pioneer Thomas Newcomen. (See pages 13-15.) He died while in the process of having a mansion built. He left an estate of $13 million.

Physically, Singer was large and imposing. Records of his personal life are incomplete. He may have married as many as five times and may have had as many as 24 children. The influence of the stage remained with him throughout his flamboyant life. Introducing a new product to the public often requires both promotional and technological skills. Singer had both in large measure.

References

The Invention of the Sewing Machine by Grace Rogers Cooper, Smithsonian Institution Press, 1968.

Sincere's History of the Sewing Machine by William Ewers and others, Sincere Press, 1970.

The Smithsonian Book of Invention, Smithsonian Institution Press, 1978.

Nuts and Bolts of the Past by David Freeman Hawke, Harper and Row Publishers, 1988.

Henry Bessemer

Surprising as it may sound, steel is the world's most precious metal. Not in terms of value per pound, but in its usefulness. Steel has more than a million uses. It is strong, inexpensive, and easily shaped. No metal has been more important to world progress. Automobile industries use the most steel, with the construction industry closely following. Steel is used in many consumer products like appliances, furniture, and household tools. Even paper clips and staples are made of steel.

Steel is an alloy of iron and very small amounts of carbon, usually 0.20 to 1.00 percent. The amount of carbon has a great effect on steel's strength and hardness. Cast iron has about 4 percent carbon and is much weaker than steel. The key to making steel lies in removing most of the carbon in iron. Some early technologists developed complicated methods that produced only limited amounts of imperfect steel. Henry Bessemer patented a technique in 1855 that literally burned out most of the carbon in iron along with other impurities. Blasting air into hot molten iron caused the carbon to burn. Bessemer's first experiment produced such a huge and unexpected shower of sparks that everyone nearby ran for cover. The Bessemer converter was the first to produce large quantities of steel.

Bessemer was born in a small village near Hitchin, 40 miles north of London. His father was a successful inventor with a productive technical career. The elder Bessemer spent time in Holland building that country's first house-sized steam engine. Then, he traveled to Paris and became a member of the French Academy of Science. He moved back to England just before the French Revolution and opened a profitable business casting metal type for printers.

The Science Museum / Science & Society Picture Library

Born:
January 19, 1813, in Charlton, England

Died:
March 15, 1898, in London, England

The younger Bessemer was educated at a local school but did not want to attend college as his father desired. Bessemer was proud of his father's accomplishments and the young man wanted to work with him. His father bought him a slide rest lathe that Henry Maudslay (1771-1831) had made very popular. (See pages 43-45.) Bessemer used it to make a small working model of a brick-making machine. But, most important, he learned how to alloy different metals for casting.

At the age of 17, Bessemer moved to London and opened his own business making specialty metals for sculptors and graphite for pencils. Bessemer was a prolific lifelong inventor. He had an exhibition at the Royal Academy when he was only 20. He married Anne Allen the same year. His major inventions at that young age were for casting metal type and making pencil graphite. Bessemer accumulated a lifetime total of 114 patents. They included such diverse appli-

This Bessemer converter was installed at a factory in north-western England in 1864. It was an experimental unit and cast the first steel from the Barrow-in-Furnace Works. It has seven blow holes in the bottom, each of which accepts a plug with seven smaller holes.

Bessemer converter outside a Sheffield museum. It dates from around 1910 and probably had a capacity of 15 tons. The vertical hydraulic actuator at the left controls the tipping, a technique patented by Henry Bessemer.

cations as the manufacture of steel, glass, sugar, and ships.

Before getting involved with steel making, Bessemer invented a process to produce high-quality bronze powder. The powder was a tint used by artists. Bessemer's work in casting sculptures introduced him to the expensive German product. He worked on an improved method using a bar of solid brass in 1843. Bessemer did not patent his process but kept it a secret. He had been disappointed after inventing a metal stamp for the British government in 1833. Designed to prevent fraud, the stamp saved the government an estimated £100,000 a year. Britain used it for 45 years. But officials had not kept promises they had made to Bessemer and he was wary of the government after that.

Bessemer cut tiny flakes on a lathe and processed them into a powder that was better and cheaper than any other available. He designed and built all the machines involved. Only his three brothers-in-law knew how the special procedure was carried out. The powder process was the basis of Bessemer's personal fortune and that of other family members. His share was enough to finance his experiments for the rest of his life.

The 1853-1856 Crimean War between Great Britain and Russia encouraged Bessemer to seek a better way to make iron. He was already a well-established inventor and was looking for another challenge. Cast-iron cannons were only strong enough to fire 12-pound projectiles. Bessemer hoped to use 24- and 30-pound projectiles. While testing new methods for converting iron to steel, he found two small pieces of processed iron that had almost all the carbon removed. He correctly guessed that air had accidentally burned it out. He used that idea to construct the first Bessemer converter.

Bessemer's 1855 experimental steel-making invention was a cast iron metal pot with an inside height of about four feet. It was open at the top and lined with heat-resistant bricks. It had six small pipes on the sides for pressurized air. Seven hundred pounds of molten iron were poured inside and Bessemer forced air through the pipes at 15 psi. The process created a massive shower of sparks and mild explosions as the carbon burned out of the metal. No fuel was added—the carbon in the iron provided all that was necessary. That first converter spewed white flames, sparks, and molten slag. Bessemer wrote, "I had in no way anticipated such violent results . . . the apparatus becoming a veritable volcano in a state of active eruption. No one could approach the converter to turn off the blast. . . . However, in 10 minutes more the eruption had ceased, the flame died down, and the process was complete."

Bessemer never expected the complex metallurgy that accompanied steel making. He sold patent rights to iron makers who could not duplicate his high-quality steel. British iron contains phosphorous, which made for brittle steel. Bessemer had unknowingly obtained his iron from one of the few places in Britain where the iron had almost no phosphorous, South Wales. Robert Mushet (1811-1891) identified the problem and a cure was eventually found. Using a specific type of firebrick in the converter removed the phosphorous. Mushet developed the first alloy steel by adding

Model of an 1865 Bessemer converter. The original could process about five tons of iron into good-quality steel.

controlled amounts of manganese and other materials.

Like Thomas Edison (1847-1931), Bessemer turned his inventions into commercial enterprises and made a fortune. He founded a steel plant in Sheffield and first used phosphorous-free iron imported from Sweden. He patented the pear-shaped, tipping-type converter shown in the photograph in 1860. He also found it more convenient to force air up from the bottom. Sparks often shot 30 feet into the air. Bessemer controlled the carbon content by shutting off the air at just the right moment. A 5-ton charge of liquid cast iron would usually have about 700 pounds of carbon, silicon, and other impurities.

In 1870, Britain produced 215,000 tons of Bessemer steel. That was half the world's output. Steel's first major use was for railroad rails. During an 1862 test on the busy London and North Western Railway, steel rails lasted seven times longer than iron ones.

The first American Bessemer steel was made by Alexander Holly at Troy, New York, in February 1865. Holly used two small 2-ton units. His converters produced a total of 118 tons of steel per month. Just a few months later, the Pennsylvania Steel Company built two five-ton units in Harrisburg. By 1880, those two converters were producing 14,000 tons of steel per year.

William Kelly (1811-1888) discovered the same process in 1851 in Eddyville, Kentucky. He received a U.S. patent in 1857 with a legal comment that his patent superseded

Bessemer's. Much early American steel was manufactured under Kelly's patent. But Kelly went bankrupt in the Panic of 1857 and lost all chance to profit from his pioneering work. The air-blast method for carbon removal is occasionally called the Bessemer-Kelly process.

Bessemer became unusually wealthy. Licensing his patented steel-making process earned him about £1 million in royalties. He was knighted by Queen Victoria and received many awards from British professional organizations. He retired at 59 and devoted the rest of his life to astronomy, solar furnaces, and his grandchildren. Bessemer had a long and happy life with his wife, Anne. They were married for 64 years and had three children. She died less than a year before he did in 1898.

The availability of large quantities of steel added a new technical dimension to the world. It allowed for longer bridge spans and larger ships. Tall skyscrapers followed the introduction of steel. Steel rails paved the way for faster and heavier trains. Machine tools could be made stronger and lighter. Bessemer's steel-making process has had a further reach than he planned.

References

Engineers, Inventors and Workers by P. W. Kingsford, Edward Arnold Publishers, 1964.

Sir Henry Bessemer F.R.S.: An Autobiography by Henry Bessemer, Engineering Publishers, 1905. [FRS = Fellow of the Royal Society]

Great Lives from History edited by Frank N. Magill, Salem Press, 1987.

The Making of the Modern World by Neil Cossons, Science Museum Publications, 1992.

Bessemer converter process. (1) Converter is tipped and about 10,000 pounds (five tons) of molten iron is poured into the brick lined container. (2) It's rotated to the vertical position and air is blown in from the bottom, burning out carbon and other impurities. (3) Converter is tipped again and about 9,300 pounds of steel is poured out and cast in molds.

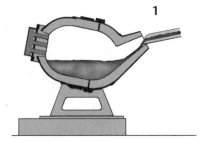

1

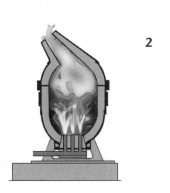

2

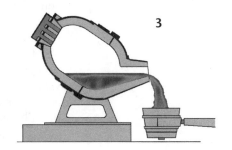

3

Erastus Brigham Bigelow

Born:
April 2, 1814, in
West Boylston,
Massachusetts

Died:
December 6,
1879, in Boston,
Massachusetts

The Jamestown Festival Park in Virginia commemorates Jamestown as the 1607 site of America's first permanent English settlement. Replicas of ships, native-American dwellings, and James Fort are the main attractions. Inside the fort are small houses that resemble those used by early European settlers. All have dirt floors. Almost as hard as concrete but not very flat, dirt floors were common in early-American homes for many years. Wooden floors were rare because cutting logs into boards with hand-operated pit saws was very labor intensive. It was also an unpleasant task, and breathing the sawdust often caused health problems.

Once water power began operating circular saws, wooden floors became more readily available. Many floors were painted a solid color, but others had intricate designs that rivaled expensive European mosaic floors. As standards of living improved, people wanted coverings for their rough wooden floors. Carpets manufactured by hand were expensive because the work took great skill and tireless fingers. The best carpets in the nineteenth century consisted of wool fibers individually knotted through a fabric backing. The more knots a carpet had, the greater its value. Some expensive Oriental rugs contained more than 2,000 knots per square inch. The word *carpet* comes from the old French *carpite*, which referred to a type of cloth. The word *rug* comes from the Swedish *rugg*, which meant "animal fur."

Like clocks, metal cooking stoves, and well-finished furniture, carpeting was a status symbol. All four items offered practical value and beauty, and average early Americans wanted them in their houses. People like Eli Terry (1772-1852) worked to provide one of those items at an affordable

Mohawk Industries, Inc.

price. (See pages 46-48.) Terry established the world's first clock factory in 1800. His clocks sold for as little as $7.50. Erastus Bigelow saw a similar need for floor coverings and invented the first power loom to make inexpensive carpets. The loom was a complex invention because carpeting is not usually woven like textiles. Threads were forced through a heavy fabric backing and held firmly in place. Bigelow's 1837 patent served as the foundation for an entire industry.

Bigelow was the second child in a farming family that lived just north of Worcester. His father earned extra income making wagon wheels and chairs. That was how Bigelow received his introduction to hand tools. He and his older brother, Horatio (1812-1868), attended the local elementary school. At 13, Erastus went to work at a nearby cotton mill. He was an inquisitive youngster and the job allowed him to study power machinery. Barely making a decent living, his father decided to take a chance with establishing a cotton factory around 1830. The effort succeeded and he employed Horatio as plant manager. Erastus briefly worked at the plant but soon wanted to be independent.

Through the age of about 23, Bigelow worked at many different jobs. He was proficient with the violin and played in an orchestra. He once worked as a clerk for telegraph inventor Samuel Morse (1791-1872), at Morse's retail store in Boston. He published a small pamphlet for stenographers that he hoped would prove profitable, but it didn't. He wanted to earn enough to enter Harvard University's medical school. His early jobs also included teaching penmanship and working in a twine factory. In hope of attending college, Bigelow taught himself Latin and paid his own way at the Leicester Academy preparatory school for a year.

Bigelow once slept under a heavy, handmade quilt that resembled carpeting. He wondered whether such material could be manufactured on a power loom. His desire for money to continue his education encouraged him to investigate the possibility. Since his father and brother worked with textiles, Bigelow had a general appreciation for the field. He first focused on a power loom that would produce lace for trimming upholstery in stage coaches. The material was only a few inches wide. After long hours of design work, Bigelow applied for a patent on a steam-powered loom when he was only 23. He received an early patent, Number 169, for the first power loom suitable for carpeting. A Boston manufacturer agreed to use his patent but went out of business. Bigelow convinced his brother to become a partner in a small factory near their home. They started operations in 1838, calling the business the Clinton Company after a favorite New York hotel. A town developed in the region and assumed the name of the company.

Bigelow's loom was complicated and product demands were strict. The lace had to match the designs then made by hand. It had to have a smooth, even face and perfect sides. The Bigelow loom did it all—and at an initial production rate more than twice as fast as hand looms. The business grew steadily and attracted new investors. The company soon employed 100 people and

made about $150,000 worth of lace each year. But production of carpeting was always at the back of Bigelow's mind.

Inexpensive nineteenth-century floor coverings came in several different forms. Matting made from long grass, corn husks,

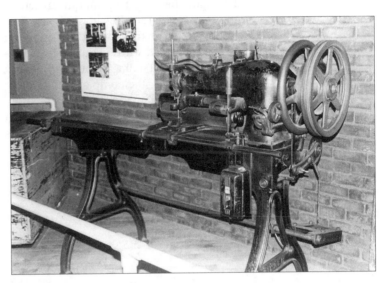

This heavy-duty sewing machine was used in a textile mill constructed about 1900. Although not specifically designed for carpeting, it sewed heavy canvas.

or strips of bark was popular. The 1909 Sears, Roebuck and Company catalog offered matting for 20 cents per square yard. Oval braided rugs were also common. They were made from rags braided into a strong rope-like yarn. The raw material usually came from textile manufacturers. Small fragments had little commercial use and were sold to rug braiders or individuals. The colorful yarn was wrapped in a flat oval and sewed together. A 1927 Sears catalog offered a 3 foot by 4 foot rug for $3.98.

Mats, braided oval rugs, and heavy textiles were all used on floors. But the most desirable coverings were traditional, large area rugs and carpets. Much nineteenth-century carpeting was made from *worsted* wool, named for the city of Worstead, England. It is a smooth, strong, twisted yarn made from long wool fibers. Bigelow often made worsted Brussels carpeting. Invented in 1710 in Brussels, Belgium, this fabric has a looped pile formed over a wire.

Bigelow's steam-powered lace loom encouraged him to think about a loom that would produce wider material. His patent already had most of the major features of a carpet loom. In 1839, Bigelow invented a power loom for producing two-ply ingrain carpeting 36 inches (one yard) wide. *Ingrain* was a desirable carpet in which the design on the face was reproduced in reverse colors on the back. Until almost 1900, more than half the carpeting manufactured in America was ingrain. Bigelow's power loom produced 25 square yards of carpeting per day as compared with 8 square yards produced by a hand loom. During the first six months of 1850, the company made more than 90,000 square yards of carpet.

The power loom not only allowed great leaps in production capacity with less skilled labor, it solved many quality problems. The length of the pattern repeats was more consistent and the carpet edges were

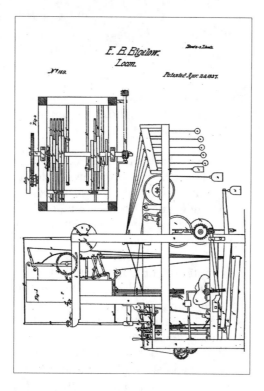

This sheet from Bigelow's 1837 power loom patent shows 2 of his 12 patent drawings. Its official title was "Power-Loom for Weaving Coach-Lace and Other Similar Fabrics." The five circles at the lower right in the drawing are weights that maintain tension in the warp fibers.

cleaner and more even. Patterns were easily changed with a Jacquard attachment. Joseph-Marie Jacquard (1752-1834) of France developed a punched card mechanism for selecting complicated weaving patterns. Wilton carpeting has a velvety cut pile face and is made on a Jacquard loom. Its name came from Wilton, England. Bigelow operated two major production facilities. His original one in Clinton had four acres under roof and he had another at Lowell, Massachusetts.

Bigelow was a founder and first president of the National Association of Wool Manufacturers. He held more than 50 patents and displayed some of his carpeting at the 1851 Crystal Palace Exposition in London. It was the first world's fair. A committee judged Bigelow's Brussels carpeting to be better and more perfectly woven than any hand-loomed goods at the fair.

Bigelow married Susan King three years before she died in 1841. He then married Eliza Means in 1843. He had one daughter and seemed happiest when his home was filled with his many friends. He bought a large estate in the White Mountains of New Hampshire that he named Stonehurst Manor. It is now a guest house and restaurant. Bigelow died at 65.

More than just an inventor and industrialist, Bigelow was an accomplished economist who wrote two books about imports and tariff problems. He ran for a Congressional seat in Washington, D.C., in 1860. He was narrowly defeated. In 1861, he became a member of the 21-person committee that established the Massachusetts Institute of Technology. The company he established with his brother is no longer in business.

References

Broadloom and Businessmen—A History of the Bigelow-Sanford Carpet Company by John S. Ewing and Nancy P. Norton, Harvard University Press, 1995.

American Rugs and Carpets by Helene Von Rosenstiel, William Morrow Publishers, 1978.

Samuel Colt

Born:
July 19, 1814, in
Hartford,
Connecticut

Died:
January 10, 1862,
in Hartford,
Connecticut

Nature's young have always found protection under the watchful eyes of their parents. An infant robin, fox, or deer cannot fend off potentially dangerous interlopers. Their parents use instinctive skills to scare away unwanted visitors. People react similarly, and defense of family and self was one reason for the development of weapons. As people moved into unfamiliar regions, achieved financial success, or asserted their legal rights, they often felt an increased need for self-defense.

Such conditions existed in nineteenth century America. Farmers, merchants, and other productive workers began to populate unmapped regions of an infant nation, barely 50 years old. When pioneers pushed west across the Mississippi River, they wanted assurance they could protect themselves. Unfamiliar and potentially dangerous animals had to be considered. There were also people with unknown motives. That environment encouraged the development of a popular handgun known as the *revolver*. No product of the era required more manufacturing skill or attention to detail than a firearm. And no handguns were better known than those made by Samuel Colt. He patented the first successful revolver in 1836.

Colt was born into a family with one sister and three brothers. His father made a good living operating several textile processing facilities. Colt's mother died when Colt was only six and his father remarried. During this stressful time, Colt was partly raised by his father's sister. He was a problem child and often rebelled against authority in small ways. Over a period of time, he secretly obtained several single-shot pistols. Colt took them apart and reassembled them until he had every part memorized.

Colt's father sent him to Ware, Massachusetts, at the age of 10 to work in one of his dying and bleaching businesses. He hoped the experience would help reshape Colt's life. But the youth led a casual life, attending school only irregularly. His exasperated father decided to send him to a college preparatory school in Amherst. But while there, Colt set off explosives on a raft in a pond. His action damaged school property and greatly concerned local citizens. With nowhere else to turn, Colt's father sent him to sea in 1830 on a small two-masted cargo ship. It was a year-long experience with a minor event that changed Samuel Colt's life.

Ship life was dull and uneventful for an ordinary deckhand. Colt's experience with handguns encouraged him to think about making a practical pistol that could fire several shots. Others had tried multiple barrels, or a magazine that held shells like a rifle, or a revolving cylinder. Colt noticed that when a sailor turned the ship's wheel, a ratchet and pawl mechanism caused it to lock in place. No matter which way the wheel spun, a spoke came in line with a clutch that could be set to hold it in place. Colt used that idea to whittle a wooden handgun with a revolving cylinder.

An early Colt pistol shows the chambers that had to be individually loaded with gunpowder and a lead bullet. The lever under the barrel pushed a ramrod into each chamber to secure the bullet. A percussion cap at the back of the chamber ignited the gunpowder through a small hole, when the cap was struck with the hammer.

Colt returned home and convinced his father to help pay for a five-shot prototype pistol. A Hartford gunsmith built the handgun, but it blew up when tested in a bench vise. Colt had not properly separated the chambers. When he fired the first bullet, it caused the gunpowder in the other chambers to fire at the same time. With no place for all the bullets to go, the pistol blew to pieces.

Somewhat dejected, Colt briefly worked for his father. He also made public demonstrations with nitrous oxide, which was commonly called "laughing gas." Colt was a stage entertainer much like a magician or hypnotist. He performed in community buildings as "Doctor Coult." This proved a profitable activity and Colt saved his money. He paid Baltimore gunsmiths to make two revolvers in 1835. They worked well— not perfectly, but well

enough for a patent. Colt took one out in 1836. The patent helped him obtain financing to open a factory in Paterson, New Jersey. He called it the Patent Arms Manufacturing Company.

Colt made handguns and rifles with revolving cylinders. Each cylinder had six holes, called *chambers*. The chambers were individually loaded with a precise amount of gunpowder. A tightly fitting lead bullet went on top. A lever under the gun's barrel moved a ramrod that forced the bullet into the chamber. A small percussion cap was placed outside at the rear of the chamber. Colt's first handguns did not use cartridges like modern weapons. The cylinder rotated as the pistol's hammer was pulled back. Each chamber lined up with the barrel and locked in place. Pulling the trigger released the hammer, which struck a percussion cap. The cap ignited the gunpowder and forced the bullet down the barrel. Partitions between the chambers eliminated crossover firing. The handgun was popular among individual purchasers. One satisfied customer was Captain Samuel Walker of the Texas Rangers. But the U.S. Government was not interested. The government had an adequate supply of single-shot pistols and worried about the reliability of Colt's complex mechanism. His factory had started with only a small investment and needed large orders to stay in business. The lengthy economic depression called the Panic of 1837 also worked against Colt. After making about 5,000 revolvers, Colt's company went bankrupt in 1842.

For the next several years, Colt worked on early torpedoes and underwater tele-

This disassembled early pistol shows some key parts in Colt's design. The removable cylinder is at the top and the lever-operated ramrod is at left center.

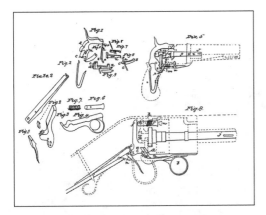

graph cables between New York City and Long Island. The 1847 Mexican-American War caused General Zachary Taylor to re-evaluate Colt's revolver. At the encouragement of Captain Walker, the government ordered 1,000 six-shot revolvers in January 1847 at $28 each. Colt was unprepared for that turn of events. He had no factory and contracted with Eli Whitney, Jr., to make weapons for him in Whitneyville, Connecticut. Ever mindful of marketing techniques, he had each of the Walker revolvers engraved with, "Address Sam'l Colt New York City." Colt did not credit the factory where the pistols were made and knew everyone was familiar with New York. He hoped that would improve sales. After five years away from gun making, Colt wanted to return to the business permanently. Always mindful of leaders in decision-making positions, Colt often gave such individuals specially engraved and boxed handguns. He once presented the Sultan of Turkey with two ornamental revolvers and soon received an order for 5,000.

Colt rented an abandoned textile factory in October 1847 to make some of his own weapons for the Mexican-American War. He bought 200 acres of land in Hartford in 1852 for a new facility. When completed three years later, the factory had cost $2 million and was the world's largest private arms factory. It eventually had 1,400 machine tools, with many designed by plant superintendent Elisha K. Root. He personally saw to the installation of 400 machine tools. Without Root's manufacturing insight, the

factory might not have succeeded. Root became president of the Colt Patent Arms Company following Colt's early death.

Colt married Elizabeth Jarvis in 1856 and they had one son who lived to adulthood. A man of astonishing vigor and energy, Colt traveled extensively to demonstrate and sell his product. He went often to Europe and South America. During trips to Russia and Turkey, he was influenced by Ottoman architecture. He built a large blue onion-shaped dome on top of his factory. After suffering from a rheumatic condition for several years, he died at the age of 47.

Colt was an enlightened employer who paid his best workers $5 per day, an amount far higher than prevailing wages. His factory was brightly lit with gas lamps, steam heated in the winter, and ventilated in the summer. He also provided the workers with modern houses and a recreation center. Colt even had a retirement fund, revolutionary for the time. In return, he expected punctuality, loyalty, and good work. His employees rarely disappointed him.

But Colt did not provide accurate enough machine tools to reach his goal of interchangeable parts. It would have taken years to pay for a machine that accomplished what a skilled worker could do with a file at the workbench. Colt's pistol parts were uniform in size and shape, but not directly interchangeable. Like others before him, Colt often claimed to have attained that goal, but he never did. David Freeman Hawke wrote in *Nuts and Bolts of the Past*, (1988) "Colt and those who adopted his production methods turned out articles with uniform, but not interchangeable, parts."

References

Yankee Arms Maker by Jack Rohan, Harper and Brothers Publishers, 1935.

A History of the Colt Revolver from 1836 to 1940 by Charles T. Haven and Frank A. Belden, Bonanza Books, 1940.

Samuel Colt's Submarine Battery by Philip K. Lundeberg, Smithsonian Institution Press, 1974.

Colt's 1836 revolver patent was not particularly complex, but the parts required precise machining.

Ada Byron Lovelace

Born:
December 10,
1815, in London,
England

Died:
November 27,
1852, in London,
England

For years, computer professionals have been viewed as a colorful group. Today, they often dress more casually than others. They sometimes use words and sentence structure that seem peculiar to the average person. They have added some unusual acronyms to our language. (An *acronym* is a word formed from the first letter of a series of words.) Computer professionals used the word ENIAC in 1945 as the name of the first electronic computer. The letters stand for Electronic Numerical Integrator and Computer. The letters in the 1964 computer language BASIC stand for Beginners All-purpose Symbolic Instruction Code. The ubiquitous CD-ROM stands for Compact Disk-Read Only Memory. The American Department of Defense (DOD) introduced the country's first government-sponsored computer language in 1980. It is named Ada. But unlike ENIAC, BASIC, CD-ROM, and many others, "Ada" is not an acronym. It was the first name of a real person who played an important role in computer development.

It is sometimes hard to identify precise origins in technology and people often add qualifying words to particular accomplishments. Examples include the first *practical* incandescent lamp, or the first *patented* automobile, or the first *dependable* fountain pen. No such restriction is necessary for Ada Lovelace, one of the most picturesque individuals in computer history. She was without question the world's very first computer programmer. Lovelace wrote a program in 1843 for Charles Babbage's computer.

Although now known as Countess Ada Lovelace, her name at birth was Augusta Ada Byron. Her birthplace overlooked London's Green Park. Called Ada by her family, she was the only child of the noted British poet Lord Byron (George Gordon) (1788-1824)

The Science Museum / Science & Society Picture Library

and his wife Annabella Milbanke Byron. Both of her parents came from financially comfortable backgrounds, though their temperaments differed considerably. Lord Byron remains Britain's most loved romantic poet and his large ancestral home of Newstead Abbey near Nottingham is open to the public. Local newspapers still print feature articles that deal with his poetry and his modern descendants. But Lord Byron led a carefree life and ignored his responsibilities. In contrast, Annabella was a calm and collected person, far more reliable than her husband. The couple's marriage lasted just one year. The future Ada Lovelace's parents separated shortly after her birth, and she never met her father.

Byron left Britain, never to return, in part because his lifestyle was unacceptable to almost everyone in his native country. But he remembered his daughter and always displayed a picture of her. He occasionally sent her presents. In his poem "Don Juan," Byron describes a woman whose "favourite science was . . . mathematical. . . . [S]he was a walking . . . prodigy." He knew that was what Annabella expected her daughter to become. Byron wandered restlessly throughout Europe and did not remarry. He died in Greece when his daughter was eight years old.

Lovelace was a gifted youngster who displayed both artistic and intellectual abilities. Her mother was an amateur mathematician who did not want her daughter to follow in her father's footsteps. So she encouraged Lovelace to pursue interests in science and mathematics, rather than poetry. Lovelace was educated by tutors, as was the custom among the upper-middle class in the nineteenth century. She received special instruction from the prominent mathematician Augustus de Morgan. He was partly responsible for developing modern algebra and had a high opinion of Lovelace's abilities. She became a close friend of Morgan's wife, Sofia.

Lovelace's mother did not neglect her instruction in the arts. She learned French and entertained guests by playing the harp and violin. She particularly enjoyed ice skating and horseback riding. But she suffered from lifelong poor health and could not always actively pursue her interests. Lovelace had severe headaches at age 7 and starting at age 13 could not walk for three years. This may have resulted from measles. Lovelace spent much of that period studying mathematics and music, two surprisingly related subjects.

In the mid-1820s, Lovelace became a close friend of popular scientific writer and mathematician Mary Somerville (1780-1872). Cambridge University used Somerville's texts and Somerville College at Oxford is named in her honor. Lovelace saw Somerville as a role model and hoped to emulate her achievements. Somerville and Sophia Morgan introduced Lovelace to Charles Babbage (1791-1871) at a dinner party in 1833. (See pages 67-69.) They often accompanied her on visits to discuss calculations by machine. It was a topic that others found dull because they could see no use for it. Not so Lovelace. She saw it as a future mathematical concept. She also saw in it the potential for achieving the type of acclaim that Somerville had. Babbage was 24 years older than Lovelace, but the two developed a close personal and professional relationship.

Somerville introduced Lovelace to William King, whom she married in 1835. He was elevated to the title of Earl of Lovelace three years later and Ada became the Countess of Lovelace, or Ada Lovelace. In spite of domestic obligations involving their three children and her own continued poor health, Lovelace expanded her mathematical skills through self-study. Her husband encouraged her interests. She regularly corresponded with electrical scientist Michael Faraday (1791-1867), astronomer John Herschel (1792-1871), and more often with Babbage. Although she was generally a shy and private person, she was friends with author Charles Dickens (1812-1870) and telegraph inventor Charles Wheatstone (1802-1875). (See pages 64-66 and 76-78.) Babbage made

The only two packs of surviving punched cards for Babbage's never-built computer. Lovelace planned for the smaller ones to be used for mathematical operations such as multiplying. The larger cards specified numbers to be used in the calculations.

technical presentations in Turin, Italy, in 1842 about the computer he was designing. An account of the lectures was published in French by L. F. Menabrea, an Italian mathematician and ambassador to France. Babbage had always encouraged women to develop scientific interests and he asked Lovelace to translate his lectures into English. Lovelace did so and, after lengthy discussion with Babbage and others, she added many of her own insights to the document.

Lovelace described the repeated use of a set of punched cards that were similar to subroutines in modern computers. She in-

Commemorative plaque inside St. Mary Magdalene Church in Hucknall, England, where Lovelace is interred next to her father, Lord Byron.

cluded mathematical examples to demonstrate Babbage's machine's calculating ability. She hinted at the possibility of programming. When published in 1843, the document she produced was three times the size of the original. It showed that Lovelace completely understood the principles of computers and computer programming.

Some people feared that a machine like Babbage's might think as people do. The modern term *artificial intelligence* mirrors that thought. For those individuals, Lovelace insightfully wrote, "For this machine is not a thinking being but simply an automaton [robot] that acts according to the laws imposed upon it." She suggested that a computer might be used to compose complex music, to produce graphic images, and for scientific calculations. Those are all modern applications. Babbage gave her great praise when he wrote that she had "entered fully into almost all the very difficult and abstract questions connected with the subject." It is only from Lovelace's publication that the modern world knows the details of Babbage's computer.

Babbage then asked Lovelace to use her skills to write a computer program. She worked out a procedure to calculate Bernoulli numbers, a complicated chore. The numbers are for specialized calculus opera-

tions. The activity of writing the program was only a mathematical exercise. Since Babbage's computer was never built, it was not possible to test Lovelace's program at the time. But it would have worked. When used with modern computers, her program gives the expected values.

Lovelace and Babbage wanted to bring the computer to the world. But it would have required a huge amount of money to make its 200,000 close-tolerance parts and carefully assemble them. The British government showed no desire for financial involvement in such a project. Unfortunately, Lovelace and Babbage then decided to use their mathematical skills in a risky manner. They developed a gambling system that they thought could provide the required funds. But their horse-racing system did not work and both lost large sums of money. Lovelace died heavily in debt in 1852, at only 36 years of age.

Lovelace's father, Lord Byron, had also been only 36 when he died. At her request, she is buried next to him in the Byron vault inside the St. Mary Magdalene church in Hucknall. Lovelace was the last Byron to be so honored.

Air Force Colonel William Whitaker served as chair of the 1970s committee that investigated a new computer language for the DOD. Its temporary name during development was DOD-1. Before approving the permanent name Ada, Whitaker asked permission from one of Lovelace's descendants. The Earl of Lytton readily agreed and noted that Ada was "right in the middle of radar."

Lovelace's work was generations ahead of her time and few others of her era understood it. It has long been reasonably speculated that had the Lovelace and Babbage horse-racing system made money, the world might have had computers in the mid-nineteenth century.

References

Ada—Enchantress of Numbers, narrated and edited by Betty Alexandra Toole, Strawberry Press, 1992.

The Computer from Pascal to von Neumann by Herman H. Goldstine, Princeton University Press, 1972.

Lord Byron's Wife by Malcolm Elwin, Harcourt, Brace & World, 1962.

Werner Siemens

Born:
December 13, 1816,
in Hannover,
Germany

Died:
December 6,
1892, in Berlin,
Germany

All inventors hope their inventions will be successful. But potential success might be influenced by specific circumstances. One problem could be timing. Charles Martin Hall (1863-1914) invented an improved aluminum-processing method in 1886. The lightweight metal had no significant use at the time. Only with the introduction of metal airplanes in the 1920s did it find a large market. Another problem could relate to national security. Howard Aiken (1900-1973) and his associates built the first modern digital computer between 1939 and 1943. But it was used by the U.S. Navy during World War II, and Aiken could not patent it until 1952. Still another problem could be economic. Steamboat inventor John Fitch (1743-1798) operated America's first scheduled steamboat service in 1790, but it did not attract enough customers to succeed financially.

Germany in the 1800s had some problems like that. Its economy was less prosperous than those of other European countries. Germany's Industrial Revolution did not begin until the 1830s, about 70 years after that of Great Britain. But by the 1850s, Germany was developing rapidly. And the country's many independent kingdom-states united in 1870 to form a politically unified country. It was in this dynamic environment that Werner Siemens's technical abilities blossomed. Siemens is Germany's best known inventor. His most significant invention was his 1866 self-excited dynamo (dc generator). He also improved on telegraphy, locomotive transportation, printing, and other areas of technology.

Siemens was born near Hannover into an educated but fairly poor farming family. His father and mother leased a country estate and grew much of the food for their large family. They had 14 children. Siemens was one of four technically successful sons who lived to adulthood. His father was partly educated in agriculture at the University of Göttingen. His mother's father was a chief official at a town near Hannover. The family lacked financial security partly because of the many children. In spite of their economic problems, Siemens's parents instilled a desire for education in and a sense of community among the family members. It certainly worked. Siemens was close to all his siblings, and particularly so with three brothers: Wilhelm (1823-1883), Friedrich (1826-1904), and Karl (1829-1906). That kinship played a prominent part in all of their technical successes.

Werner Siemens wanted to major in engineering at the world-famous Berlin Academy. Technology and engineering were essentially identical career paths during the nineteenth century. He was the first family member to have such aspirations, but his father told him he could not afford it. He

The world's first electric locomotive was built by Siemens in 1879. It operated for four months as a demonstrator at the Berlin Trade Exhibition. A large brass plate on the side says "Siemens & Halske, Berlin."

decided to apply for officer training in the Prussian Army Engineering Corps. His station in structured German life gave him only a slight chance of acceptance. Siemens walked the 160 miles to Berlin to take the challenging tests. He passed. And as luck would have it, his instructors also taught at the Berlin Academy. Army life was rigorous, but he received an excellent technical education over three years. Siemens met Georg Ohm (1789-1854) and received his commission in 1838. (See pages 58-60.)

Siemens's parents died within six months of each other in 1840. As the eldest son, he assumed the responsibility to support and educate his siblings. He even brought Wilhelm to live with him in Magdeburg to help him finish high school. Siemens's education and army work encouraged his interest in the emerging technology of electricity. In his free time, he worked on designing a new dynamo to use for electroplating gold and silver. He took out his first patent and the dynamo proved surprisingly successful. Siemens sold the German patent rights. Wilhelm went to England with it in 1843 and sold rights for its use in that country for £1,600. About that time, Friedrich and Karl moved in with Werner to finish their educations. They had previously lived with other relatives.

Still in the Army, Siemens transferred to a military administrative unit. The change in duties resulted in his meeting Hermann Helmholtz (1821-1894), one of Germany's greatest electrical scientists. Their discussions prompted Siemens to work on an in-

dicating telegraph, similar to that of Charles Wheatstone (1802-1875) in England. (See pages 76-78.) Working with instrument maker Johann Georg Halske (1814-1890), they developed a serviceable system. In 1845, the partners established the Siemens and Halske Telegraph Company. They devised a new method for forming insulation and soon became the leading supplier of telegraph wire. Siemens invented a press in 1847 that applied a seamless insulating coating of *gutta percha*. The insulation was rigid and strong, and nothing could match it. It started out as tree sap from the jungles of Malaya. Its name evolved from the Malaysian words for "gum" and "tree," *getah* and *perca*.

Siemens's military work helped him get contracts to develop telegraph systems for northern Germany. His first major line linked Berlin with Frankfurt, about 300 miles apart. It was Europe's first long-distance telegraph line. Siemens left the army in 1848, on his way to many technical triumphs. He invented a duplex circuit in 1854 that sent two signals on the same line. He patented an improved dynamo armature in 1856. In 1870, his company completed an incredible London-Berlin-Tehran-Calcutta telegraph line. The 6,000-mile line, some of it underwater, was completed in a series of stages. Younger brother Wilhelm designed the cable-laying ship *Faraday* in England. In 1875, it laid the first transatlantic cable to directly connect Europe with the United States.

Wilhelm became a British citizen in 1859 and handled the company's electrical interests in that country. He was the first to use an electric arc to melt steel in 1878. Wilhelm developed a highly successful open-hearth furnace for making steel. Friedrich moved to England to help with the Siemens-Martin open-hearth steel process. Pierre Martin (1824-1915) had independently discovered the same method in France. Youngest brother Karl moved to St. Petersburg, Russia, and helped establish that country's telegraph system. They purchased 75 indicating telegraphs for the St. Petersburg to Moscow telegraph line. In spite of the great distances that separated the four brothers, they traveled often and regularly saw each other.

Siemens spearheaded a company that connected much of Europe and Asia with telegraph lines. But in terms of his inventiveness, his most important discovery was the self-excited dynamo. Early dynamos used expensive and inherently weak permanent magnets. They provided the magnetic field necessary to generate electricity. But electromagnets held the potential for producing more powerful dynamos. In 1866, Siemens found a way to use the small amount of magnetism that remained in the iron core of an electromagnet. It was enough to start electricity production when a dynamo began turning. Electromagnets then took over.

Siemens investigated electricity for transportation as well as communication. He built the world's first electric locomotive in 1879. It was a demonstration model meant to show the feasibility for its use in mines. The small train operated over four months, carrying a total of 90,000 passengers around a 300-meter circular track. Electricity came from a small steam-powered dynamo. The locomotive picked up the electricity from a flat center rail.

So internationally accepted were the Siemens brothers that three of them received titles of nobility in different countries. Werner was raised to the Prussian (German) peerage by Kaiser Friedrich III. Karl received Russian nobility status from Czar Alexander III. Wilhelm, who had changed his name to William, was knighted in England by Queen Victoria.

Unlike many hands-on technologists, such as Thomas Edison (1847-1931), Werner Siemens valued scientific inquiry. He often published his research results in the *Proceedings of the Berlin Academy of Sciences*. He helped establish universal standards of measurement. Siemens's financial assistance in 1887 resulted in the formation of the government-operated Physikalische Technische Reichsanstalt research organization. Helmholtz was its first director. He married Mathilde Drumann in 1852. After she died in 1865, he married Antoine Siemens, a distant relative. Siemens had six children. His two oldest sons, Arnold and Wilhelm, succeeded him in operating the family business in 1884. Self-made and one of the wealthiest people in the world, Siemens died a week before he turned 76.

The Siemens industrial organization had 6,533 employees when Siemens died in 1892. About 100 years later, it had 250 manufacturing sites in 42 countries. Siemens's companies had 46,000 employees in the United States alone. That massive corporation began because Christian and Eleonore Siemens made sure their 14 children respected and helped each other.

References

Werner von Siemens—Inventor and International Entrepreneur by Wilfried Feldenkirchen, Ohio State University Press, 1994.

Siemens Gmbh homepage [on-line]. Available: http://www.siemens.com/usa

Timetables of Technology edited by Bryan Bunch and Alexander Hellemans, Simon and Schuster Publishers, 1993.

Siemens's 1866 self-excited dynamo used residual magnetism in the iron core of the electromagnet. This meant it was no longer necessary to use weak and expensive permanent magnets.

George Henry Corliss

Born:
June 2, 1817,
in Easton,
New York

Died:
February 21,
1888, in
Providence,
Rhode Island

Steam engines were so important to the Industrial Revolution that people often directly link them. James Watt's (1735-1819) improvements made steam engines powerful and efficient enough for use in factories. (See pages 25-27.) After Watt's inventions, people continued to work on new possibilities for improving engine performance. Some tried huge cylinders. In 1769, at least eight had cylinder diameters larger than 60 inches. Others worked on a more regular supply of steam by installing multiple boilers for each cylinder. One 1763 engine in England had a 76-inch-diameter cylinder served by four boilers.

As useful as such modifications may have been, they were not design changes. One American invented a totally new form of steam valving that resulted in an engine with almost constant speed, which was particularly important for factories. Machine tools and production equipment in nineteenth-century factories were driven by a belt from one steam engine. Watt-type engines could not quickly respond when loads were added or removed from the main power shaft. George Corliss's 1849 improvements were so universal that an entire class of engine was named after him. The technical world ranked him as Watt's professional equal.

Corliss was born into an intellectual, well-to-do doctor's family in eastern New York State. His father practiced medicine until he was 80, and both parents took a strong interest in their son's education. The family moved to Greenwich, New York, in 1825 so that the younger Corliss could attend good schools. He showed particular interest in mathematics and mechanics, and he developed a strong interest in mechanisms. After completing school at 14, he set

From the Collections of the Henry Ford Museum and Greenfield Village (Negative Number B51749.)

out to make his own way in the world.

America was an agricultural nation and Corliss could not find a job that allowed him to pursue his mechanical interests. So he started his working career as a clerk at the company store of William Mowray and Sons, a cotton manufacturer. Corliss learned the basics of bookkeeping, sales, customer relations, and other business pursuits that would help him later in life. He remained with Mowray for four years, until he felt he had learned all he could there. At 18, Corliss asked his father to send him to a Vermont preparatory school for additional schooling. After three years at the school, he returned to Greenwich where he opened a general store. At this stage in his life, Corliss was still not sure what professional path he would take.

The minor incident that launched his amazing steam engine career concerned customer complaints about the stitching in leather boots that Corliss sold. Corliss assumed responsibility for the problem. Because of his interest in mechanical things, he invented a machine to sew the boots back together. It passed two needles and thread through the heavy leather in opposite directions at the same time. The machine worked reasonably well. Corliss received a patent for his sewing machine in 1842 and went to Providence, Rhode Island, in search of a company to manufacture it. Corliss's mechanical abilities impressed people at a company that made machine tools and steam engines. Fairbanks, Bancroft, and Company agreed to help him with his machine if he would work for them as a designer/drafter. Corliss moved his wife, Phoebe, and their two children to the city where he would spend the rest of his life. He soon dropped his sewing machine pursuits, and within a year he had designed the basic valve mechanism that would revolutionize construction of steam engines.

Watt-type steam engines used a sliding valve to send high-pressure steam to one side of a double-acting piston while removing spent steam from the other side. A simple flapper in the main steam line from the boiler controlled engine speed. The sliding valve and flapper arrangement worked adequately, but the engine could not respond quickly to load changes.

Corliss's improved design placed separate rotary inlet and outlet valves at each end of the double-acting piston, four per cyl-

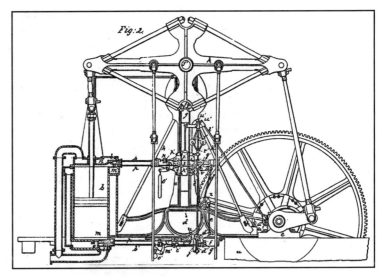

inder. The valves resembled round metal rods with a flat section on one side. A rocker arm quickly opened and closed the valves by rotating a small portion of a turn. Both intake and exhaust could be regulated. The valve action was so rapid that people referred to the valves as *cutoff valves*. All Corliss's improvements allowed his engines to maintain an almost constant speed, usually within 1 percent. That made them particularly popular with textile mill owners whose cotton thread quality depended on constant spindle rotation.

Corliss left Fairbanks in 1848 to form a new company that would be based on his technical innovations. Others found financial backing for Corliss, Nightingale, and Company. In 1849, that firm constructed the first Corliss engine that used his revolutionary rotary valves. The engines were large and turned slowly, making them more suitable for factories than for locomotives or motor cars. Operating from 100 psi steam, they typically had a 60-inch stroke with a single horizontal 24-inch diameter piston and a flywheel about 15 feet in diameter. They often developed 300 horsepower and turned at 70 to 100 rpm. Cities like Fall River, Massachusetts, became textile centers because of Corliss engines.

Corliss saved his money and purchased a controlling interest in

Some Corliss designs were beam engines as in this 1849 patent drawing. The huge piston on the left pushed and pulled the horizontal beam at the top of the engine. That action was transferred to a crank that rotated the geared flywheel at the right.

A patent model for an 1859 Corliss rotary valve mechanism is displayed at the Smithsonian Institution's National Museum of American History. Unlike those engines with long-throw sliding valves, only a slight motion was required to open or close rotary valves. The engine turned the offset circular disk at the left, which controlled valves that do not appear in this model. The plaque at the middle right says: "George H. Corliss, Providence."

his company. He had a new factory built in 1856 and changed its name to the Corliss Steam Engine Company. Within a few years, it had expanded to almost 1,000 employees with Corliss as president. It was the world's largest manufacturer of stationary engines. In spite of his business success, Corliss never forgot his technical roots. He personally devised almost all the continued improvements in his company's engine mechanisms. His first came in 1850, his next in 1852, then in 1858, 1867, and 1875. Two more significant improvements followed in 1880. In all, Corliss held 69 patents.

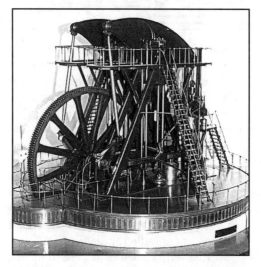

A model of the huge twin-cylinder Corliss steam engine used at the 1876 American Centennial Exposition in Philadelphia. Almost four stories tall, it provided mechanical power for all the displays in Machinery Hall.

After their introduction in 1848, many purchasers would consider only Corliss steam engines. Corliss had so much confidence in his expensive engines that he sometimes sold them under an unusual contract. Instead of an outright purchase price, the customer paid only the money saved in fuel over a two-year period. Corliss engines were about 25 percent more efficient than comparable Watt engines. One customer signed such a contract for a Corliss engine, then, to his surprise, saw a huge reduction in fuel bills. The customer soon negotiated to pay twice the market value of the engine. He would have paid much more than that had the contract not been renegotiated.

Corliss's most dramatic engine was the twin-cylinder steam engine that provided all the power for Machinery Hall at the 1876 Centennial in Philadelphia. It was the largest steam engine in the world and developed 1,500 hp. The disassembled 700-ton Centennial Engine arrived in 65 railroad freight cars. The pistons were 40 inches in diameter with a 10-foot stroke and operated at a steam pressure of 80 psi. The 56-ton, 30-foot diameter geared flywheel was the heaviest ever made. The engine turned at only 36 rpm and operated almost noiselessly. The 14-acre exhibition hall had 8,000 machines used for demonstrations such as rifling gun barrels, printing, veneering, brick making, and water pumping. A main power shaft under the floor meshed with the geared flywheel and extended toward all the exhibits. Six-foot-diameter bevel gears on the shaft allowed turns of 90 degrees. After the Centennial closed, George Pullman (1831-1897) bought the engine for his railroad car factory near Chicago. It operated until 1910 and was cut up for scrap in 1914. Standard-sized Corliss engines still exist in museums, which occasionally operate them for the public.

Corliss's name was as closely identified with steam power as any other in history. By 1876, he had built engines with a combined power of five million horsepower. He was recognized throughout the world. His biggest award was an 1870 Rumford Medal, a precursor to the Nobel Prize. Corliss was a man of high integrity and personally paid for the $100,000 Centennial Engine. Among the most honest of all American technologists, he was completely trusted by everyone. His handshake was a binding contract that he never failed to honor.

References

The Power of Steam by Asa Briggs, University of Chicago Press, 1982.

Great Inventions, edited by Charles Greely Abbot, Smithsonian Scientific Series (Vol. 12), 1934.

Heat Engines by Charles N. Cross, Macmillan Company, 1930.

"Harris-Corliss Steam Engine, Randall Brothers, Inc.," *Regional Historic Mechanical Engineering Landmark,* American Society of Mechanical Engineers Publications, October 16, 1985.

Maria Mitchell

It would be hard to find a science that predates astronomy. Early people probably stared at the heavens and wondered about the many bright spots in the sky. Astronomers in the sixteenth century took measurements with naked-eye instruments and tried to explain the motion of the stars and planets. Galileo Galilei (1564-1642) was the first to aim a telescope toward space, in 1609. Isaac Newton (1642-1727) did his best work when he explained the motion of the planets around the sun. His 1686 *Mathematical Principles of Natural Philosophy* is viewed by many as the greatest scientific work ever published. One of Newton's most impressive comments was, "If I have seen further than others, it is because I have stood on the shoulders of giants." Astronomy was an important early science because it captured everyone's imagination. People have always been curious creatures. Technologies that attempted to answer ancient questions often attracted the best minds.

That curiosity extended into the twentieth century. American taxpayers supported the first manned landing on the moon in 1969 by Neil Armstrong (1930-) and Edwin Aldrin (1930-). The space shuttle *Discovery* launched the Hubble Space Telescope into earth orbit in 1990. Named for astronomer Edwin Hubble (1889-1953), it was designed to see farther into space than had previously been possible. Those projects built on information gathered by giants of the type Newton described. One such person was America's first important woman scientist, Maria Mitchell. (Her first name was pronounced "mah-rye-ah.") She rose to prominence with her discovery of a comet in 1847. But perhaps her greatest technical contribution was her unorthodox teaching style. Instead of lecturing to students, she

Born:
August 1, 1818, in Nantucket, Massachusetts

Died:
June 28, 1889, in Lynn, Massachusetts

involved them in her work. Mitchell's graduates became excited about their futures and helped guide American science, engineering, and technology.

Mitchell was 1 of 10 children whose father enjoyed astronomy as a hobby. Her father held a variety of jobs on Nantucket Island. Among them, he was a barrel maker, a school headmaster, and a bank officer. Mitchell was a particular favorite of her father, probably because she had a natural aptitude for mathematics and science. Only one other sibling went on to a technical career. Mitchell's brother, Henry, was a hydrographer and studied navigable waters like oceans, rivers, and lakes.

Mitchell and her father often observed the night sky from the roof of their home. Like many other houses in the community, theirs had a fenced-in flat area on the roof known as a "widow's walk." The name suggested women looking out sea for husbands on returning ships. The Mitchell father-daughter team often adjusted chronometers (clocks) for the whaling ships by checking them against stellar observations. Ship navigation was such a way of life in Nantucket that practically every home had a sextant. Thus, Mitchell grew up in a community where people were generally in the habit of observing the night sky.

Mitchell's father felt that young women should be as well educated as young men

The Maria Mitchell Observatory at Vassar College is no longer used for celestial observations. However, classes are still held in the beautiful red-brick building.

and encouraged his daughter as best he could. He lost some of his family's wealth in a business deal. At 18, Mitchell went to work as a librarian at the new Nantucket Antheneum. She was there for 20 years and used the library's resources to improve her knowledge of mathematics, astronomy, and foreign languages. Her father's financial condition improved and he obtained an expensive four-inch telescope. He made star position observations for the Coast Guard, assisted by his daughter. They recommended particular sightings to determine a ship's longitude and latitude. During their thousands of observations together, Maria often recorded data for her father and made calculations. But he strongly encouraged her independent use of the telescope.

The last of the great naked-eye astronomers was Tycho Brahe (1546-1601) of Denmark. To commemorate Brahe, Denmark's King Frederick VII offered a gold medal to

anyone who discovered a comet using a telescope. To that point, all comets had been discovered by nontelescopic observation. As luck would have it, Mitchell located a comet in 1847. It was five degrees above the North Star. Mitchell reported her findings to Harvard University and astronomers there confirmed the sighting. Mitchell received the medal and the comet was named Miss Mitchell's Comet in her honor. The discovery was Mitchell's springboard to recognition. She was elected to membership in the American Academy of Arts and Sciences in 1848, the first woman so honored. The United States Nautical Almanac Office asked Mitchell to develop tables of positions for the planet Venus. She began traveling to scientific conventions and met other astronomers. Over the next several years, she also wrote articles for professional publications.

Mitchell spent a year in Europe beginning in 1857. She met many famous international scientists, including mathematician and science writer Mary Somerville (1780-1872) of England. Somerville had been a close associate of Ada Lovelace (1815-1852), the world's first computer programmer. (See pages 106-108.) Mitchell was a house guest of Sir John Herschel (1792-1871), a world-renown British astronomer. She also met with German meteorologist Alexander von Humboldt (1769-1859) and others. After her return to America, a group of women combined their resources and presented her with a five-inch refracting telescope. It had been made by Alvan Clark (1832-1897), an internationally famous American telescope maker.

When her mother died in 1861, Mitchell moved with her father to a married sister's home in Lynn, Massachusetts. Shortly afterward, a successful brewer, Matthew Vassar (1792-1868), established a women's college in Poughkeepsie, New York. Vassar College was the first with equipment and resources to equal men's colleges. Matthew Vassar was eager to hire a female astronomer of Mitchell's stature. He offered her free reign and a modern observatory with a 12-inch refracting telescope, the third-largest in the country. Mitchell had not attended college and feared that she might not be prepared for a teaching career. But Vassar's offer included everything Mitchell had ever wanted. She moved to Poughkeepsie in 1865 and became professor of astronomy and director of the college's observatory. She held the position for the next 23 years. Mitchell lived in the observatory, which is now a National Historic Landmark. Although it is no longer used for telescope observations, regular classes still meet in the building.

Mitchell was always a forthright and honest person. With no background in higher education, Mitchell taught her students the only way she knew how, by involving them in her astronomical observations. Her colleagues considered her methods radical, but she produced excellent graduates. Mitchell's work included observations of solar eclipses and changes in planetary surfaces, particularly Jupiter and Saturn. She pioneered the daily photography of sunspots. She observed that sunspots are whirling vertical cavities and not clouds as many astronomers of her day believed. A remarkably encouraging teacher, Mitchell inspired her students by sharing all her research work with them.

Mitchell was often abrupt in her dealings with people. But she was kind and helpful to interested students, some of whom even purchased their own telescopes. Mitchell had graduates like Ellen Swallow Richards (1842-1911), the first woman to receive a technical degree from an American univer-sity. Richards graduated from the Massachusetts Institute of Technology in 1873 with a degree in chemistry. Another of Mitchell's students, Christine Ladd-Franklin (1847-1930), earned an international reputation for her work with optics. The book *Who's Who in America* listed 25 of Mitchell's graduates.

Mitchell was an early member of the American Association for the Advancement of Science. She helped organize the Association for the Advancement of Women in 1873. She was the first woman elected to membership in the American Philosophical Society. Founded by Benjamin Franklin (1706-1790), it was the country's first scientific society.

Mitchell's health began failing in 1888 and she tried to retire from Vassar. The college trustees refused her request. Instead, they gave her an indefinite leave of absence at full salary. Mitchell returned to Lynn, planning to continue her work with a small observatory she had built there. But she died the next year at the age of 70.

Mitchell is often described as America's most prominent nineteenth-century scientist. Vassar College has a faculty position called the Maria Mitchell Chair in Astronomy. But Mitchell also had interests related to social conditions. She hoped to see technical methods used to solve such problems. Besides being active in traditional scientific organizations, she also served as a vice president of the American Social Science Association. Mitchell led a productive life of discovery that influenced and encouraged generations of young women.

References

Sweeper in the Sky: The Life of Maria Mitchell by Helen Wright, Attic Studio Press, 1997.

Women Scientists in America by Margaret W. Rossiter, Johns Hopkins University Press, 1982.

Women in Science in Nineteenth-Century America, Smithsonian Institution Press, 1978.

Notable American Women, 1607-1950 edited by Edward T. James, Harvard University Press, 1971.

Pictorial History of Women in America by Ruth Warren Crown Publishers, 1975.

This bust of Mitchell is at the observatory at Vassar College. The bronze plaque commemorates the building's status as a National Historic Landmark.

Jerome Increase Case

Born:
December 11,
1818, in
Williamstown,
New York

Died:
December 22,
1891, in Racine,
Wisconsin

It has often been written that agriculture was the most important development in the history of the world. It allowed people to stay in one place, giving them the opportunity to establish other technologies that would improve living standards. Agriculture has been particularly important to America. Because of its geological conditions, the United States has large areas of fertile soil. Adequate rainfall helps support modern agricultural production. America's 465 million acres of cropland are the most of any country in the world.

But in the early 1800s, there was no efficient way to use that abundant natural resource. People still farmed and processed crops much as their ancestors had centuries earlier. Farming methods changed dramatically with the invention of mechanical equipment. Practical reapers invented by Obed Hussey (1792-1860) in 1833 and Cyrus McCormick (1809-1884) in 1834 allowed farmers to quickly cut wheat from large fields.

However, before the wheat could be milled into flour, the grain had to be separated from the plant. Earlier farmers used sticks to beat out the grain, a process called *threshing*, from the word *thrash*. Wheat production became completely mechanized in 1869 when Jerome Increase Case introduced a steam-powered thresher, or threshing machine. He was the first manufacturer of a steam engine used exclusively for agricultural purposes.

The youngest of four sons, Case was born into a farming family near Syracuse, New York. Originally from the eastern part of the state, his parents were among the pioneers in the Syracuse area. Case actively participated in clearing land and cultivating, threshing, and selling wheat. He had

Courtesy of Case Corporation

little formal education and went to school only when he could, but he showed a strong aptitude for work with machinery. Taking advantage of his son's ability, the elder Case purchased a small horse-operated thresher and put the teenager in charge of the family's thresher service.

Early wooden threshers required two sections. One was the power unit with one or two horses that provided rotation by walking on a slanted treadmill. A long circular leather belt connected the treadmill's output shaft to a cylinder on the second section, the thresher. Farm workers fed dried wheat into the rotating cylinder, which forced the grain from the stalk. The process was much like that of cracking open a peanut shell. A fan blew away unwanted plant matter as the grain fell into cloth sacks.

Case enjoyed his work and was soon regarded as an expert thresher. His good reputation and ability would stay with him all his life and would contribute to the success of his future company. Seeing the growth in reaper use, the young man decided he'd find his future in America's western farmland. For Case, that meant the state of Illinois and the nearby territories of Wisconsin and Minnesota. He purchased six small threshers on credit and loaded them onto a boat at the Lake Ontario port of Oswego.

Arriving in Chicago in late summer of 1842, Case unloaded his cargo. The city had a population of fewer than 15,000, and manufactured products were sometimes difficult to obtain. Case easily sold five of his threshers to finance the trip. He kept the sixth to use in a thresher service, as he had done for several seasons in New York. Sensing that the Wisconsin Territory was on the verge of a wheat boom, Case headed north. He spent a night at a hotel near Racine, where he learned that there was, indeed, a need for his services.

Case settled in the area and traveled among the local wheat farms. The thresher Case brought with him was originally patented in 1784 by Andrew Meikle (1719-1818) in Scotland. It was the first to use a cylinder, and other inventors had 267 patents granted for what they claimed were improvements. Case's thresher was described as having a "ground-hog" design. It was about the size and shape of a small desk and only generally resembled the one shown in the photograph here. Horses powered a belt connected to a cylinder that separated the wheat and chaff. A fan blew away the lighter chaff, and wheat fell down the chute and into cloth bags. Case could process 150 to 200 bushels of wheat per day. In the off season, he built an improved thresher that could process 300 bushels of wheat per day. He used a cylinder with short pegs that revolved rapidly between other pegs. The dynamic action shook grain from the stalk. It was a modification of a method pioneered in 1834 by American brothers John and Hiram Pitts. When farmers saw Case's thresher in action, they all wanted to purchase one.

After securing financing in 1844, Case built a factory to manufacture affordable, low-capacity threshers. He made improvements to each model under construction. Three years later, he constructed a modern three-story, water-powered factory that became the hub of his farm equipment business. His horse-powered threshers sold for

$290 to $325. Just 10 years later, Case was manufacturing 1,600 machines a year in the largest factory west of Buffalo. Much of his company's success was attributed to Case being an outstanding thresher himself. He gave many personal demonstrations at customer's farms. Seeing the company president deftly operating farming equipment helped promote brand loyalty.

After the Civil War, Case saw that continued westward expansion would require steam-powered agricultural equipment. He built a steam engine in 1869 that he unimaginatively named "Number 1." It looked like an early locomotive with a seat attached to its fold-down smokestack. The 70 psi steam engine developed 8 hp, but it was not self-propelled. Horses pulled it to the threshing site. Fueled with coal or wood, the boiler was about 2 feet in diameter and 10 feet long. It consumed over a cord of wood per day. The engine's single cylinder had a diameter of 7 inches and a 12-inch stroke. It turned the 42-inch flywheel at 180 rpm. A leather belt around the flywheel went to the thresher, located about 25 feet away for fire safety. Case matched the engine to an improved thresher he named *Eclipse*, which was much larger than his original ground-hog unit. The combination could thresh 1,000 bushels of wheat per day.

At a time when wheat brought a dollar a bushel and yields were about 25 bushels per acre, the J. I. Case Company sold few of the $1,000 engines. Only 75 left the factory in 1876 during the company's first year of steam engine production. Demand slowly grew as farmers saw the benefits of the new threshing method. It was typical for several farming families to make a joint purchase that they could share. By 1886, Case had sold a total of 3,163 engines of various power ratings. At the time, the company was the world's largest manufacturer of steam

Small model of an eighteenth-century, hand-operated threshing machine. Unprocessed wheat went into the top hopper. Unwanted chaff was blown away by the fan and grain fell down the chute into a cloth bag. The thresher resembles one used by Case when he was a teenager.

Case's original 1869 single-cylinder agricultural steam engine is on display at the Smithsonian Institution. The fold-down smoke stack has a seat for a person who guided the horses that pulled the engine from farm to farm.

engines. It would ultimately manufacture 36,000.

Case was so well respected that even those who disliked his gruff personality would comment on his honesty. After a dealer and plant mechanic could not repair a thresher in Faribault, Minnesota, in 1884, the 65-year-old Case unexpectedly arrived at the farm. He could not properly adjust the thresher either. Disgusted that such a product left his factory, Case burned the thresher to the ground. The next day, the farmer received a new machine that operated perfectly.

Case married Lydia Ann Bull in 1849, and they had four children. He was never one to live beyond his means. He used public transportation and often walked to the farms he visited. Public transportation schedules were imprecise in those days, and when they were first married, Case's wife convinced him to purchase a horse and buggy. He liked horses and kept several race horses at Glenview Stock Farm, a farm he owned in Louisville, Kentucky. In a play on his own initials, Case named his favorite and most successful horse Jay Eye See. The horse was the world's all-time champion trotter-pacer.

Case was a dynamic person who did not see technology as an end in itself. He felt an obligation to help society in other ways. He served three terms as mayor of Racine and two terms as state senator. Case was a founder of the Wisconsin Academy of Science, Art, and Letters. He used his resources to establish banks in Racine and Burlington, Wisconsin. Those banks provided loans for young people just starting out in life. In considering other people's situations, Case recalled how he had borrowed money for the six threshers he took to Chicago. His abrupt and straightforward approach to life camouflaged his concern for others.

References

Machines of Plenty by Stewart H. Holbrook, Macmillan Publishers, 1955.

J.I. Case—Historical Summary, JI Case, A Tenneco Company, Form No. 3798, undated (circa 1987).

Steam Power on the American Farm by Reynold M. Wik, University of Pennsylvania Press, 1953.

Edwin Laurentine Drake

Petroleum is an extraordinarily important liquid in the modern world. Using complicated searching, processing, and distribution networks, oil companies provide fuel for the majority of transportation systems. Petroleum also heats buildings and provides energy for many manufacturing processes. It supplies raw material for plastic, paint, medicine, and other products. Countless oil wells produce more than 20 billion barrels of oil worldwide each year.

Petroleum was known long before modern times, and the word has been in use since the Middle Ages. It comes from the Greek petra, meaning "rock", and *oleum*, meaning "oil." The foul-smelling, greenish-black liquid often oozed from the ground. In the eighteenth and nineteenth centuries, people used it as an ointment for wounds and skin abrasions. It was even taken internally in the mistaken belief it could cure certain physical problems.

Lighting was one important application. Petroleum could be distilled into kerosene for use in oil lamps. Whale oil was commonly used, but excessive hunting led to a whale shortage that began in around 1845. Some people thought that petroleum could be found by drilling for it, a novel idea at the time. The person in charge of drilling the first successful oil well was Edwin Drake. He found oil in northwestern Pennsylvania in 1859. By 1865, oil wells were producing 3.5 million barrels a year. Since all anybody really wanted was a little more kerosene, Drake's biggest problem was what to do with the oversupply of petroleum.

Drake was born into a farming family a few miles north of New York City. When he was eight, his parents moved to Vermont, where he attended public school. Drake showed no particular academic interests.

Rendering by Tim Harmon

Born:
March 29, 1819, in Greenville, New York

Died:
November 8, 1880, in Bethlehem, Pennsylvania

Eager to strike out on his own, he left home at 19 and worked as a clerk on a Lake Erie steamboat that traveled between Buffalo and Detroit. Drake was a very pleasant person who got along well with everyone. He later held clerk jobs in a hotel and dry goods store.

Drake moved to Springfield, Massachusetts, when he was 26 and got married. For five years, he worked at the Boston and Albany Railroad station. He left to become a conductor for $75 a month on the newly opened New York and New Haven Railroad. The job would give him a chance to travel and meet more people. Drake was never a healthy person. He was slender, frail, and prone to catching colds.

In 1854, Drake's wife died and he moved to a New Haven boarding house. James Townsend, the president of a local bank, also stayed there and the two struck up a casual friendship. Townsend was an organizer of the Seneca Oil Company. Seneca owned about 100 acres of land near Titusville, Pennsylvania, and had an oil spring on the property. Due to Townsend's encouragement, Drake purchased $200 worth of stock.

Drake contracted a back disorder that made it difficult for him to walk up and down the aisles of moving railroad cars. This forced his resignation from railroad work in 1857 and he had no other employment possibilities. Officers of the Seneca Oil Company had never been to Titusville, and soon they agreed to hire Drake to inspect and develop the property. Townsend suggested obtaining oil by drilling into the ground. Drake had never drilled a well and had little technical knowledge of any kind. His only qualifications seemed to be that he owned some stock, was available, and lived in the same boarding house as Townsend.

Smithsonian Institution Photo No. 91-13294

Drake's drilling equipment included a steam engine in the building at the left. It lifted a heavy drill bit to the top of the derrick, then released the bit to drop into the ground. Drake is at the right, wearing a top hat, with friend Peter Wilson, the town pharmacist, to the left.

Drilling into the ground was not a new idea. As America expanded west, people drilled wells to bring up salt brine to use in preserving foods. The crude oil that frequently came with the brine was an unwanted liquid, but it could be processed into lamp oil (kerosene), spot remover (naptha), candles (paraffin), and salve (petroleum jelly). During his trip by train, Drake stopped and studied salt-well drilling operations in Syracuse and Pittsburgh.

The slim, frail Drake arrived in Titusville in April 1858 with his second wife and infant daughter. The family stayed at the American Hotel where their room and board was $6.50 a week. Drake had married Laura Dow the previous year, and they eventually had four children. To impress local leaders, Seneca sent him letters addressed to "Colonel Drake." Although Drake never earned that rank in any military organization, the title stuck and he used it for the rest of his life. Working with a $1,000 budget, Drake hired several workers at $1 a day to dig for oil using shovels around known seepage areas. The method was slow, required much physical effort, and did not prove very productive. In the spring of 1859, Drake met William Smith, a salt-well borer who had an excellent reputation. Drake hired him at $2.50 a day.

Smith built a 50-foot derrick of pine and placed a strong pulley at the top. He fashioned a crude drill bit with a convex chisel point about four inches wide, the same diameter as the well hole. The drill bit resembled a garden trowel and was about 12 inches long. It screwed into the end of a narrow 20-foot-long square iron bar that weighed 500 pounds. A rope tied to the other end of the iron bar passed over the pulley and was attached to a drum connected to a 6 hp steam engine. The steam engine slowly rotated the drum, wrapping up the rope, and lifting the iron bar to the top of the derrick. Releasing the drum from the steam engine dropped the drill bit into the well hole. The heavy metal bar provided the weight that forced the drill into the ground. The drill did not rotate as happens with modern equipment. Smith selected a drilling site near a stream called Oil Creek because it always had a film of oil on its surface.

On August 27, after two weeks of drilling, Smith located crude oil at 69 feet. This was the first time that petroleum had been located at its source. The well was not a gusher. The oil rose in the well and was transferred to storage tanks with small buckets. The first day, the well produced eight barrels, but production reached 22 barrels two months later. By pure chance, Smith had located his well where high-quality oil was close to the surface. The Oil Creek field turned out to be the shallowest and most productive ever discovered.

Drake had difficulty storing the sudden large quantity of petroleum. He found all the empty whiskey barrels he could and built two large vats. He soon had 300 barrels of petroleum at the well site. By the end of the next year, 74 wells in the Oil Creek area were producing 1,165 barrels a day.

Oil wells soon used open derricks like those shown in this reproduction oil boom town in Beaumont, Texas, near the 1901 Spindletop oil field.

Since there were few refineries and no established pipelines, storage became a real problem.

Drake made a sales trip to the industrial centers of Chicago, Cincinnati, and Pittsburgh. He hoped to interest companies in petroleum's possibility as a lubricant. Traditional lubricants were refined from animal products. Drake had little success with his odorous and impure liquid. It took several years before the infrastructure was available to process and distribute petroleum efficiently. The first small crude refineries were quickly constructed in 1860. They had capacities between 1 and 100 barrels a day. About 75 percent of the petroleum was converted into profitable kerosene.

Drake remained in Titusville for four years. He could have taken out a patent on his drilling operation, but did not. He also ignored the advice of friends who suggested he become an oil developer in the area. Drake had an established reputation, was well liked, and could easily find people to work for him. He unwisely chose to remain a paid employee for Seneca. When it came to financial matters, Drake often lacked common sense. He saved $16,000, then moved to New York and lost all his money in oil speculation. He lived out his years with gifts from Titusville friends and an annual pension of $1,500 from the state of Pennsylvania. He and his wife moved to Bethlehem, Pennsylvania, in 1870 and he died 10 years later at 61.

Drake was one of those people who happened to be in the right place and at the right time. He had no technical background and did little innovative work. Yet he will always be remembered as the first person to drill a successful oil well. Few other technologists have been in a similar position.

References

"Petroleum: What is it Good For?" by Paul Lucier in *American Heritage of Invention and Technology*, Fall 1991.

"Sitting on a Gusher" by Hildegarde Dolson, *American Heritage*, February 1959.

Modern Petroleum, by Bill D. Berger et al., The Petroleum Publishing Co., 1978.

Lord Kelvin (William Thomson)

Born:
June 26, 1824, in Belfast, Ireland

Died:
December 17, 1907, in Largs, Scotland

The history of human achievement in science and technology can often be read in measurement units. One common temperature scale was named for Gabriel Fahrenheit (1686-1736). He was born in Gdansk, Poland, and made the first successful mercury thermometer in 1714. Another everyday scale was named for Anders Celsius (1701-1744), a Swedish astronomer. In 1742, Celsius was the first to describe a temperature scale based on water's boiling and freezing points. He recommended dividing the scale into 100 increments.

Both the Fahrenheit and Celsius scales were based on the characteristics of water. Technologists often find absolute temperature scales more useful. Absolute zero is the lowest temperature that can be reached. It occurs at -460° Fahrenheit. A scale that begins at that temperature is the Rankine scale, named for William Rankine (1820-1872) of Scotland. He was one of a handful of people who created engineering as an academic discipline. In the nineteenth century, engineering and technology were identical career paths.

A person who spent almost all his life in Scotland also has a temperature scale named for him. He is Lord Kelvin. Absolute zero on the Kelvin scale occurs at -273° Celsius. Lord Kelvin, born William Thomson, worked with heat and energy, or *thermodynamics*. He also invented a special galvanometer that made the 1865 transatlantic telegraph cable practical.

Kelvin was the fourth of six children in his family. He was born in Ireland, where his father was a professor of mathematics. His father had some ideas about education and personally taught Kelvin and his brother James at home. Kelvin's mother died in 1830 and his father assumed personal responsibility for raising all the children. When he ac-

Lord Kelvin—An Account of His Scientific Life and Work by Andrew Gray, J. M. Dent Publishers, 1908

cepted a position as head of mathematics at the University of Glasgow, he moved them all to Glasgow. Kelvin's father deserves some credit for providing the environment that placed his son on the path to technical discovery and international recognition. Kelvin was a child prodigy who entered the university at the age of 10, along with his older brother who was 12. Kelvin was usually at the top of his classes and James was often second. James eventually became a professor of engineering at the University of Glasgow.

Kelvin published the first of his 661 technical papers when he was only 17. It dealt with advanced mathematics. He continued his studies at Cambridge University and Paris University in France. Kelvin specialized in mathematics, optics, and astronomy. He received an advanced degree from Cambridge in 1845. A professorship in physics became available at Glasgow and the 22-year-old Kelvin accepted the position. His father's influence was obvious. But the professorship was a good fit for Kelvin. He stayed at the University of Glasgow for the next 53 years.

Scottish technical education was the best in the world at the time and Kelvin worked to keep up the standards. One of his early projects involved the construction of Britain's first physics laboratory. His department offered the most advanced physics instruction available in Britain through the 1840s. Early on, Kelvin had an interest in determining the age of the earth. In an era before the discovery of measurement methods based on radioactivity, this was a difficult determination. Kelvin based his analysis on estimated cooling rates when compared with the sun. That work pointed him toward studies with heat and energy, two subjects that fascinated him for the rest of his life.

Famous Men of Science by Sarah Knowles Bolton, Thomas Y. Crowell Co. Publishers, 1889

Lord Kelvin, later in life

James Joule (1818-1889) lived in England at the time and published papers on heat and energy that were all but ignored. Kelvin met Joule and was impressed with his work. He saw to it that Joule's technical efforts received proper attention. The two men became close professional allies, working together between 1847 and 1852.

Joule and Kelvin noticed that as gas cooled, its volume decreased by 1/273 for every degree Celsius below zero. Others had also observed that characteristic. But Kelvin proved that the volume would not drop to zero at -273° Celsius. Instead, all molecular motion stopped and the temperature could not go any lower. Kelvin called this the "absolute zero of temperature" and the name *absolute zero* stuck. He developed a temperature scale with its zero point at absolute zero. That made the freezing point of water 273° Absolute and its boiling point 373° Absolute. The absolute temperature scale was later renamed the Kelvin scale.

The results of Kelvin's and Joule's work became the basis for liquefying gases, such as liquid hydrogen. Liquid hydrogen fueled the

Saturn 5 rocket that took people to the moon in the 1970s. The fuel is also used by the Space Shuttle's three main central engines.

During the late 1800s, much of Britain's technical community was working on the theory and practical aspects of a transatlantic telegraph cable. The American Cyrus Field (1819-1892) laid the first cable in 1858, but it operated only briefly. He established a permanent cable in 1865, followed by others over the next few years. Kelvin sailed on I. K. Brunel's (1806-1859) huge ship *Great Eastern* during its first successful cable-laying crossing in 1865.

There were problems associated with receiving weak electrical telegraph messages from an underwater cable that was 2,300 miles long. The person in charge of signal transmissions was E. O. W. Whitehouse. He thought the underwater cable should operate at high voltage to ensure that the electrical current was pushed along the entire distance. Kelvin quietly thought otherwise. During a test,

Kelvin's 1876 tide predictor was the first machine to predict high and low tides. It combined 10 tidal components, with one pulley for each. During operation, the smallest wheel spun at 1,600 rpm.

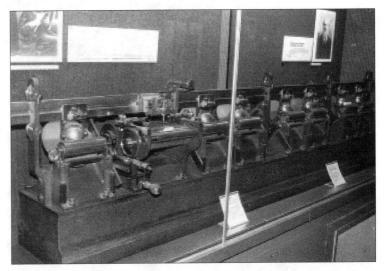

This 1878 harmonic analyzer interpreted graphical records of atmospheric temperature and pressure. Part of the mechanism was invented by Kelvin's brother James.

Perhaps Kelvin's most significant invention, this mirror galvanometer of the 1860s saved the undersea transatlantic telegraph cable project. This one was used at the North American end of the Newfoundland-to-Ireland cable.

Whitehouse's high voltages damaged the cable, but did not put it out of service. After some years of legal and technical arguments, Kelvin installed a low-voltage readout system that used his highly sensitive mirror galvanometer. About the size and shape of an ordinary flashlight battery, Kelvin's galvanometer had a tiny magnet and mirror hanging by a silk thread inside a coil of fine copper wire. Weak transatlantic signals arriving at the coil caused the mirror to rotate slightly. A beam of light reflected by the mirror multiplied the effect. Motion of the light beam to the left or right indicated Morse Code dots or dashes. Kelvin's mirror galvanometer had rescued a huge financial investment. Queen Victoria knighted him for his effort in 1866. He patented the invention and held a total of 70 patents.

Kelvin was the first scientist to grow wealthy through his many discoveries. His estate totaled £160,000 in the early 1900s. He married Margaret Crum in 1852. She became ill soon after the wedding and was an invalid until she died in 1870. Kelvin married Frances Blandy in 1874. They built a large house in Largs, near Glasgow, to complement one he had in London. Kelvin had no children from either marriage. He and Frances traveled to America in 1884, where he delivered a series of lectures in Baltimore and other places.

Kelvin once fell on the ice while playing curling, a Scottish game similar to shuffleboard. His leg was badly broken and he walked with a limp for the rest of his life. He enjoyed sailing and bought a yacht in 1870

that he named *Lalla Rookh*. He was an accomplished navigator and spent much of his spare time on the yacht. Kelvin's lifelong interest in sailing encouraged him to develop new instruments for that venture. His nautical inventions include new types of compasses, depth gauges, and a tide predictor.

After many years of productive scientific work, Queen Victoria conferred on Kelvin the title of Lord Kelvin in 1892. He was the first person to be so honored for technical achievement. He chose the name from the small River Kelvin that flowed through Glasgow. Kelvin received almost every honor a person could. He was president of the prestigious Royal Society from 1890 to 1894. He received awards from 250 organizations. He retired in 1899 and immediately signed up for university classes as a special student. He died eight years later at the age of 83.

Kelvin did much work with theoretical analysis. His inventions included navigation and communication equipment, electrical circuits like the Kelvin Bridge and the Kelvin-Varley voltage divider, and a huge mathematical analyzer that was larger than a metal lathe. An extraordinary and likable scientific genius, Kelvin once called himself a failure because he could not "fit physical science into the engineer's concept of nature." Britain accorded him its highest final honor. Kelvin is buried next to Isaac Newton in London's Westminster Abbey.

Author Dennis Karwatka during a visit to the University of Glasgow, where Kelvin worked for 53 years.

References

Famous Men of Science by Sarah Knowles Bolton, Thomas Y. Crowell Company Publishers, 1889.

The Making of the Modern World edited by Neil Cossons, Science Museum Publications, 1992.

Francis Marion Roots

Rendering by Tim Harmon

Born:
October 28, 1824,
in Oxford, Ohio

Died:
October 25, 1889,
in Connersville,
Indiana

The success of a technology often depends on the success of an important but sometimes obscure component. One example might be a simple fan for moving air, which is critical to a building's environmental control system. It forces cooled or heated air to all the rooms. Eliminating this piece of air-handling equipment would severely compromise heating and cooling systems. Processing steel also requires forced air for blast furnaces. It helps to provide the high temperatures necessary to melt metal. Two-stroke-cycle diesel engines use a blower to force burned gases from the cylinders. Invented by Charles Kettering (1876-1958) in 1929, a blower is crucial to the engine's design.

Francis Roots operated a water-powered textile factory in Indiana with his brother Philander (1813-1879). While experimenting with wooden models for a more efficient waterwheel, Roots invented a unique twin-lobe, continuous-flow blower. It was not a chance discovery since both water and air are elements of fluid power. The first evidence of what Roots had invented occurred in 1859. A belt-operated test unit, roughly the size and shape of a coffee table, forced air out the top and blew Philander's hat off his head.

Roots was born into a middle-class manufacturing household. His mother was Sylvia Yale, a member of the family that established Yale University in 1701. His father operated a profitable woolen mill that produced cloth for blankets, shirts, and dresses. Roots had four older brothers and a younger sister. He was educated in local schools and briefly attended Miami University, which was in his hometown. The Roots family was not wealthy and Roots's father made sure that he and his brothers gained familiarity

with every department in the mill. They all worked in the factory during the summer months. Roots quickly developed technical insights and made several small production recommendations to his father.

When he was 21, Roots's father sent him with a wagon to sell woolen goods in the more sparsely settled areas of Indiana, Illinois, and Iowa. He traveled the region for about a year and found a particularly good location to build a water-powered factory in Connersville, Indiana. Potential factory builders had to purchase rights to use the water. But the city wanted to bring in new industry and charged little for use of the Whitewater Canal. Another advantage was that the area had several sheep farms for raw material. Roots recommended that his father build a woolen mill in the city. A four-story brick building was constructed on what is now 6th Street. The new factory went into operation around 1847. Roots and his older brother Philander supervised fac-

A small modern Roots blower partially disassembled to show the interior lobes. This one is driven by an electric motor to provide a continuous flow of air.

Comparing this drawing from an 1881 patent with the disassembled modern Roots blower shows how little the basic design has changed. In the patent drawing, the upper lobe turns clockwise and the flow of air is from the left to the right.

tory operations.

Roots met Esther Pumphrey and wanted to marry her. But he felt he was not in a financial position to do so. Answering to the California Gold Rush, he went west in April 1849 to seek his personal fortune. He teamed up with others near Sacramento and they made a claim in an area called Scorpion Gulch. It was a successful location and everyone in the small group of partners profited. Roots returned home via the Isthmus of Panama with a considerable amount of money. He married in 1850 and he and his wife moved in with Philander and his wife for about a year. Roots had a large Victorian two-story house built and invested some of his new wealth in the woolen mill. He and his wife had a total of six children.

Francis and Philander Roots were close and worked well together. Their relationship paralleled that of the Stanley twins who established America's first successful automobile factory in 1897 and the Wright brothers who constructed a successful airplane in 1903. It was around 1859 when the Roots brothers considered building a new waterwheel to replace the overshot wheel that powered the factory. They made test models out of carefully crafted wood, but none worked particularly well. One of their designs consisted of two wooden lobes, each in the shape of a figure eight. The lobes revolved in opposite directions when water was forced past them. Their rotation turned a shaft to provide power for the textile machinery. During long-term tests, the wood swelled from the moisture and the device was not successful as a potential replacement for their waterwheel.

To refine the design, the brothers ran the device after it dried out. While rapidly spinning it during a balance test, the unit sent out a blast of air that blew Philander's hat from his head. The air discharged upward

from a 10-inch-diameter outlet. That minor event caused the brothers to consider using the unit as a blower for foundries. Their device had two long impellers mounted on parallel shafts. As one rotated inside the other, it provided a constant flow of air. It was not a compressor but what is technically known as a *positive displacement blower*.

The Roots brothers spent several years refining their design before deciding to go into limited production. They converted the basement of the woolen mill into a small factory for making blowers. They sold their first ones to blacksmiths and foundries. The continuous air blast onto burning coal or coke produced higher temperatures for working metal. Roots blowers were also used for factory and mine ventilation. Several went to the gold and silver mines at the Comstock Lode in Virginia City, Nevada.

Most blowers were fairly large. The Roots brothers made seven sizes in 1869, numbered one through seven. Their Model 5 blower, for example, was five feet long, 3-1/2 feet in diameter, and weighed 6,200 pounds. When rotating at a typical speed of 300 rpm, the blower could provide 5,800 cubic feet per minute (cfm) of air flow. The pressure was about 7 to 10 psi. The Model 1 blower, the smallest, produced 900 cfm

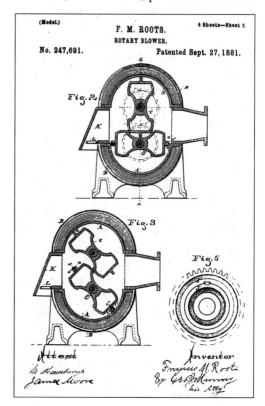

at 300 rpm. The largest, Model 7, produced 9,300 cfm. For comparison, a modern portable air compressor provides about 4 cfm of air flow.

The Roots brothers purchased an old Connersville foundry in 1864 and expanded their production. The business grew so rapidly that they sold the woolen mill. Other industries that found use for the Roots blower were refineries, food processors, chemical plants, water-treatment plants, and paper mills. Its most colorful application was for an experimental pneumatic subway in New York City in 1867. A Roots blower

This Roots blower was a supercharger on an early experimental aircraft. Navy Lt. Apollo Soucek flew it to several world altitude records with his XF3W-1 Apache fighter in 1929 and 1930.

literally blew a passenger car to its destination, one block away. The blower then reversed its rotation to create a partial vacuum and pulled the car back to its original location. The huge blower used impellers that were 16 feet long. It turned at 60 rpm and produced an air flow of 100,000 cfm.

The brothers worked well together and accumulated at least 15 patents. In general terms, Francis was the innovator and Philander was the manufacturing manager. Although neither would have characterized their contributions in quite that fashion. Francis Roots enjoyed traveling. He made several successful marketing trips across the United States and to Europe, sometimes accompanied by his family. The brothers' blower won many prizes at international

trade fairs that were popular at the time. They won top awards for design, workmanship, and efficiency at the 1867 Paris Exposition, at an exposition in Vienna in 1873, and in Philadelphia at America's 1876 Centennial Exposition.

Francis Roots was active in the civic affairs of his community. He was half-owner of the Connersville Hydraulic Company, which assisted water-powered factories on the Whitewater Canal. He served as its president for 14 years. He also served as president of the Connersville Furniture Factory. Roots helped establish the First National Bank of Connersville and was its president for seven years. A deeply religious man, he died at the age of 64.

The beautiful 1955 Ford Thunderbird is often considered a classic. Frederic Chopin's (1810-1849) exciting *Polonaise* is also a classic. And so is the timeless design of the Roots blower. First manufactured more than a century ago, it is essentially unchanged and still in production. Considering how technological improvements have altered so many products, that is a remarkable testimony. The National Aeronautics and Space Administration uses Roots vacuum blowers to help simulate conditions in outer space. All major NASA space facilities use them for chamber evacuation. The field of technology is not separated from the rest of the world and some technical innovations are clearly classic. Look carefully and you may find others that share that distinction with Francis Roots's 1859 blower.

References

History of Fayette County Indiana–Her People, Industries, and Institutions edited by Frederic Irving Barrows, B.F. Bowen and Company Publishers, 1917.

Welcome to Roots–Home of the Original Rotary Lobe Blower, undated company pamphlet, Dresser Industries, Connersville, Indiana, circa 1990.

Patent Force Blast Rotary Blower, PH and FM Roots Co. trade catalog, circa 1904 (copy in Henry Ford Museum Archives, Dearborn, MI).

Biographical and Genealogical History of Wayne, Fayette, Union, and Franklin Counties, Indiana, Lewis Publishing Co., 1899.

Joseph Wilson Swan

Born:
October 31, 1828,
in Sunderland,
England

Died:
May 27, 1914, in
Warlingham,
England

It is not uncommon for the popular press to introduce a historical technologist as the "inventor of the bicycle," for example. They might also refer to the "inventor of the fountain pen" or the "inventor of the milling machine." What they really mean is *an* inventor of a bicycle, or *an* inventor of a fountain pen, or *an* inventor of a milling machine. Few inventions have emerged fully formed from the mind of one person. Practically all evolved through the efforts of many people. One of the best examples is Thomas Edison (1847-1931), who is often characterized as the "inventor of the electric incandescent lamp." However, Edison did not invent the lamp and never claimed that he had. He grew tired of explaining that he only made it practical.

Edison tested his first successful prototype lamp in October 1879. Early records are unclear, but it appears that Scottish schoolmaster James Bowman Lindsay may have produced light from electricity in 1835. The first patent for an incandescent lamp may have been one taken out in 1845 by J. W. Starr (1822-1847) in Cincinnati, Ohio. The lamps proved unsuccessful because of the difficulty in removing all the air from the glass envelope. One electrical pioneer might have supplanted Edison in people's minds if only he had patented his lamp in a timely manner. Joseph Swan operated several incandescent lamps using filaments of carbonized paper in 1860. That was almost 20 years before Edison succeeded with a similar filament.

Swan was born in an industrial seaport community in the northeast of England. His father sold metal fittings for ships, but he did not profit from the business. He wasted much of the family income on impractical inventions and poor business practices. The

The Science Museum/Science & Society Picture Library

younger Swan attended the Hendon Lodge School until money ran out when he was 14. But he said the out-of-school education he received was invaluable in establishing his life's work. Sunderland was a bustling city of iron works, ship building, coal mining, and supporting industries. The inquisitive teenager liked to visit the shops with his brother John and to talk with the workers.

Although Swan had hoped for another career, circumstances led him to an apprenticeship with a druggist in nearby Newcastle. But the druggist was pleasant and provided an environment for Swan to work with the new technology of electricity. He experimented with coils, condensers, and batteries. At 17, Swan heard a lecture on electricity by W. E. Staite in Sunderland. From then on, he applied much of his free time to making a practical incandescent lamp. Doing so was not a new idea and had been discussed in technical circles for more than 30 years. While growing up, Swan's long winter evenings were spent in rooms poorly illuminated by candles and oil lamps. The dim light hampered his studies after sundown and he hoped to find a better lighting method. By 1848, he had developed an incandescent lamp using a carbon filament that glowed in a vacuum in a glass envelope. But the vacuum was not good enough nor were the batteries powerful enough. Swan turned to other areas of technology.

Swan left his apprenticeship to enter the pharmacy and chemistry business with John Mawson, who would soon become his brother-in-law. Mawson was a respected businessman and introduced Swan to many community leaders. Swan was a pleasant person, who soon emerged as head of the local literary society. The charming young man met people who provided financial backing for his early experiments. His first profitable venture was in the field of dry photographic plates.

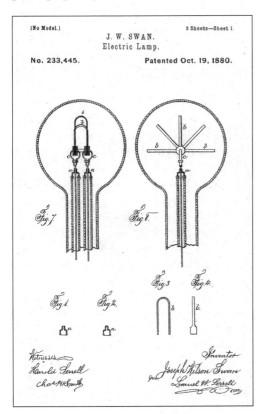

This drawing is from an American patent Swan took out in 1880. Although still using a thick high-current filament, the glass envelope and connections clearly show the influence of Edison's globe design. Swan's earlier lamps had resembled short bananas with electrical connections at each end.

Swan saw his first photograph in 1850 and became fascinated by the process. Early photographs were exposed on glass plates that carried a wet, sticky emulsion. Swan was the first person to produce a truly dry plate. He found that if a silver bromide solution was carefully heated, its sensitivity increased. Mawson helped finance the construction of a factory in 1856. Swan's work with photography led him to investigate light-sensitive photographic paper. He patented a bromide process in 1878 and paper like his is still used to print negatives.

While manufacturing his photographic printing materials and specialty chemicals, Swan continued his lamp experiments. He paid particular attention to filament quality and high vacuum. His 1860 lamp used a quarter-inch-wide filament heated by a 50-cell battery. But the arrangement proved impractical. Swan became a well-established inventor and ultimately earned more than 70 patents. He won several awards at the Paris Exposition of 1867. Then, that same year, several tragedies befell him.

His close friend Mawson died in an accident while trying to destroy some nitroglycerine explosive. Swan's wife of five years, Frances, and their twin boys died of a mysterious fever. To escape from a house marked by sad memories, he moved a few miles away to Gateshead and stayed in virtual seclusion for more than two years. In 1871, he married Hannah White, returned to his factory in Newcastle, and was soon working again at his normal pace. Swan had a total of nine children.

The development of efficient dynamos (dc generators) and improved vacuum pumps encouraged Swan to make another serious try at the electric lamp. He did well enough financially to hire a full-time assistant, Charles Stern, to help produce light from a glowing carbon filament. Swan's lamps operated at low voltage and relatively high cur-

The shorter building in the middle is the site of Swan's early work with incandescent lamps. He was a partner in a chemistry business with John Mawson at 13 Mosley Street in Newcastle. A circular plaque at the left of the entrance door commemorates the building's technical heritage.

Swan presented
this lamp to
London's Science
Museum in 1908.
The history of
this particular
lamp is unclear.
It may have been
made in 1878
and used as an
exhibit in an
important
demonstration in
1888.

rent. A typical 20 watt Swan lamp driven by a 5 volt dynamo required a current flow of 4 amperes. His lamp filaments had to be fairly thick to carry that much current. In addition, it would have been impractical to distribute low voltage dc and Swan seems never to have thought of doing so.

Swan made a public demonstration of a successful incandescent lamp to the Newcastle Chemical Society in December 1878. But his work with the electric lamp was primarily a hobby and Swan was slow to patent his invention. Less than a year later, Edison operated his lamp and immediately applied for a patent. He received it in January 1880.

Swan had seen his mistake and applied for a related American patent, which was granted in October 1880. He opened the Swan Electric Light Company in Britain in 1881. It proved quite successful. His lamps were the first used in the House of Commons of the British Parliament and in the British Museum. Swan once received an order from America for 25,000 lamps. Edison brought a patent infringement suit against Swan and they settled out of court in 1883, forming the Edison and Swan United Electric Company. The company dominated Great Britain's electric light manufacturing business for many years.

Swan's fertile mind also produced the first synthetic thread and fabric. While looking for a suitable filament, he produced a thick nitrocellulose solution. It was a dangerous mixture, chemically related to nitroglycerine. Swan developed a method of forcing the hot, viscous solution through a grid of small holes into a cooling liquid. He carbonized the resulting threads and used them as lamp filaments. In a related activity, Swan treated the threads to eliminate their flammability, wove them into fabric, and made some small textile articles. He called the material *artificial silk* but did not exploit the idea. The fabric was related to rayon, which was first commercially produced in about 1910.

Swan was a dignified and modest man who lived a long and productive life. He received many honors and served as president of three professional organizations. He

moved to London to be near his company and then to the south of England during his declining years. He was a poetic writer of lively letters and well liked by everyone. He died of a heart ailment at the age of 85.

In spite of the closeness of the incandescent lamp race, most historians credit Edison with inventing the first practical one. His lamps were more efficient and used hair-thin high-resistance filaments instead of the thicker low-resistance strips used by Swan. Edison also supported his invention with an electrical generating and distribution network. His 1882 Pearl Street Station in New York City was the world's first light plant. It produced 110 volts of dc electricity for 59 original customers.

Edison regularly researched the technologies with which he worked. He almost certainly knew about Swan's earlier experiments. And Edison was quick to settle his 1883 patent infringement suit. Without Swan's pioneering work, Edison might not have succeeded with his own incandescent lamp.

References

Sir Joseph Swan and the Invention of the Incandescent Electric Lamp by Kenneth R. Swan, Logmans, Greene, and Company Publishers, 1946.

Inventions That Changed the World, Reader's Digest Association, Ltd., 1982.

George Herman Babcock and Stephen Wilcox

Few forms of early eighteenth-century energy have had the staying power of steam. In modern America, steam generates more than 85 percent of the nation's electricity and drives many large ships. It heats homes and buildings and provides rotational power and heat energy for manufacturing. Its importance to the modern world cannot be overemphasized. And perhaps the most important component in any steam system is the boiler. Far from being simple containers, boilers are sophisticated devices that safely develop large volumes of high-pressure steam. But they are largely overlooked. Modern room-sized boilers are often located in inaccessible parts of buildings.

Early haystack-shaped boilers produced relatively small quantities of low-pressure steam for engines built by Thomas Newcomen (1663-1729). (See pages 13-15.) His huge beam engines operated at a leisurely six cycles per minute to pump water from English coal mines. Later steam engines provided mechanical power for machine tools in factories and generators in electrical plants. Those engines were the first portable *prime movers*, devices that took energy as it occurred in nature and used it to drive machinery.

John Stevens (1749-1838) developed an improved boiler in Hoboken, New Jersey. (See pages 31-33.) His small vertical boiler for an experimental 1825 locomotive had several internal tubes. The tubes provided a larger heating area, which produced more steam. Partners George Babcock and Stephen Wilcox built on that design with their first patented safety boiler in 1867. The high-pressure, high-capacity, multi-tube boilers they built for power-generating sta-

Smithsonian Institution
Photo No. 91-13726

Smithsonian Institution

tions would not explode if a problem developed. Their design now dominates boiler construction.

Babcock was born in eastern New York state. His father, an inventor, had several successful textile factory innovations to his credit. They included a weaving loom that made plaid fabric. His mother's relatives were also inventors. Babcock's family moved to Westerly, Rhode Island, when he was 12. Five years later, while still in high school, he met Stephen Wilcox. The two developed a friendship that lasted the rest of their lives. After finishing secondary school, Babcock took up the new technology of daguerreotype photography. He enjoyed the challenges provided by early photography and was a well-respected amateur photographer for the rest of his life.

Babcock started the *Westerly Weekly* newspaper when he was 19 and continued in the printing business until 1854. He and his father invented a successful color printing press that was manufactured in fairly large numbers. It won a medal in its class at the 1851 Crystal Palace Exposition in London, the first world's fair.

George Babcock
Born:
June 17, 1832 ,in Otsego, New York

Died:
December 16, 1893, in Plainfield, New Jersey

Stephen Wilcox
Born:
February 12, 1830, in Westerly, Rhode Island

Died:
November 27, 1893, in Brooklyn, New York

A model of Thomas Edison's Pearl Street Station shows two cutaway Babcock and Wilcox slanted-tube safety boilers on the ground floor. The world's first electrical generating plant, it had a total of four boilers with six steam engine/dynamo combinations on the second floor.

Wilcox was born at the south end of the Connecticut–Rhode Island border. His father was a banker and he had a comfortable early youth. Wilcox received his primary education in local schools and developed an aptitude with tools. Excited about technical work, he began early on to try to invent things. When he tried to patent them, he found most had already received patents.

Much like John Ericsson (1803-1889), the builder of the Civil War steamship *Monitor*, Wilcox was influenced by what he read about hot-air engines. He built one to operate fog horns for the U.S. Lighthouse Board. This project met with limited success, and Wilcox turned his attention to a related technology: steam. In partnership with D. M. Stillman, he patented a tube-type boiler with angled tubes in 1856. The invention provided the germ of the idea that would result in the world-renowned Babcock and Wilcox boiler.

Meanwhile, Babcock had moved to Brooklyn and worked for a patent lawyer while attending night school to learn technical drawing. With the outbreak of the Civil War, he took work at the Mystic Iron Works in Connecticut. He designed explosive shells and machinery for various steamships. Shortly after the war ended, he joined forces with Wilcox to open a factory in Elizabethport, New Jersey, to construct large stationary boilers. Banking contacts through Wilcox's family helped the men locate financing. They later built factories in Bayonne, New Jersey, and Glasgow, Scotland.

Babcock and Wilcox boilers were not simple containers for water, such as tea kettles. The boilers had many wrought-iron tubes inside the firebox to increase the heating surface. More surface area meant more steam pressure and a larger continuous quantity of steam flow. Thomas Edison (1847-1931) used four Babcock and Wilcox boilers in the world's first electric plant near Wall Street in New York City. Each developed 120 psi of pressure and could convert almost 3,000 gallons of water per hour into steam. Together they powered six 125 hp, single-cylinder Porter and Allen steam engines. The engines were connected directly to six generators with an electrical output to operate 7,200 incandescent lamps. The system went on line in 1882. Babcock and Wilcox boilers were the ones most frequently used by communities building their own electrical power plants. They found applications all over the world.

Babcock and Wilcox's 1867 tube-type, or *safety*, boiler was the first really successful design of its type. It had a working pressure of 200 psi. If 1 of the 12 to 28 tubes burst, only a small amount of steam would immediately leak out. The boiler would not explode like a shell-type (tea kettle) boiler. The slanted, individually replaceable tubes inside the huge firebox were only half filled with water, providing a space to collect steam. Coal was the typical fuel, although some boilers were later converted to fuel oil or natural gas. The fire directly heated the water near the top of the slanted tubes. Furnace baffles then forced the heat to pass down through the lower end of the water tubes. The design became very popular because it effectively eliminated serious boiler explosions.

Babcock and Wilcox tried building steam engines, but by 1878 the boiler business took all their time. Steam engine inventor George Corliss (1817-1888) recommended Babcock and Wilcox boilers, as did other manufacturers such as Armington and Sims, Westinghouse, and the Ames Iron Works. (See pages 112-114.) Chicago's Fisk Street power station installed 24 Babcock and Wilcox boilers in 1903. It was the first utility to use steam turbines exclusively for electric power production.

Wilcox was the real inventor, while Babcock acted as company president and salesman. Wilcox did much of his design work while sailing on his yacht *Reverie*. He was often assisted by his nephew William Hoxie. That experience may have influenced Hoxie, who went on to perfect a marine version of the Babcock and Wilcox boiler. A 250 psi model was standard in the British Royal Navy in 1900.

In all, Wilcox had 47 patents, several held jointly with Babcock. Wilcox was liked by everyone and seemed unaffected by his fame and influence. Although he was the one who developed the boiler, he agreed to have Babcock's name appear first in the company's title. Public spirited and generous with his wealth, Wilcox provided money to build a park, library, and high school in his home town of Westerly. He married Harriet Hoxie in 1865 and died at the age of 63 at his home in Brooklyn.

To help promote the business, Babcock wrote technical papers, gave lectures at professional meetings, and busied himself with the company's financial matters. He was a charter member and president of the American Society of Mechanical Engineers.

Babcock was also a visiting lecturer at Cornell University for eight years. He served on the board of trustees of Alfred University and contributed liberally to the school. He also provided financial support to expand the public library in the city of Plainfield, New Jersey, where he lived for many years. Babcock was a quiet, considerate, devoutly religious, and cultured individual, who taught himself French at age 60. He was married four times and died only 19 days after Wilcox.

Babcock and Wilcox were close friends who achieved a technical milestone. The field of technology has seen several similar partnerships. Joseph Brown and Lucien Sharpe had a machine tool company that grew to international prominence based on their 1862 universal milling machine. William Harley and Arthur Davidson provided the most enduring nameplate in motorcycle history with their first motorcycle in 1903. Leopold Mannes and Leopold Godowsky invented the first color slide film, Kodachrome, in 1935. Today, some people may be so focused on the technology alone that they might overlook the importance of human relationships to the field. But, the world would be noticeably diminished had not George Babcock and Stephen Wilcox been close friends.

One-eighth-scale model of Babcock and Wilcox's patented 1867 boiler.

This operational Babcock and Wilcox boiler is at the electrical generating plant at Greenfield Village in Dearborn, Michigan.

References

"Driving the Dynamo" by John Bowditch, *Mechanical Engineering*, April 1989.

A Short History of the Steam Engine by H.W. Dickinson, Cambridge University Press, 1939.

"Pearl Street Station—Fiftieth Anniversary" (C540-C556), c. 1922, General Electric Hall of History Archives, Schenectady, NY.

Alexandre Gustave Eiffel

Born:
December 15,
1832, in Dijon,
France

Died:
December 27,
1923, in Paris,
France

Sometimes a person's name becomes so closely identified with a particular item that it is hard to separate the object from the individual. Charles Wheatstone (1802-1875) used an electrical resistance bridge circuit named after him, though he did not develop it. (See pages 76-78.) His major claim to fame was a telegraph system that established the first intercity communication link. It connected Liverpool with Manchester, England, in 1839. George Ferris (1859-1896) constructed his 264-foot Big Wheel for the 1893 World Columbian Exposition in Chicago. But he also constructed several bridges, three over the expansive Ohio River. Ferris was an expert in the properties of structural steel and established the profession of materials testing. In a similar fashion, the name Eiffel brings to mind a tall tower in Paris, France. It was built in 1899 and remained the world's tallest structure for more than 30 years. Considered unsightly by many people at the time, it has come to symbolize Paris. Gustave Eiffel and his staff designed the tower and he supervised its construction. But Eiffel also had other accomplishments to his credit. Like Ferris, Eiffel had been a competent bridge builder with an impeccable reputation. He also worked on the early stages of the Panama Canal, was a pioneer in the field of aeronautics, and designed the supporting framework for the Statue of Liberty.

Eiffel's youth was typical of others in his region of east-central France. His father was an officer who served with Napoleon's Imperial Army and was twice wounded. He was posted to the garrison in Dijon where Eiffel was born. Eiffel was the oldest child in the family and had two younger sisters. Eiffel managed the difficult achievement of graduating from high school with a double

emphasis in literature and science. In 1852, he entered the Ecole Centrale des Arts et Manufactures, one of Europe's first great engineering schools. Eiffel majored in chemistry, because he expected to go to work in a vinegar distillery owned by a favorite uncle, Jean Baptiste Mollerat. Just before graduation, however, Eiffel's parents and the Mollerat family had a disagreement over politics and stopped speaking to each other. Knowing that he could not work for his uncle, Eiffel sought a job in Paris.

Primarily because he graduated from the highly regarded Centrale, Eiffel found employment with Charles Nepveu. Nepveu's firm specialized in building railway rolling stock and other equipment. The firm merged with a large Belgian railway company and Eiffel completely abandoned chemistry for construction work. At age 25, his first big project was to supervise the construction of a bridge over the River Garonne in Bordeaux. The $600,000 undertaking was completed efficiently and on schedule.

With a well-established career in hand, Eiffel married Marie Gaudelet in 1862 and they moved into a Paris apartment. They had five children before her death from pneumonia 15 years later. Eiffel never remarried.

A French economic recession in 1864 led Eiffel to work as a free-lance consultant. He was good at his work and found many individual contracts. His hallmark was building strong, lightweight wrought-iron structures. Eiffel was among the first to design such structures by following mathematical theory rather than intuition. His company grew as he worked on bigger and bigger projects. As a specialty, he built many bridges, mostly for railways. Some historians point to a bridge near Porto, Portugal, as his most stunning. It included a magnificent 530-foot span. Opened over the River Douro in 1876, it was named Maria Pia in honor of the Queen of Portugal. Another success was the iron structural work for the Bon Marche Department Store in Paris in 1879. This was the first glass and cast-iron department store. Eiffel also designed the 110-ton dome of the Nice Observatory in 1885 that could be moved by hand.

Paris was planning for the 1889 World's Fair to commemorate the one-hundredth anniversary of the French Revolution. Organizers wanted a dramatic structure to symbolize the event and conducted a design competition that Eiffel and his company easily won. Eiffel had an international reputation, was the best in the business, and was French. The exposition committee approved Eiffel's tower design in 1887 and construction began shortly afterward.

Often called the "Thousand Foot Tower," it was designed to be 300 meters tall. It would be the tallest structure in the world and Eiffel was breaking new ground with its construction. Some of Paris's citizens were outraged by a structure they viewed as an unpleasant-looking rude industrial imposition. Signed complaints came from writers, painters, architects, and ordinary citi-

zens. They took some comfort in the knowledge that the tower was scheduled to be dismantled 20 years after the fair closed.

Eiffel had evaluated potential wind loading in his

From *Gustave Eiffel* (Electra Editrice, Milano, 1957)

bridges and his tower design shows that influence. It is a cross-braced latticed girder that offers little wind resistance. The sway at the top is less than five inches. Perhaps its most unusual

The Eiffel Tower under construction in mid-1888

design features are the huge hydraulic cylinders under each of its four legs. Each cylinder was rated to lift 900 tons. Eiffel was unsure whether he could build such a large structure to be perfectly level. When he completed the first platform, 190 feet from the ground, Eiffel checked the structure. He added water to all the huge cylinders and permanently pinned them in place with iron wedges once he was satisfied the floor was level.

The Eiffel Tower took only 26 months to complete. One reason for the rapid construction was that Eiffel had many of its parts built off site. Another was that the parts were standard bars, angle brackets, or plates. The only non-French item on the tower was the inclined elevator be-

Photograph by Gina Karwatka

From *Gustave Eiffel* (Electra Editrice, Milano, 1957).

One of Eiffel's most dramatic bridges is the Maria Pia over the River Douro on the coast of Portugal.

tween the first and second levels. America's Otis Elevator Company built it, since no French manufacturer wanted that particularly difficult job. The 12,000-piece, 7,000-ton, wrought-iron tower was completed in April 1889, with a perfect safety record. No worker had died from an accident on the scaffolding. There are many statistics associated with the Eiffel Tower. Among the most interesting is that painting it takes 60 tons of cinnamon-colored paint.

Eiffel was also involved with the early stages of constructing the Panama Canal. The effort was initiated by Ferdinand de Lesseps (1805-1894), whose Suez Canal had opened in 1869. Work on the Panama Canal started in 1881 but was soon fraught with mismanagement and allegations of fraud. Lesseps hoped that Eiffel's impeccable reputation would ease the problems. Incredibly, Eiffel agreed in 1887 to do what he could. It was a bad decision and only served to tarnish Eiffel's image and embroil him in the legal system. Unfamiliar with jungle construction projects, Eiffel estimated that he could finish the canal by 1890. It was an impossible task and France effectively ended construction in 1891 and completely abandoned the project in 1904. The United States paid $40 million for the French interests in the Panama Canal.

With the advent of balloons and aeronau-

tical gliders, Eiffel renewed his interest in the power of wind. When he was 73, he had an apartment and laboratory on the tower and was quite unwilling to retire. He experimented by sailing scale models 377 feet down from the tower on thin wire. That activity evolved into a wind tunnel Eiffel built in 1909 at the foot of the tower. A 70 hp electric motor drove a fan that blew air over carefully constructed test models. By 1911, he had carried out more than 5,000 experiments. The low-speed wind tunnel still operates but is now used to evaluate automobile shapes.

Eiffel published 31 books and received medals from many countries. He was always formal in his bearing. His grandchildren rarely saw him without a suitcoat and he typically maintained an air of aloofness. He remained vigorous until well into his 80s, when he taught his grandchildren to swim and fence. Eiffel distributed five mimeographed copies of his memoirs to his children two months before he died at 91. Eiffel quickly became identified with the tower during its construction. It remained the world's tallest structure until the 77-story Chrysler Building opened in New York City in 1930. Eiffel's writings suggest that he was mildly astonished by the publicity his tower generated. It was not taken down in 1909 as originally scheduled. The new technology of radio made the tower an ideal broadcasting location. And much like the Gateway Arch in St. Louis, the tallest monument in America, the Eiffel Tower has become its city's most immediately recognized landmark.

References

The Tallest Tower by Joseph Harriss, Regnery Gateway Publishers, 1975.

Modern Marvels, "The Eiffel Tower," a video presentation of the History Channel, written and produced by Andy Thomas, circa 1994.

Alfred Bernhard Nobel

When individuals are particularly good at what they do, organizations and institutions often try to recognize their accomplishments. Public schools acknowledge the academic performance of various students. Seasons of sporting activities often end with an awards ceremony. In the field of technology, as well, organizations honor people who make significant contributions. Sir Geoffrey Copley's generosity inspired the first major international award in science and technology. He gave the Royal Society of London £100 of seed money in 1709 to carry out scientific experiments. The society decided in 1736 that "a medal or other honorary prize should be bestowed on the person whose experiment should be best approved." Since that date, winners of the Copley Medal have included Georg Ohm in 1841, Josiah Willard Gibbs in 1901, and Albert Einstein in 1925. (See pages 58-60 and 145-147.)

Benjamin Thompson (1753-1814) established technical awards in America and Britain. (See pages 34-36.) He carried the title Count Rumford and provided $10,000 in 1796 to establish Rumford Medals in both countries. The American Academy of Arts and Sciences selects the American winners. Winners have included John Ericsson (1862), George Corliss (1869), Josiah Willard Gibbs (1880), and Thomas Edison (1895). (See pages 79-81 and 112-114.) The Copley and Rumford Medals acknowledged outstanding achievement of an international character. They provided a way to honor the world's best technical professionals.

Both awards remain active. Rumford Medals were given in the 1980s and Copley Medals in the 1990s. Alfred Nobel used Copley's and Thompson's frameworks to establish awards over a broader spectrum

Courtesy Deutsches Museum, München

Born:
October 21, 1833, in Stockholm, Sweden

Died:
December 10, 1896, in San Remo, Italy

of human achievement. He used the fortune he made by inventing dynamite to sponsor the awards. Since 1901, Nobel Prizes have been granted annually in physics, chemistry, medicine, economics, literature, and world peace. They continue to represent the pinnacle of human achievement.

Nobel was the third of four sons born into a family with an enviable technical heritage. His great-great-great grandfather on his mother's side was Olaf Rudbeck (1630-1702). A botanist, Rudbeck was one of Sweden's greatest seventeenth-century scientists. Nobel's father was a successful industrialist and inventor in his own right. Nobel received his early education in Stockholm. Then, Russia had purchased the production rights to an underwater mine invented by Nobel's father. The Russian government wanted him to supervise its manufacture and, at the age of nine, Nobel moved to St. Petersburg with his family. This provided an enlightening opportunity for

The Nobel Diploma for the 1964 Nobel Prize in physics for Charles Townes. Townes, an American, shared the award with two Russians for their combined work with lasers.

Nobels no longer had a market for their products. They declared bankruptcy in 1859 and returned to Sweden.

Nobel's father had often experimented with black powder, or gunpowder. The ancient Chinese or Arabs invented this mixture of crushed charcoal, sulfur, and saltpeter. Roger Bacon (1214-1292) first recorded the formula for the western world in 1242. A relatively weak explosive, gunpowder was used primarily for fireworks or to propel rockets and small shells. Nobel's father wanted to develop an explosive more powerful than gunpowder. He hoped to see it used to clear paths for road, rail, and canal transportation. Alfred became interested in a new oily explosive liquid invented in 1847 by Ascanio Sobrero (1812-1888) in Italy. It was nitroglycerine, an incredibly powerful—and equally dangerous—substance.

The term *high explosive* is sometimes casually used to cover all forms of explosive materials. Professionals in the field use the term more precisely. During combustion, a high explosive changes to a gas in an extremely short time. The rate of burning can be as fast as 20,000 feet per second (13,600 mph). Nitroglycerine, TNT (tri-nitro-toluene), and some other materials are classed as high explosives. There are also *low explosives*, such as gunpowder. They have burning rates of 300 feet per second or less.

When used in mining or road construction, nitroglycerine was imperfectly detonated by heat. It was particularly sensitive to mechanical shock. A small jolt could easily make it explode, often with disastrous

the young Nobel. Poor health from a spinal ailment kept him from attending school. Swedish and Russian tutors taught him at home, and Nobel and became fluent in Russian. He began serving an apprenticeship in his father's factory at 15. Nobel's parents encouraged their sons to be as inventive as they possibly could.

At the age of 17, Nobel began a long study trip that took him to Germany, France, Italy, and America. He had a lifelong interest in chemistry and used the opportunity to improve his knowledge and experimental skills. While in America, he received the bulk of his technical education working with Swedish-born John Ericsson (1803-1889). Ericsson made ship equipment of his own design. He is best remembered as the designer of the 1862 iron-clad Civil War vessel *Monitor*. Nobel developed excellent language skills during his travels. Along with Swedish and Russian, he also mastered German, English, and French.

Nobel's father became an important arms manufacturer. He helped produce equipment for Russia during the 1853-1856 Crimean War. Nobel returned to St. Petersburg and worked for his father during that period. He received his first patent in 1854 at the age of 21. It was a gasometer, a device for containing and measuring gas. To meet production schedules, Nobel's father built large manufacturing facilities. The Russian government canceled its contracts after it lost the war to the allied forces of Great Britain, France, and Turkey. The

A Nobel medal in physics

results. Nobel used that characteristic in 1863 with his first important invention, a detonator. Made from mercury fulminate, its small explosion caused a more predictable detonation of the nitroglycerine.

Nobel's father had found a comparatively simple way to manufacture nitroglycerine. The family constructed the world's first true factory for producing it in 1865 in an isolated region outside Stockholm. Nobel served as the company's director and plant manager. These responsibilities marked his beginning as an inventor and industrialist. Unexpected explosions caused by nitroglycerine's dangerous unpredictability had resulted in several deaths, including that of Nobel's younger brother Emil in 1864. Nobel's first major work was to make the substance a safer explosive.

Nobel focused on absorbing the liquid into a suitable solid. He found that diatomite worked best. A chalky earth material, it absorbed three times its weight in nitroglycerine, while still appearing to remain almost dry. The resulting product was dynamite, the most valuable industrial explosive ever discovered. Nobel took out an American patent in 1868. He coined the word *dynamite*, from the word *dynamic*. He packaged it in waxed-paper cylinders so it could be inserted into holes drilled in rock, coal, or mineral ores. Dynamite is a relatively safe and solid form of nitroglycerine that requires a blasting cap for ignition. Ordinary mechanical shocks do not cause it to explode. Now, more than 1,000 tons of dynamite are used every day on construction projects in America.

Nobel had interests in electrochemistry, optics, and biology, and he held 355 patents. His pioneering work helped later inventors make artificial rubber and silk, synthetic precious stones, and other products. But it was the manufacture of dynamite that made him wealthy. He had factories in 12 countries, which included operations in New York and San Francisco. Dynamite was still a dangerous product that led to many accidental deaths.

Nobel thought that dynamite would be used exclusively in civil engineering projects, but it was soon applied to weapons of war. In a mistakenly printed obituary, one newspaper called him a "merchant of death." Nobel was so disappointed by the turn of events, that he decided to use his wealth to fund annual prizes for those who "conferred the greatest benefit on mankind." His will provided $9.2 million for the awards following his death at 63.

Nobel never married and he traveled almost constantly. He had no permanent home during his adult years. A reserved, shy, and generous person who disliked publicity, he had few close friends. After his death, potential beneficiaries contested Nobel's one-page handwritten will. Such valuable prizes had never before been offered for professional achievement. Nobel's potential heirs considered him crazy for making such a bequest. Following several years of fierce litigation, the courts decided otherwise. The first Nobel Prizes were offered in 1901. Those first prizes carried both supreme international recognition and large cash awards of $30,000. Winners still receive their awards on December 10, the anniversary of Nobel's death.

Nobel received an American patent in 1868 for dynamite, which he called "Improved Explosive Compound." He had received a similar British patent in 1867.

References

Nobel—The Man and His Prizes, edited by the Nobel Foundation and W. Oldberg, University of Oklahoma Press, 1950.

Alfred Nobel—The Loneliest Millionaire, by Michael Evlanoff and Marjorie Fluor, Ward Ritchie Press, 1969.

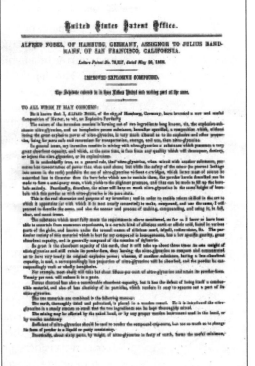

Ferdinand von Zeppelin

Born:
July 8, 1838,
in Konstanz,
Germany

Died:
March 8, 1917, in
Charlottenberg,
Germany

Some transportation devices are so large that it is hard to comprehend their size. The massive 1912 steamship *RMS Titanic*, with a length of 882 feet, was once the largest ship in the world. But another transportation device of the same era rivaled it in size. The huge cigar-shaped rigid airships of the early 1900s were the biggest flying machines ever constructed. The enormously successful *Graf Zeppelin* safely flew more than a million miles between 1928 and 1937. It was 775 feet long, well over three times longer than a Boeing 747 jumbo jet.

French brothers and paper makers Joseph Montgolfier (1740-1810) and Jacques Montgolfier (1745-1799) were the first to operate a lighter-than-air (LTA) craft. They filled a large paper-lined linen balloon with hot air in 1783. It floated in the air for 10 minutes and landed 1.5 miles from the launch site. Other people experimented with lifting gases. French scientist Jacques Charles (1746-1823) operated the first hydrogen-filled balloon, also in 1783. Through the 1800s, all such LTAs were unsupported gas bags, like children's small balloons. Ferdinand von Zeppelin had a different idea. He built rigid airships with aluminum frames that had significant advantages over gas bags. They could be fitted with several engines for propulsion and could readily carry passengers. His first, LZ-1, for *Luftschiff Zeppelin* ("airship Zeppelin"), lifted from a floating dock in southern Germany in 1900. Zepplin gave his name to a series of LTAs.

Zeppelin was born near Lake Constance in southern Germany. His family had a long history of government service, and he had a comfortable and supportive childhood. Zeppelin had a close relationship with both parents. His mother died when he was 13 and Zeppelin considered becoming a missionary. He graduated from a military high school in

The Science Museum/Science & Society Picture Library

1857 and studied engineering at the University of Tübingen. Zeppelin was commissioned as a first lieutenant in the Corps of Engineers after graduation.

To gain firsthand military experience, he went to America as a German observer during the Civil War. Arriving in 1863, he received a letter of introduction from President Abraham Lincoln that allowed him to move freely among the Union Army forces. He participated in several campaigns in Virginia and grew interested in flights made by hydrogen-filled observation balloons. Zeppelin took notes during experiments with tethered balloons at Fort Snelling in St. Paul, Minnesota, and personally flew to 700 feet. From then on, he constantly considered various forms of LTAs.

Zeppelin returned to Germany and, while briefly hospitalized after a horse-riding injury in 1874, he outlined his plans for a rigid-frame airship. He visualized a metal structure of vertical rings held in place by long rows of horizontal girders. Individual gas bags filled with hydrogen would provide lift. The entire airship would be covered with fabric. Zepellin did more than just speculate. His diary included a plan that had a gas capacity of 706,200 cubic feet in 18 gas cells and room for 20 passengers. A bold design, it lacked a proper power source. Zeppelin would have to wait until the four-stroke-cycle internal combustion engine was invented by his countryman Nickolaus Otto (1832- 1891) in 1876.

Zeppelin reached the rank of brigadier general and retired from military service in 1890. He had received at least two medals for bravery in battle. Zeppelin sought funding to design and build the first fully controllable rigid airship. The conservative German government considered his design too radical. He approached individual investors, who gave him 90 percent of the $250,000 necessary for the project. Zeppelin provided the remainder. Much came from his wife, Isabella, who drew from a family inheritance. The couple had one daughter. Work on the LZ-1 began in Stuttgart in 1898. Made with a structure of scarce aluminum, the elongated airship would measure 420 feet long and 38 feet in diameter. It was a difficult shape to work with. The LTA had to be rigid, strong, and lightweight—and have a large volume for the gas cells and crew.

Hydrogen was the most efficient lifting gas, but it was potentially explosive, not unlike the gasoline used in modern automobiles. To eliminate leakage, the pressure in the 17 drum-shaped gas cells was a very low 0.40 inch of water pressure, or about 0.014 psi. The gas provided a gross lift of 14 tons. The entire airship was covered with treated cotton fabric. Propulsion came from two four-cylinder 14 hp Daimler gasoline engines turning three-

bladed propellers. One engine was in a gondola under the airship about 100 feet from the rear. The other was in a gondola about 100 feet from the front. Five crew members, including Zeppelin, rode in the open gondolas. Ascent and descent were controlled by a 250 kg weight that moved along the bottom. When placed near the rear, the airship lifted its nose and the engines propelled it upward. Sliding the weight forward assisted in a controlled descent. A concept airship, the LZ-1 was not designed to carry passengers.

Because it was a prototype, partly constructed by trial and error, it took two years to build. The airship's first 18-minute flight in July 1900 began from a floating hangar on Lake Constance. With Zeppelin at the controls, the LZ-1 reached an altitude of 1,300 feet and showed some minor design flaws. Following modifications, it made flights in October that lasted more than an hour at speeds up to 17 mph. Thousands of spectators lined Lake Constance and saw the successful liftoffs. But the government decided not to spend public money on future airships. Zeppelin's company went out of business. LZ-1 was dismantled and sold.

While trying to keep the airship idea alive, Zeppelin often flirted with bankruptcy. But he never wavered in his strong faith in his LTA design. He had a steely quality and the photograph on the preceding page shows that characteristic. Zeppelin's determination and drive, especially at his advanced age, stunned the German public. But some described him as stubborn and

This model of the *Graf Zeppelin* shows how the control car near the front melded into the envelope. The construction shed was about 114 feet tall and the airship was 113 feet in diameter. The actual clearance was only 15 inches.

The *Graf Zeppelin* was scrapped in 1940. This nose section at Germany's national museum in Munich is one of the largest complete pieces remaining.

unwilling to accept technical advice from more knowledgeable persons. Nonetheless, his popularity soared. Ordinary people saw him as a visionary and a patriot who spent his own money to help the government with a worthwhile project. The German Parliament eventually voted to subsidize airship construction, and airships became a popular and safe means of transportation. Zeppelin's next was the 1906 LZ-2, and the LZ-3 came later in 1906. Many zeppelins had both names and an LZ designation. The 1911 LZ-10, for example, was named *Schwaben*, for a district in Germany. It was the first successful commercial zeppelin.

The tragic fate of the *Titanic* was still fresh in people's minds in 1912. Airships helped restore confidence in transportation technology. By 1914, the start of World War I, nearly 34,000 passengers had been safely transported in 1,588 airship flights. Zeppelin saw 92 airships constructed before he died at 78. Many had been built for use in the war.

Completed in 1928, the LZ-127 *Graf Zeppelin* (*Graf* means "count") was a long-range craft, built to connect Germany with New York City and Buenos Aires, Argentina. The airship flew around the world in 1929. With 3 million cubic feet of hydrogen, its heavy 94-ton gross weight easily lifted from the ground. It was propelled by five Maybach 550 horsepower engines turning propellers and had a cruising speed of 73 mph. Germany considered it the national flagship and it flew to many European destinations.

The success of the *Graf Zeppelin* encouraged Germany to build an even larger airship, the *Hindenburg*. Named for Paul von Hindenburg, the country's president between 1925 and 1934, it first flew in 1936. It was only 12 feet longer than the *Graf Zeppelin* and the largest airship ever built. Like the *Titanic*, the *Hindenburg* was destroyed under dramatic circumstances. It mysteriously exploded in 1937 while docking to a support pole at Lakehurst, New Jersey. A total of 36 people lost their lives, far fewer than the approximately 1,500 who went down with the *Titanic*. Until then, no airship passenger had ever even been injured.

But the images of the explosion and fire caused people to thereafter distrust airship travel. That was very unusual. People did not generally distrust ship, rail, or car travel following serious disasters. And airplane travel has always been popular, even during the 1930s when air crashes were not uncommon. The *Graf Zeppelin* was broken up for scrap in 1940.

References

Hindenburg by Rick Archbold, Barnes and Noble Books, 1997.

Giants in the Sky by Douglas H. Robinson, University of Washington Press, 1973.

Airship by Patrick Abbott, Adams and Dart Publishers, 1973.

The Great Dirigibles by John Toland, Dover Publications, 1972.

Lighter-Than-Air Flight edited by LTC Carroll V. Glines, USAF, Franklin Watts Publishers, 1965.

This drawing from an American patent taken out by Zeppelin bears a close resemblance to LZ-1. It was launched about three years after the patent was filed. The item marked "b" is a movable weight that assisted ascent and descent.

Expensive zeppelin flights were often partly financed through the sale of special stamps. This enlarged museum display stamp shows one issued by Madagascar. It depicts the *Graf Zeppelin* flying over New York City and includes a photograph of Ferdinand von Zeppelin.

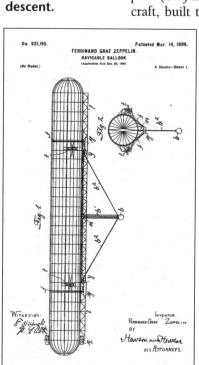

Josiah Willard Gibbs

Born:
February 11, 1839,
in New Haven,
Connecticut

Died:
April 28, 1903, in
New Haven,
Connecticut

The United States is the seat of enormous technical innovation and has been the home country of many great technologists, engineers, and inventors. These are the people who make things happen. Science is a related profession that explains *why* things happen. Science and technology are so closely connected that dictionaries often use the word *science* in their definition of the word *technology*. The relationship between the two fields is exemplified by the discovery of the basic principles of heat engines by French scientist Sadi Carnot (1796-1832) in 1824. His discovery did not find practical application until nineteenth-century technologists took his theoretical ideas and developed useful steam, gasoline, and diesel engines.

In spite of the importance of science, the United States has produced fewer scientists than some other countries. British-born Isaac Newton (1642-1727) and German-born Albert Einstein (1879-1955) are two examples of world-renown theoretical scientists. Only one native-born American was in the same league professionally as Newton and Einstein. Josiah Willard Gibbs led a quiet life while discovering the principles of *physical chemistry*, which are now used in all industrial processes.

Gibbs and his four sisters were born into a family with an academic heritage that stretched back six generations. His father taught religion at Yale University and his mother's father was a trustee at Princeton University. The children were raised in an intellectual and cultured middle class environment. Gibbs had the same name as his father so the family called him Willard or Will. He contracted a severe case of scarlet fever as a youngster that left him in frail health for the rest of life. He received his primary education in neighborhood private schools and attended Yale around the time that universities were establishing advanced technical degrees. He was the first American to receive a Ph.D. in engineering. Earned in the field of mechanical engineering in 1863, his doctoral thesis showed Gibbs's roots in technology. It concerned methods of machining gear teeth and was titled "On the Form of the Teeth of Wheels in Spur Gearing."

For three years after graduating, Gibbs tutored students in physics and Latin at Yale. Not knowing if he wanted to follow a technology or science career path, he also worked on steam turbines and engine governors. He also patented an inertia-operated railroad brake in 1866. Gibbs and his older sisters, Anna and Julia, received small inheritances after their parents and two other sisters died. This was enough to send all three to Europe in 1866. One sister, Julia, returned in 1867 to marry, but Gibbs spent

Gibbs's house on the Yale University campus once stood where this stone fence now stands. A chiseled plaque on the fence states, "Here Stood the Home of Josiah Willard Gibbs (1839-1903). Class of 1858. Professor of Mathematical Physics."

three years studying science and technology at universities in Paris, Berlin, and Heidelberg. His father had taught him basic French and German and he spoke both fairly well. The journey was a turning point in Gibbs's professional life and he decided on a career as a scientist. He returned to New Haven and Yale hired him as a professor in the field of mathematical physics. He was the first person in America to have such a teaching assignment and served in that capacity for the next 32 years. Other than his trip to Europe, Gibbs lived his entire life at the same address, 121 High Street in New Haven. The house, which no longer exists, was located a block away from his college building. A plaque on a stone wall directly across from Yale's library commemorates its location.

Gibbs taught upper-level classes like advanced mathematics, the wave theory of light and sound, and the theory of electricity and magnetism. But he also had time to consider technical research. Gibbs started with recently discovered laws of thermodynamics. He performed few, if any, experiments. But with an education grounded firmly in technology, his first investigations concerned the most practical subject of the time, steam engine efficiency. He went from analyzing steam engines to analyzing steam and then to analyzing the state of water.

Gibbs described in scientific terms how ice becomes water, how water becomes steam, and how steam becomes hydrogen and oxygen. He showed how every process in nature is one of change, and Gibbs was the first to discover those laws of nature. Newton discovered the science of mechanics, and Einstein developed the theory of matter and energy. Gibbs's great discovery was the science of physical chemistry. Where most scientists only contribute to a field of study, Gibbs created one.

Although not a particularly good teacher, Gibbs was an outstanding researcher. He soon developed an international reputation. The recently established Johns Hopkins University in Baltimore tempted him with an 1884 employment offer of about $5,000 a year. Gibbs had held his position at Yale without a salary, while living on his small inheritance. Yale then belatedly provided him with a salary. Only two-thirds of what Johns Hopkins had offered, it was enough for Gibbs. He lived simply and spent little money.

Gibbs's work was so advanced that only a handful of people could understand his ideas. His publications are still hard to read and follow. Some European contemporaries even found it easier to rediscover his results than read his papers. Scientific publishers accepted Gibbs's work on faith alone—they, too, were unable to understand it. Within 50 years of the publication of Gibbs's work, chemistry had invaded all the world's industries. Making metal alloys was a chemical process, as was manufacturing photographic film, soap, lamp filaments, and thousands of other products. Production methods in those fields are based on reactions that have been analyzed, corrected, or predicted using the results of Gibbs's work. His discoveries made possible many new industrial methods. One was the production of new steels from precisely combining carbon and iron. Another involved the industrial synthesis of ammonia from nitrogen and hydrogen. Of all the great nineteenth-century technical theories, Gibbs's is the only one that still stands with-

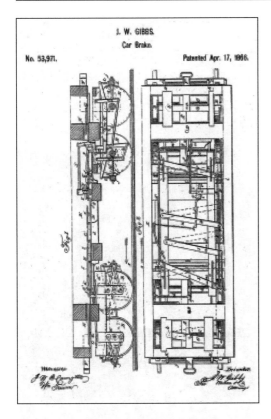

J. W. GIBBS.

Car Brake.

No. 53,971. Patented Apr. 17, 1866.

out major modification. All his publications were translated into French and German during the 1890s. In the half century after his death, Nobel Prizes were awarded four times for accomplishments based on some aspect of his work.

Gibbs was a quiet bachelor who lived in the family home with two of his sisters. Julia had married Yale's librarian, Addison Van Name, and had a family of her own. Relatives knew Gibbs as the gentle and understanding Uncle Will. They had no knowledge of his achievements. Even his students were often unaware of them. Gibbs once received a carefully crafted three-dimensional plaster model representing a relationship now known as a Gibbs Function. Coming from James Clerk Maxwell (1831-1879), the world's most influential scientist, it was an unsurpassed compliment. He merely told his students the model came from "a friend in England." Triode vacuum tube inventor Lee De Forest (1873-1961) was a student of Gibbs. In his diary, De Forest referred to Gibbs as "a great man."

While proud of his accomplishments, Gibbs had no aggressive pride nor did he advertise himself to the world. He was a simple, unaffected gentleman who had a quiet sense of humor and was highly ap-

proachable. Absorbed in his work, he was also content in his home environment. He enjoyed long walks and ice skating. He spent occasional summer vacations in the mountains of New Hampshire.

Besides physical chemistry, Gibbs also created two other fields of physics. They were *vector analysis* and *statistical mechanics*. Ordinary people did not understand Gibbs's work. But as was the case with Newton and Einstein in their times, most knew he was a great thinker. He received 19 awards and honorary degrees, including the highest international prize for scientific achievement, Britain's Copley Medal, in 1901. Gibbs also received America's Rumford Medal in 1880. Both awards were precursors to the Nobel Prize. Gibbs did nothing to invite such honors and even his closest friends never knew the full list of his achievements until they read his obituary in 1903. A slender bearded man with piercing blue eyes, Gibbs died from a brief intestinal ailment at the age of 64.

The outward life of J. Willard Gibbs was quite uneventful. He experienced no adventures other than those common to most other Americans of his time. He did not participate in any events of historical importance. He neither sought nor occupied positions of influence even in the field of science. He was never in the public eye. Yet his inner life was one of high adventures such as few ever know. He explored the far horizons of the unknown scientific world, traveling farther than any of his counterparts. America produced its only great theoretical scientist during the country's most materialistic years. There was never one before Gibbs and there has not been one since.

References

Willard Gibbs by Muriel Rukeyser, Doubleday, Doran & Co. Publishers, 1942.

Josiah Willard Gibbs—The History of a Great Mind by Lynde Phelps Wheeler, Yale University Press, 1951.

The Engines of Our Ingenuity, "J. Willard Gibbs," Number 119, by John Lienhard, University of Houston, circa 1993, broadcast by National Public Radio.

Drawing from an 1866 patent for a brake, the only patent Gibbs ever took out. An inertia brake designed for use on railroad boxcars, it was not a successful invention.

Karl Benz

Born:
November 25, 1844, in Pfaffenrot, Germany

Died:
April 4, 1929, in Mannheim, Germany

The modern automobile is so universal it seems its inventor's name should be common knowledge. But there is some uncertainty as to who that person was. Some people point to the bulky, wooden, three-wheeled, twin-cylinder, steam-powered carriage made by Nicholas Cugnot (1725-1804). It briefly took to the roads of France in 1769. Others cite the Austrian Siegfried Marcus (1831-1898). He mounted a gasoline engine on a crude chassis that may have operated as early as 1875. Both of these were landmark vehicles, but neither progressed beyond its initial successes. Each existed as a single point in technology and not as part of the continuous evolution of the motorcar.

Experimental motorcars of the late 1800s were called *horseless carriages*. Early technologists often used high-wheeled, horse-drawn wagons as the basis for their chassis designs. But Karl Benz did not. He designed a unique three-wheeled frame to carry his equally unique gasoline engine. Benz's patented 1885 motorcar was the basis for the first production automobile offered for sale to the public. He opened a factory in Mannheim, Germany to produce his Velo motorcar, named after the Latin word *velox* for "swift." He sold a total of around 1,200. Many people describe Benz as the person who designed and built the first workable motorcar driven by an internal combustion engine.

Benz was born in the Black Forest region of southwest Germany. His relatives had been blacksmiths for generations and his father started his career in that trade. Benz's father died at an early age from a job-related incident after becoming a railroad mechanic. Benz's mother had little income but great hopes for her son's future. The

The Science Museum/Science & Society Picture Library

two moved to the industrial city of Karlsruhe where she could earn money as a cook. Benz graduated from the local high school with a desire to continue his family's technical heritage. He entered the Karlsruhe Polytechnic Institute in 1860. There, Ferdinand Redtenbacher was one of his favorite teachers. Redtenbacher pointed out the need for a power source more portable than a bulky steam engine. That suggestion had a lasting effect on his teenaged student. Benz had some practical experience with internal combustion. The school purchased an engine fueled with natural gas and asked him to help with the installation. It used a two-stroke cycle, which Benz later used in a production engine he manufactured.

Benz was not a worldly man and preferred to stay close to his home region of Germany. After graduation, he worked at a locomotive shop in Karlsruhe. Although he received valuable experience, he saw no future at the company. Over the next several years he worked as a drafter, designed

weight scales, and helped build bridges. None of these were satisfying occupations. Benz met Bertha Ringer in 1870 and wanted to marry her. To become more financially secure, he and mechanic friend August Ritter opened a small machine shop in Mannheim.

The men specialized in making metal bending machines for other companies, but their business was barely successful. Ritter soon backed out and Benz's future bride came to the rescue. She partly funded the venture with her dowry and she and Benz married in 1872. The couple had a happy 57-year life together and raised five children. But the business was not so fortunate. Benz often flirted with bankruptcy.

Thinking he needed an innovative product, Benz wanted to build an internal combustion engine for general use. He had not yet considered the motorcar. Nikolaus Otto (1832-1891) had patented a four-stroke-cycle gasoline engine in 1876. Otto's horizontal, single-cylinder, five-horsepower bench engine weighed about 750 pounds. It was the size of a metal lathe. Had Benz made a four-stroke cycle engine, he would have had to pay a large royalty to Otto. He decided to work on a two-stroke-cycle design.

After a year of experimentation, Benz had his engine running for the first time on New Year's Eve 1879. At his wife's suggestion, the two spent that evening in the machine shop working on the balky engine. It started and Benz later wrote that they "both listened to it run for a full hour." Benz patented the engine and people began investing in his company. Benz was soon making bench engines for agricultural use, pumps, dyna-

mos, and other stationary applications. His engines produced 5.6 hp at 153 rpm and sold well. In his free time, he began to consider making an engine for a motorcar.

The courts declared Otto's patent invalid in 1886. Benz had hoped for such a turn of events and was working on a small four-stroke engine. Planning to use it for road transportation, he completed his first motorcar in late 1885 and operated it in the factory's courtyard. Other experimenters often fastened an engine to an existing carriage. Benz designed both the engine and the frame assembly. The car's tubular frame and wire-spoke wheels were influenced by bicycle design. The vehicle weighed 580 pounds and, to simplify steering, had just three wheels. The single-cylinder, liquid-cooled engine developed 0.90 hp at 250 rpm. Benz invented electric ignition to replace the flame ignition used in earlier engines. Although the vehicle had a clutch, there was no gearbox. The engine transmitted power directly to the rear wheels through a differential and two chains. Its top speed was 10 mph.

With Benz's stationary engine business expanding, he employed 50 people in a 40,000 square foot factory. He sold a total of more than 1,000 engines by 1894. He used some factory space and employees to help him construct and assemble his early motorcars. Initially, the vehicles were more of a hobby than a serious business venture. Benz built two others of slightly different design. He soon thought of the vehicle in terms of production and displayed one at the 1888 Munich En-

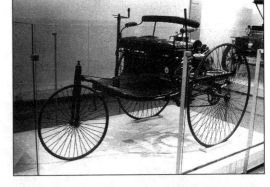

The original three-wheeled 1885 Benz motorcar is on display at the Deutsches Museum in Munich.

gineering Exposition. Although he won a gold medal, few people wanted to order an automobile. It was a new product, never before offered to the public, and they were unsure. Benz exhibited his motorcar at the 1889 World Fair in Paris and contracted

The first Benz production motorcars had three wheels. Benz's agent in France, Emile Roger, sold this model in 1888. It is displayed at London's Science Museum as the oldest motorcar in Great Britain.

with Emile Roger to act as his sales agent in France. A few orders began to trickle in.

Benz did not start a real production line until 1890. All his early cars were three-wheeled vehicles. He had sold only 69 cars up to that time, but that kept him in business. Benz started manufacturing his four-wheeled Velo in 1893. With a selling price of around $650, it had a 2 hp engine that averaged 15 miles per gallon. A Velo participated in America's first road race in 1895. Run between Chicago and Waukegan, Illinois, the race was won by Frank Duryea (1869-1967) driving one of his own cars. (See pages 178-180.) Duryea and his brother made America's first production cars in Springfield, Massachusetts. The only other vehicle to finish the 10-hour race was the Velo.

The Velo remained in production and sold well until 1900. Its estimated total sales were 1,200 and the number of company workers increased to 250. Benz built the first bus in 1895. It resembled a stagecoach, had a 5 hp engine, and carried eight passengers.

Benz was the world's largest manufacturer of motorcars. With the success of his company assured, he retired from active participation in the company in 1903, rejoining as a member of the board of directors. Benz enjoyed a long retirement and often traveled with his ever-supportive wife. They were frequently photographed together. Benz died at the age of 84.

This dramatic 1934 Mercedes Benz 500K was one of only 354 constructed. The display card explains that with its eight-cylinder 160 hp engine, the car "can still silently attain a top speed of 160 km/hr."

Emil Jellinek was a wealthy Austrian who distributed automobiles made by Gottlieb Daimler (1834-1900). He purchased a specially built Daimler motorcar in 1899. Jellinek entered a race in France, registering under the name Mercedes, after his young daughter. Jellinek won and was so pleased with the car that he ordered 36 for his dealership. It was a large order. His only requirement was that the model be called Mercedes. The Daimler Motor Works officially adopted the name in 1902. The com-

This 1897 Velo is shown with a mannequin at the steering tiller. The motorcar weighed 280 kg and its single-cylinder engine developed about 2 hp.

panies founded by Daimler and Benz merged in 1926, long after each had made his contribution to motorized transportation. The first Mercedes-Benz motorcar was sold in 1927. Production that year was 7,918 and the company remains the world's oldest automobile manufacturer.

References

The Star and The Laurel—The Centennial History of Daimler, Mercedes, and Benz by Beverly Rae Kimes, Mercedes-Benz of North America, 1986.

The Complete Encyclopedia of Motor Cars, second edition, edited by G. N. Georgano, E. P. Dutton and Co., 1973.

Internal Fire by C. Lyle Cummins, Society of Automotive Engineers Press, 1989.

The Engines of Our Ingenuity, "First Auto," Number 125, by John Lienhard, University of Houston, circa 1993, broadcast on National Public Radio.

Otto Lilienthal

The Science Museum / Science & Society Picture Library

Born:
May 23, 1848, in
Anklam, Germany

Died:
August 10, 1896,
in Berlin,
Germany

When considering early heavier-than-air flight, many people think first of Wilbur and Orville Wright. Their 1903 flight in North Carolina was the first that involved a controllable and powered aircraft that carried a person. But there were other types of heavier-than-air apparatuses. These included kites, engine-powered model airplanes, and full-sized gliders. Early flight technologists experimented with all of them.

The Chinese or Greeks were probably the first to fly kites and may have done so as early as 200 B.C. Even the Wright brothers flew kites to evaluate their biplane designs. John Stringfellow (1799-1883) flew a twin-propeller steam-powered model airplane along a horizontal wire in London in 1848. With a 10-foot wingspan, Stringfellow's was the first airplane to fly using mechanical power. Although not a complete success, his design showed the way for others. But it was the people-carrying, full-sized gliders of the late 1800s that most captured everyone's imagination. Otto Lilienthal of Germany was the first person to fly in a heavier-than-air machine. He made his initial glider flight in 1891 and established aerodynamic principles that led directly to the Wright brothers' flights. The Wright brothers frequently and unhesitatingly credited Lilienthal with making their early successes possible. Lilienthal was the first person to design and systematically test aircraft that were heavier than air.

Lilienthal was born near the Baltic Sea in 1848. As children, he and his younger brother Gustav (1849-1933) spent time watching storks on the marshland near their home and noted the arched shape of their wings. They then built small model gliders to test youthful theories. Some worked well, but not all. With great enthusiasm, the two brothers built strap-on wings for their arms in 1867 and jumped from low rock formations. Made from wooden frames covered with linen, the wings did not lift them from the ground. The brothers then laboriously fastened feathers to the wings with tar. Although they failed in their early attempts to fly, both kept their excitement for technology and what it might offer for the future.

Gustav went on to become a successful architect, but Otto wanted to work more directly with technology. He attended the Potsdam Technical School for his early education. He then went to the Berlin Technical Academy, before joining the army in the 1870s. Returning to civilian life in 1880, he convinced others to invest in a machine shop business he wanted to establish in Berlin. He made small ship equipment, small steam engines, and other items he had invented. Lilienthal's most significant invention was an 1885 portable steam boiler for which he received German patent No. 1603. Lilienthal sold many of the boilers and the proceeds helped to finance his aeronautical experiments.

This full-size replica shows Lilienthal in his most popular Type 11 glider. He primarily supported himself with his forearms in short padded tubes, while he held a wooden crossbar with his hands.

Lilienthal soon realized that his first attempts at flight failed because he had not properly analyzed the problem. So he experimented with wing surfaces and was the first to prove conclusively that an airfoil shape produced more lift than a flat surface. One of his test devices resembled a small oil derrick. A weight attached to a rope slowly fell from the top, causing a wheel to spin. The wheel had various airfoil shapes attached to it, which Lilienthal evaluated. He used the results to make charts of lift pressure and drag force for the various shapes. These were included in his 1889 book *Bird Flight as the Basis of Aviation*. It was the most complete book on aerodynamics ever published and all aviation pioneers used it. The book included many excellent illustrations that the

One-seventh-scale model of Lilienthal's 1874 airfoil test apparatus. He used it to measure the properties of various airfoils by whirling them around a circular track. Lilienthal built several such devices while searching for a suitable airfoil. The biggest, reconstructed here, was 4 meters high and had a diameter of 7 meters. The two 200 kg weights descend and provide energy for the rotation. Lift and drag could be measured as soon as the rotation rate had stabilized after the release of the weights.

talented Lilienthal personally drew and painted. For research, he kept four young storks and hand fed them to tame them. He studied their flying and gliding techniques.

There was no technical information on flight, so Lilienthal now made his first tentative attempts from low rocks as he had done as a youngster. He then found that running down a hill into the wind was safer and more controllable than jumping off a high point. His first two experimental glider designs did not succeed, but his Type 3 showed real promise. It roughly resembled a modern hang glider. It had a frame of peeled willow rods covered with cotton fabric. Lilienthal covered the fabric with several coats of wax. The glider's wing span was 23 feet, it had a surface area of 151 square feet, and it weighed 40 pounds. Lilienthal attached himself to the glider with a leather harness and stood erect. He braced his arms in padded tubes and held a crossbar with his hands. In flight after flight, the confident and courageous Lilienthal lifted from the ground for short distances. He briefly sailed though the air with his legs hanging free. The glider had no flight controls because aviation theory did not exist in 1891 and Lilienthal did not know what was required. He controlled the flight by shifting his body over a limited range to alter the glider's center of gravity.

During the next several years, Lilienthal made more than 2,000 glider flights. He was the first person photographed in flight and images of him inspired would-be flyers all over the world. Wilbur Wright once wrote, "Of all the men who attacked the flying problem in the nineteenth century, Otto Lilienthal was easily the most important. . . . No one equaled him in fullness and clearness of understanding of the principles of flight." Most of Lilienthal's glider flights went no more than a few hundred feet and lasted 12 to 15 seconds. But several went more than 1,000 feet. Lilienthal learned to ride air currents and often flew higher off the ground than his point of departure. Lilienthal also influenced the American glider pilot Octave Chanute (1832-1910), who became one of the Wright brothers' most trusted advisors.

To provide a convenient launch site for his experiments, Lilienthal had a cone-shaped hill built in a field near his home with a shed on top for glider storage. The

50-foot-tall hill was made with dirt excavated from a canal under construction. No matter which way the wind blew, Lilienthal could run down hill into the wind. The hill provided a perfect launch pad for his experiments.

Lilienthal made 18 different types of gliders, 15 monoplanes, and 3 biplanes. The planes weighed between 33 and 55 pounds. One biplane had a shorter 18-foot wingspan and a 200-square-foot wing area. Lilienthal hoped the design would make it easier to control. He was wrong. He sold production copies of his gliders to others. His maneuverable Type 11 monoplane was the most popular and sold in great numbers. It easily folded up and could pass through an ordinary doorway.

Perhaps too strongly influenced by his observation of birds, Lilienthal briefly worked on powered flight that used a wing-flapping method. He had little faith in propellers because none existed in nature. Successful fixed-wing gliders like Lilienthal's could have been built at any time in human history. But earlier technologists were overly influenced by the sight of birds' flapping wings. However, they certainly observed how birds glided with wings held still, and modern historians have trouble coming to terms with this apparent oversight. Even the highly perceptive Leonardo da Vinci (1452-1519) designed an airplane on paper with mechanically complex flapping wings.

Lilienthal considered steam power and designed a 15 kilogram engine that he never constructed. He also developed a mechanism powered by a cylinder of compressed gas. A hand-operated valve allowed the mechanism to flap the tips of an otherwise fixed wing. Lilienthal calculated that he might achieve a four-minute flight with the aircraft. But he never had the opportunity to test it.

At the same time, Lilienthal was conducting an elevator experiment with another glider that used head motion to control the airplane's pitch. He used a rope attached to his forehead. Moving his head backward would cause the glider to ascend, moving it forward would cause it to descend. On a particularly windy August day in 1896, an unexpected gust caused the glider to nose up. Lilienthal may have been confused with his new and unfamiliar control method. Within two or three seconds, the glider stalled, lost lift, and fell from 50 feet. Lilienthal did not survive the crash.

It was never Lilienthal's aim to build a flying machine according to certain flight doctrines. He just wanted to fly and would have done anything to get into the air. He showed by example that mastery of flight was possible. Lilienthal was as much an ad-

Diorama of Lilienthal gliding down from a 50-foot hill at Lichterfelde, near Berlin.

venturer as he was a brilliant experimenter. During his flights, he always wore hiking boots and the clothing of an outdoor adventurer. He lived an exciting life of danger and discovery, and would not have had it any other way. Lilienthal provided inspiration for all the flight pioneers who followed him.

References

The Conquest of the Air by C. L. M. Brown, Oxford University Press, 1927.

Aviation–The Pioneer Years edited by Ben Mackworth-Praed, Chartwell Books Inc., 1990.

Science and Technology in Nineteenth Century Germany, Goethe-Institut London and Deutsches Museum, Munich, 1982.

A page from Lilienthal's book *Bird Flight as the Basis of Aviation*. He drew these images and colored several others for the book.

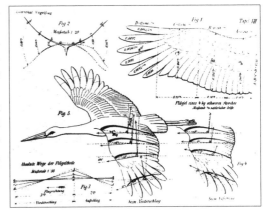

John Ambrose Fleming

Born:
November 29,
1849, in
Lancaster,
England

Died:
April 18, 1945,
in Sidmouth,
England

The late 1800s saw a rapid growth in desktop technologies like punched-card data processing, photography, and electricity. It wasn't obvious that electricity would soon have a major worldwide impact, but it was making surprising and diverse inroads. Thomas Edison (1847-1931) in America and Joseph Swan (1828-1914) in England manufactured incandescent lamps for sale in the 1880s. In 1888, Frank Sprague's (1857-1934) street cars in Richmond, Virginia, showed that electricity could be used for power applications. Telegraph systems by Charles Wheatstone (1802-1875) in 1837 and Samuel Morse (1791-1872) in 1840 showed the potential for its use in hard-wire communication. As the nineteenth century merged with the twentieth, electricity's close relative electronics began to emerge. But only those who could interpret subtle experimental results could read the signs pointing toward electronics. Heinrich Hertz (1857-1894) first sent electronic signals through the air in Germany in 1886. (See pages 76-78, 130-132, 166-168, and 169-171.) But his experiment did not immediately result in wireless communication. No one knew how to control, or *modulate*, the signal to send readable information. That task faced John Fleming. He created the field of electronics with his 1904 invention of the first practical radio tube.

Born in northwest England, Fleming was the oldest child in his family. His family moved to London a few years later and Fleming spent the rest of his life there. His father was a minister. His mother came from a family that developed the manufacture of portland cement, which was named after the Isle of Portland off England's south coast. Fleming's family was comfortable, but not wealthy. Fleming received his primary edu-

The Science Museum/Science & Society Picture Library

cation in the local schools and graduated from the University College of London (UCL) in 1870. He had to work his way through college as a clerk. In spite of limited study time, Fleming tied for top honors with one other student. Fleming continued his education by winning scholarships to various colleges. His special interest was finding methods to measure electrical resistance. He developed a bridge circuit he called Fleming's Banjo. It was used for several years at university laboratories. Fleming's academic work was of high quality and he accepted a position as "demonstrator in mechanisms and applied mechanics" at Cambridge University in 1880.

Early in his career, Fleming moved between academic and industrial positions. Incandescent lamps had just gone into production and Fleming was fascinated by the prospects of an electrical world. He briefly taught at a college in Nottingham, then resigned to become a consultant for the

Edison Electric Light Company in London. He stayed there for about three years. He was appointed professor of electrical technology at UCL in 1885 and remained there for 41 years. His early projects included transformers and high-voltage transmission, and he developed the right-hand rule. The rule relates the direction of the thumb, index finger, and middle finger to magnetic field, conductor motion, and electromotive force. Fleming worked with James Dewar (1842- 1923) on electrical effects at temperatures near that of liquid air, about -330° Fahrenheit. He was comfortable in front of groups and had a reputation as an excellent teacher.

Fleming worked on improving the carbon filaments in incandescent lamps. They had a short life and tended to darken the glass globe with use. Edison had unsuccessfully tried to improve bulb performance by placing an extra electrode alongside the glowing filament. He hoped it would absorb excess carbon, somewhat like a "getter" in later radio tubes. Edison observed in 1883 that when the extra electrode was connected to a positive voltage, a small current flowed between it and the filament. This was later called the "Edison effect." Edison saw no special value in the effect but patented it anyway. (Edison seemed to patent everything in sight.) That phenomenon was the key to the first radio tube. Bright as Edison was, he did not see the subtle secret being revealed to him.

Like some others, Fleming began serious research into the Edison effect in 1889. The merged Ediswan Electric Light Company gladly made 12 special lamps for his use. Many people had theories about the effect, but none had the answer. Fleming spent almost all his professional life in academics, but his experimental work made him the first in the field of applied electronics. He experimented with electrical conduction from glowing filaments in a vacuum. He hoped this would contribute to wireless communication, and he carried out countless experiments on transmission and reception. Fleming had a flair for linking complex electrical mathematics with their practical effect. He often made popular presentations on radio waves at the UCL, the Royal Institution, and the Royal Society. But the research seemed to lead nowhere and he left it for several years.

Fleming was a friend of Italian radio developer Guglielmo Marconi (1874-1937) and became a technical advisor to Marconi's company in 1899. He helped Marconi design a powerful wireless transmitter in southwestern England. That transmitter sent the first faint transatlantic radio signal to St. John's, Newfoundland, on December 12, 1901. It was the letter "S"—three dots—repeated over and over in Morse code.

The signal was hard to detect and Fleming looked for ways to improve radio circuitry. He faced the problem of converting a weak alternating current (ac) into a direct current (dc) to operate a receiver. Fleming realized that his 1889 Edison-effect lamp could convert ac to

One of Fleming's original diodes. This is one of the special lamps that Fleming had made in 1889 for investigating the Edison effect. He modified it in 1904 to make his diode.

Undated production Fleming valve. Its label says, "Fleming Valve. J. Ambrose Fleming developed the vacuum tube diode. This was a practical application of the 'Edison Effect' in 1904."

dc because it let current flow in only one direction. He later wrote, "To my delight, I . . . found that we had, in this peculiar kind of electric lamp, a solution." With the idea in his head, Fleming easily made up a small circuit with one of the lamps. He experimentally confirmed in 1904 that his invention could detect radio waves. The electron had just been discovered and the Edison effect was known to be caused by electronic emission from a heated filament. Fleming's discovery is generally considered to be the beginning of the electronic era.

Fleming's diode had two electrodes inside a glass globe from which the air was evacuated. One electrode, the filament, was heated and emitted electrons. Incoming radio waves were directed toward the other electrode. It accepted electrons from the heated filament when it was positive. It repelled electrons when it was negative. In other words, it allowed current to flow in only one direction. Fleming called the electrode a *valve*, since it turned on when current flowed in one direction and turned off for flow in the other direction. It worked just like a one-way or check valve in hydraulics. The electronic symbol, a triangle, was borrowed from hydraulics. Britons still call the tubes "valves," but they are known as *vacuum tubes* in America. Fleming's invention allowed more precise detection of radio waves and it was heavily used in the early years of amplitude-modulated (AM) radio communication. It was an essential part of all transmitters and receivers for more than 50 years.

Fleming was married to Clara Ripley for 30 years before her death in 1917. He married Olive Franks at the age of 84. He had no children. His favorite pastimes were wet-plate photography and painting with water colors. Like Edison, he was almost deaf and said that his poor hearing heightened his powers of concentration. Fleming was a captivating speaker. He made his last public presentation at the age of 90. He wrote three books and more than 100 technical papers. He was strong supporter of John Logie Baird (1888-1946) in his effort to establish television. (See pages 199-201.) Fleming served as president of the Television Society from 1930 until his death. He lived long enough to see his invention used in advanced electronic communication systems. A remarkably energetic individual, he remained professionally active almost all of his life. He died in 1945 at the age of 95.

Fleming's valve ushered in the field of practical electronic communication. It is still part of modern circuitry and used with television picture tubes, computer screens, and radio transmitters. But the valve was only a detector, not an amplifier. It laid the groundwork for the 1907 triode amplifier invented by Lee Deforest (1873-1961) in America. The triode made it possible to receive voice communication and opened up the world to anyone with a radio. The word *radio* was suggested as a trademark at a 1906 international convention in Berlin. It was widely adopted after 1915. No can say for sure when the Information Age began. A good argument can be made that it started with Fleming's simple electronic valve.

References

Making of the Modern World edited by Neil Cossons, John Murray Publishers, 1992.
The Principles of Wave Telegraphy and Telephony by John Ambrose Fleming, Longmans, Green, and Co. Publishers, 1910.

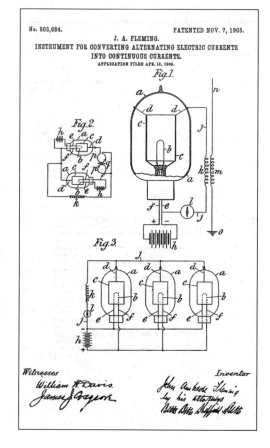

Fleming took out an American patent shortly after he applied for his British patent. Quite nicely drawn for the period, the hot carbon filament is identified as "b" and the secondary electrode as a metal cylinder "c".

Charles Algernon Parsons

Thomas Newcomen (1663-1729) showed people how to convert steam pressure into mechanical energy. He built the world's first steam engine in 1712 near Birmingham, England. James Watt (1736-1819) improved Newcomen's design and built engines more suitable for factory use. (See pages 13-15 and 25-27.) Steam engines were also used for ships, locomotives, and early electric power plants. Both Newcomen and Watt used the back-and-forth motion of pistons to develop useful work. The words *steam engine* refer to a piston-type power plant.

Technologists in the nineteenth century wanted to find a more effective way to use steam. Many worked on turbines. A turbine directly converts energy from a moving fluid to rotating mechanical motion. Water wheels and wind mills are simple examples. But high-pressure steam has much more energy than flowing water or wind, and the technical problems associated with harnessing it for use with a turbine seemed insurmountable. Every time technologists directed high-pressure steam at a turbine, it spun too fast, often destroying itself. Charles Parsons used an insightful approach in 1884. He built a steam turbine that had 30 sections, which allowed for using energy in stages instead of in one large blast. Parsons's first production units generated electricity for use on ships. Later ones powered the propellers for huge transatlantic liners like the 1907 *Mauretania*. Its Parsons steam turbines developed a total of 73,000 hp.

Few technologists have been born into a more academically talented family than Parsons. His mother, Lady Mary Rosse, was skilled at photography and photographic chemistry. She understood blacksmithing and designed much of the iron work around

The Science Museum / Science & Society Picture Library

Born:
June 13, 1854, in London, England

Died:
February 11, 1931, in Kingston, Jamaica

the family's Birr Castle estate in central Ireland. Parsons's father, the Third Earl of Rosse, was a wealthy and internationally recognized astronomer. He was the first to sketch the moon's surface and was president of the prestigious British Royal Society. His 12-ton telescope, with its six-foot diameter reflecting mirror, cost £30,000 and was the largest in the world. His academic credentials and offers of financial support attracted many scientists to Birr Castle. Parsons and his five older brothers were raised in an intellectually rich environment.

Although Parsons was born in London, his family lived in Ireland. His strong-willed father had little faith in the local public schools and hired private tutors for his children. Most of the tutors were scholars working in his observatory. Private instruction meant that Parsons rarely mixed with peers outside of his family. That may have accounted for the shy and retiring nature he sometimes displayed. He and his brother

Parsons's original 1884 steam turbine. Steam entered through the white flange in the center. It then split between 15 stages to the left of the inlet and 15 stages to the right of the inlet. It powered the alternator at the far right of the photograph.

Richard spent much of their spare time in the castle's extensive workshops. They constructed a small steam engine, a lens grinder, and an ocean depth gauge. They enjoyed taking the family yacht on voyages in the Irish Sea and the North Sea.

Parsons attended Trinity College in Dublin for two years and graduated from Cambridge University in England in 1877. His degree was in mathematics because British universities did not at that time offer a degree in engineering or technology. Parsons had little desire to follow an academic life and wanted to work with his hands. Parsons paid £500 to serve an apprenticeship at the Elswick Works of W. G. Armstrong and Company in Newcastle in northeastern England. Armstrong (1810-1900) was involved with shipbuilding and military artillery. An advanced engineering firm, his factory provided ideal practical experience. Parsons worked for Armstrong for four years. He then went to a locomotive builder in Leeds for two years before taking a position as junior partner with Clarke, Chapman and Company in 1884. That company manufactured electricity-generating equipment near Newcastle. The production of cheap electricity required driving systems with more power than piston engines. Parsons began to investigate steam turbines, a prime mover that had long been considered but never successfully constructed.

Instead of using just one spinning turbine wheel, Parsons connected several together. That allowed his turbine to take advantage of the steam power gradually instead of with one large blast. Within three months, he constructed a staged steam turbine connected to an alternator. The combination was the size of a small bed and produced 7.5 kW (10 hp). It rotated at 18,000 rpm, 15 times faster than steam engines. It was the world's first turbo alternator and the company manufactured them for electric lights on ships. The first to use one was the Tyne Shipping Company's *Earl*

The 100-foot-long *Turbinia* is the world's first steam-turbine-powered boat. It has been restored and is displayed at Newcastle's Discovery Museum.

Percy. Within just four years, about 200 were in service.

Parsons's axial-flow steam turbine showed a difference over modern designs. Capturing 100 psi steam traveling at 1,500 mph presented some technical challenges. Parsons decided to send the steam to the middle of the power plant. Half the steam turned left and went through 15 turbine stages. The rest turned right and went through the other 15 stages. This was a clever idea that worked well right from the beginning.

Patent royalties and family resources provided the money for Parsons to open C. A. Parsons and Company in 1889. He built equipment for the first public power station to use turbo alternators. The Forth Banks power station in Newcastle went on line in 1890 with two 75 kW units. Parsons then turned his attention to another potential application: turbine-powered ships.

After testing six-foot-long scale models in a pond, Parsons personally designed and had built an experimental boat 103 feet long by 9 feet wide. His slender Turbinia was a beautiful lightweight vessel that displaced 44 tons. It originally had a single propeller shaft that turned at 1,750 rpm from a 2,000 hp steam turbine. At first, *Turbinia* had disappointing results. The shaft rotated at such high speed that it produced *cavitation*. Parsons was the first to encounter this characteristic. Bubbles created from dissolved air produced large air pockets in the water and led to a drop in propeller thrust. Parsons built the first cavitation test tunnel in 1895 to evaluate the problem and design a solution.

A redesigned *Turbinia* had three propellers, each connected to a separate 700 hp steam turbine. Parsons introduced his new craft in a most dramatic manner. The British Navy was conducting sea trials near

Portsmouth in southern England in June 1897. Its 150 ships spread out over five square miles of water. In the *Turbinia*, Parsons easily steamed past the navy's piston-powered vessels at a speed of 34.5 knots (40 mph). At the time, British battleships traveled at 17 knots. High-speed destroyers could manage only 27 knots. The embarrassed naval officers did not appreciate Parsons's unauthorized demonstration and at first rejected steam turbines for their ships. But in 1906, the battleship *Dreadnought* became the first large ship to use steam turbines. The 1907 four-stack *Mauretania* was the first large commercial ship to use them. Named for an ancient Roman province in Africa, it had a top speed of 26 knots. The *Mauretania* made 269 transatlantic round-trips and held the record for the fastest crossing for 22 years.

Not all of Parsons's projects met with such success. He had an interest in the artificial

Parsons's 1895 cavitation device for evaluating propellers. Though plain in appearance, it was a sophisticated piece of experimental test equipment. It was the world's first cavitation tunnel.

production of diamonds and spent much time and money in the effort. He never succeeded in producing a diamond. Parsons also tried to market his own version of a phonograph. It was a technical success but a commercial failure.

Parsons was a pleasant man with a warm personality. He married Katherine Bethell in 1883 and they had two children. Their son lost his life in World War I and their daughter became a marine engineer. One of Parsons's pastimes was making mechanical

toys for his children and grandchildren. He made a rubber-powered helicopter and an airplane with a small engine. He later installed the engine in a small car that ran erratically around his backyard and greatly amused everyone. He always loved the sea and took voyages with his wife to such exotic places as the West Indies, South America, and Kenya. He died at 77 on the ship *Duchess of Richmond* during a visit to Jamaica.

All steam turbines trace their heritage to Parsons's original, now displayed at London's Science Museum. Steam turbines are used in ship transportation and it is not unusual to find them rated at 150,000 hp. But a more important application is generating electricity. That is how Parsons used his first turbine and his insight was astonishing. About 85 percent of America's electricity comes from fossil- and nuclear-fuel power plants that use steam turbines connected to alternators. Modern units with rotating blades and stationary vanes are amazingly similar to the one invented by Charles Parsons more than a century ago.

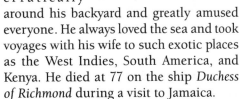

This 2,000 hp steam turbine was *Turbinia's* first power plant. It had an unusual design called an outward radial flow. It caused cavitation, was removed from the boat, and replaced with three smaller steam turbines.

References

Turbinia—The Story of Charles Parsons and His Ocean Greyhound by Ken Smith, Newcastle Libraries and Information Service, 1996.

Master Minds of Modern Science by T. C. Bridges and H. Hessell Tiltman, Books for Libraries Press, reprinted 1969.

"Turbinia, The Experiment Which Transformed the World's Navies" by Gerald James, *The Rolls Royce Magazine*, Rolls-Royce Industrial Power Group, March 1991.

Birr Castle web site.[On-line]. Available: http://ireland.iol.ie/birr-castle/science-center.html

King Camp Gillette

Born:
January 5, 1855,
in Fond du Lac,
Wisconsin

Died:
July 9, 1932,
in Los Angeles,
California

It has been said that all things can be arranged into one of two types. Successful inventions are no exception. They can be classified as "significant" or "less significant." Some inventions have been so important that they had a notable effect on world civilization. One example was Guglielmo Marconi's wireless telegraphy system. The Italian-born Marconi's most dramatic accomplishment came in 1901 when he sent the first transatlantic wireless signal. Another important invention was Robert Goddard's 1926 liquid-fueled rocket. Goddard was the only American to invent an internal combustion engine. His powerplant sends astronauts and satellites into space.

Then there are the less momentous but useful inventions. They affect day-to-day living on a more personal level. One example was Whitcomb Judson's invention of the zipper in 1893 in America. His original patent was for a "clasp locker or unlocker for shoes," but it evolved into the modern slide fastener. Another was the first practical ballpoint pen, invented by Lazlo Biro in Hungary in 1938. The ballpoint pen's tip did not break as easily as that of a pencil and it did not require refilling like a fountain pen. While not earth-shattering in importance, such inventions help eliminate some of life's minor annoyances.

One of those annoyances was endured by half the adult population at the turn of the twentieth century. Careful daily shaving with a sharp straight razor and its unprotected blade took most men about a half hour. It was such a risky morning ritual that many men grew beards or frequently visited barber shops. In 1901, King Camp Gillette invented an inexpensive safety razor that made shaving faster and safer. It

National Portrait Gallery, Smithsonian Institution

was successful right from the start and is the basis for all similar razors sold throughout the world today.

Gillette's parents both had careers and were good providers for their seven children. His father worked in business and had some minor inventions to his credit. His mother, an authority on the art and science of cooking, wrote several books on the subject. The most successful was her 1887 *The White House Cook Book*. Another book was published in five languages. In all, she may have sold three million copies of her books. The family moved to Chicago when Gillette was four and he was educated in the city's public schools.

The Great Chicago Fire occurred when Gillette was 16 and it destroyed all his family's possessions. The fire was a huge calamity for the city, burning out many neighborhoods and businesses. About 30 percent of the city's residents were left homeless. The disruption to Chicago's social systems encouraged Gillette to leave school and start working. His first job was at a wholesale hardware business at 40 Lake Street on the city's near north side. After two years there, he moved to New York City and took a similar job.

Gillette eventually became a traveling salesman for hardware firms. He made a decent living and married Alanta Ella Gaines in 1890. They had one son. Like many people, Gillette was always looking for a way to earn more income. One of his employers gave him a clue that put him on the trail of invention and wealth. William Painter had earned a fortune with a disposable bottle cap. It was similar to the modern variety and had to be removed with a small opener. His Crown Cork and Seal Company in Baltimore was once the world's largest bottle cap company. Painter told Gillette to "invent something people have to have, but which they can use, throw away, and then buy another one." His bottle cap was just such a device. Tin cans and disposable cardboard collars were others. To identify a potential product, Gillette ran through the alphabet listing items in frequent use. He stopped at the letter "R" for razor.

Men used flint or pieces of shells to shave their faces 20,000 years ago. Bronze and iron razors have been discovered during archaeological digs. Both the ancient Greeks and Romans were clean shaven. Barber shops in Athens and Rome were places of discussion and gossip. Shaving was usually for fashion or comfort, but another reason related to combat. An enemy could gain advantage by grabbing a man's beard.

Shaving with a straight razor involved scraping the face with a very sharp knife. Poor control could chafe the skin and a small slip could cause a nick. The average man often wound up with several nicks in the course of a slow and deliberate morning shave. Many men preferred to trust the job to a skilled professional barber. Barbers made lubricating shaving lather from soap at the bottom of mugs that looked like modern coffee mugs. Small brushes created the suds, which barbers applied to the customers' faces. Each customer had his own shaving mug at the barber shop. It was not unusual for a shop to have hundreds of mugs and several barbers. Barber shops

soon became male social clubs. Barber shop quartets originated there. The quartets were singing groups with four different voices. Gioacchino Rossini made the barber Figaro a prominent character in his 1816 opera *The Barber of Seville.*

Gillette had taken out some patents with his brothers Mott and George. So he had some knowledge of design and the patent procedure. Gillette thought about a razor design for several years. He carved a wooden model in 1895 that he thought could offer the convenience of a quick and easy shave at home. His safety razor used a

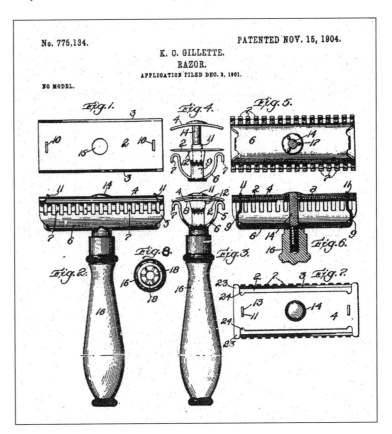

thin sharpened steel blade clamped between metal plates. A handle screwed the plates together. Enough of the blade was exposed to present an edge to the face but not enough to nick a cautious shaver. Gillette enlisted the help of Massachusetts Institute of Technology professor William Nickerson. Gillette had no formal technical training and Nickerson was an expert machinist and inventor. Nickerson's most successful invention was the push-button system for elevators.

The two men made their first double-

This is the drawing used with Gillette's safety razor patent. It was his only successful patent as a single inventor.

This 1960s-era safety razor differs only slightly from the one Gillette patented in 1905. The razor is shown with its stainless steel, double-edge blade. Stainless steel blades were introduced in the 1960s.

edged blades from steel ribbon used for clock springs. Experts told Gillette he couldn't grind a satisfactory razor edge on the paper-thin 0.006-inch-thick steel. Nickerson didn't agree and went on to make special machines that automatically made both the blades and the razors. The blades were the most difficult item to manufacture. Nickerson determined the proper size, shape, and thickness. He found a method for hardening the metal, perfected a tempering process, and designed the machinery to sharpen the edges.

But it took six long years before the idea was complete enough for Gillette to patent. He found financial backers in Boston who invested a total of $65,000. The Gillette Safety Razor Company opened in 1903 above a fish market. The company made 51 razors and 168 blades during its first year. The next year, production increased dramatically to 90,000 razors and 12.4 million blades. Customers replaced their double-edged blades when they became dull, usually after about four shaves. Within a few years, the company had several factory sites around Boston. They had a com-

bined floor area of five acres and 2,000 employees. Gillette soon had factories in Canada, England, France, and Germany. An entire generation was introduced to the safety razor when the American government issued Gillette razors to its troops during World War I. An image of Gillette appeared on every package of blades manufactured up to 1963.

Gillette became wealthy and retired from active participation in the business in 1913. He left much of the day-to-day company operations to others so he could spend more time on his main interest, social reform. He wrote four books on the subject. One, published in 1910, for example, was titled *World Corporation*. He retired to Los Angeles in 1922. Gillette spent much of his time growing fruit trees and working with real estate. He died at the age of 77.

Gillette was not the only person to become wealthy around the turn of the century. Andrew Carnegie (1835-1919) gave away much of his wealth for libraries, colleges, and other public organizations. George Eastman (1854-1932) contributed heavily to colleges and dental schools. Henry Ford (1863-1947) organized the Ford Foundation in 1936. It is the world's largest philanthropic organization. But Gillette was not in their league. For all his writing about social reform, Gillette used little of his considerable wealth to help others.

References

The Entrepreneurs—An American Adventure, by Robert Sobel and David B. Sicilia, Houghton Mifflin, 1986.

Sources of Invention, by John Jewkes et al., Macmillan Publishers, 1969

Purnell's Encyclopedia of Inventions, edited by Peter Presence Purnell Books, 1976.

Extraordinary Origins of Everyday Things, by Charles Panati, Harper & Row, 1987.

Everyday Inventions, by Meredith Hooper, Angus & Robertson Publishers (UK 609 H7665e).

William Seward Burroughs

Technical advancements have improved life, in part because people looked for easier ways to accomplish their objectives. Mathematical calculation devices are one example of this. Early civilizations used beads and stones as counters. The word *calculus* comes from Latin and means "stone" or "pebble." Ancient people once used those small markers to help them maintain numerical order while solving mathematical problems.

The first significant mechanical calculator appeared in France in 1642. It was an adding machine built by 19-year-old Blaise Pascal (1623-1662). The word *calculator*, related to the word *calculus*, describes a device that adds or subtracts, as well as one that can multiply or divide. Pascal was a very remarkable person who investigated mathematical theory and fluid power. The computer language Pascal is named after him. One of the basic rules of hydraulics is Pascal's law. It states that pressure exerted on a confined fluid is equal in all directions. The metric pressure unit of *kilo Pascal* (kPa) also comes from Pascal's name.

Pascal's adding machine was a small 12-inch-long rectangular brass box with six exposed metal wheels on top. Its seventeenth-century operator moved its wheels to add or subtract numbers. The answer appeared though small holes on top. The device, called a *Pascaline,* did not have a keyboard and did not resemble a modern calculator. Also, the crude construction methods used at the time resulted in unreliable operation. Pascal had 70 machines built for sale, but the venture did not succeed commercially.

The nineteenth century saw the introduction of adding machines with keyboards. That improvement proved both a blessing

and a curse. It was a blessing because it made the machines much easier to use. It was a curse because it required a complex mechanical design that was hard to manufacture. In 1888, William Burroughs became the first person to make a successful keyboard-operated adding machine. American financial institutions quickly adopted it for their day-to-day operation.

Burroughs was born near Syracuse in central New York State. His birth year appears variously as 1855 and 1857. A maker of patent models and other related devices, Burroughs's father had inventive talent and a thorough knowledge of mechanical devices. But he was not financially successful. The younger Burroughs was raised with two brothers and a sister. He received a basic education in the local schools but was not happy with his situation at home. Burroughs began earning his own living at

Born:
January 28, 1857 (?), in Auburn, New York

Died:
September 14, 1898, in St. Louis, Missouri

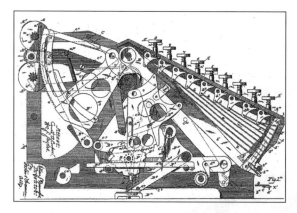

Burroughs's first calculator patent, issued in 1888, was very complex, but many of the parts were identical. Each column had nine keys, and each of those keys was connected to duplicate rods, levers, and gears. Burroughs felt that a full keyboard would eliminate input errors. He was right.

This Burroughs No. 3 registering accountant machine was made in 1898. It had a paper-tape output and the mechanism is visible at the rear. The color coding of the keys shows its financial heritage. The dark keys at right are for cents. The machine's maximum value would be $9,999,999.99.

15. He started working in the Cayuga County National Bank as a beginning-level clerk. He then moved to a retail store and a lumberyard.

By that time, typewriters and telephones had already made their appearance in modern office work. Both speeded up normal daily transactions. But bookkeepers and clerks like Burroughs had only traditional methods and struggled to keep up with the other advances. His work included mostly adding long columns of numbers. He found the activity demanding and monotonous. Burroughs had to check and recheck his work.

At age 24, Burroughs married Ida Selover. They had four children together. Also at age 24, Burroughs learned that he had tuberculosis. Medical professionals in New York advised him to find a milder climate and a less-demanding occupation. He moved to St. Louis in 1882. Burroughs found employment with a company that made woodworking machinery. His banking work and experiences with his father's patent model work led him to invest his spare time on inventing an adding machine. He met machine shop owner Joseph Boyer, who supported himself by making specialty items for other companies. Boyer was impressed with Burroughs and offered him shop space and equipment for Burroughs to test his ideas for a practical adding machine. He even provided an assistant and supported Burroughs financially from time to time.

Boyer became president of Burroughs's company in 1902. Recalling the early years, he wrote, "There was Burroughs, greater than any of us could fully appreciate. . . . I used to leave him at his bench in the evening and find him still there in the morning." The accuracy demands of a mechanical adding machine presented a daunting challenge. Burroughs's plans had to be so precisely developed that he used an industrial needle to draw on polished sheets of copper and zinc. It was his method for drafting full-size plans for a prototype device that he knew must be minutely accurate. Although using paper would have been more convenient, its size changed with the temperature and humidity. Obsessed with accuracy, Burroughs located hole centers under a microscope.

Burroughs made some parts himself, but more were made by his assistant, Alfred Doughty. After months of work, he had a working prototype in 1885 and filed his first patent application. The keyboard had nine columns of nine keys and worked as follows: Suppose a person wanted to enter $3,701,584.92. The operator would press "3" in the first column, "7" in the second column, no key for the third column, "1" in the fourth column, and so on. Each key stayed down after it was pressed. The operator then pulled a handle on the right side and the number was mechanically stored in the adding machine. Subsequent numbers were added each time the operator pulled the handle. Subtractions were also possible. Burroughs first targeted bankers and accountants with his new machine.

Burroughs contacted three local business people for financing. Thanks to Boyer's recommendation, he had no difficulty establishing the American Arithmometer Company with capital of $100,000. The adding machine's many parts had to be precisely manufactured and the Boyer machine shop workers had the required skill. Boyer's shop was on the first floor of an industrial building. The Arithmometer company was on the second floor. Although the adding machine operated well under Burroughs's hands, not all operators had his skillful touch. It was 1887 and trouble was just over the horizon.

Burroughs had sold about 50 adding ma-

chines when customers discovered a disturbing problem. The machines did not give repeatable results. Pulling the handle too vigorously interfered with the calculator's accuracy. The new company appeared to be headed toward bankruptcy. Burroughs quickly solved the problem by inventing a dashpot governor to control handle movement. It consisted of a small cylinder filled with oil that contained a plunger. The design was similar to an automobile shock absorber. It automatically regulated the adding machine's calculating mechanism no matter how the handle was pulled. Burroughs's company prospered as banks and other businesses converted to the new reliable adding machines. Within 10 years, the company had its own factory and 65 employees.

Burroughs continued to make improvements. In 1892, he patented an adding machine that automatically printed the input numbers and total. It was the world's first practical adding and listing machine. The invention was a success, even though it cost $475. By comparison, Karl Benz's Velo motor car cost $650 in 1893.

One innovative feature of Burroughs's machine was its ability to work on two sets of figures at the same time. A handle on the left side allowed the operator to shift from one set of numbers to another. Burroughs called them *registers*. Modern electronic calculators use the term *memory* or *memory channel*. Burroughs had sold more than 1,000 adding machines by 1898 when he invented an automatic ribbon reverse. It soon became a standard feature on all typewriters.

Unfortunately, Burroughs's tuberculosis did not improve. He retired from the company and moved to the quiet southern Alabama community of Citron-elle. Alfred Doughty, his assistant from the earliest days of experimental work, took over as company president. The Franklin Institute in Philadelphia awarded Burroughs its coveted John Scott Medal in 1897. He died the next year at the age of 41, while visiting friends in St. Louis.

Burroughs's first adding machine was complex and required careful evaluation by the patent examiners in Washington, D.C. His patent application had nine pages of highly technical text and seven pages of drawings. Most late nineteenth-century patents took only a few months to process. Burroughs's first calculator patent was filed in January 1885 but not issued until August 1888. His precise design coupled with careful fabrication of parts resulted in a successful machine. It would have sold in larger numbers if people had been less concerned about losing jobs to the new device. In spite of that perception, sales in 1907 alone were greater than the combined sales of all competitors during all the years of their existence. The company was renamed the Burroughs Adding Machine Company.

References

The Making of the Micro by Christopher Evans, Van Nostrand Publishers, 1981.

Total to Date, The Evolution of the Adding Machine: The Story of Burroughs by Bryan Morgan, Burroughs Adding Machine Ltd., 1953.

Burroughs History, Unisys Corporation, Burroughs Division, circa 1960.

Encyclopedia of Computer Science edited by Antony Ralston and Edwin D. Reilly, Van Nostrand Publishers, 1993.

The label for this Burroughs machine states, "Adding machine used in the head office of Barclay's Bank (London) from 1897 to 1913." It was probably used for two different sets of numbers. The color coding of five columns on the left follows the same pattern as the five columns on the right.

This is a replica of Blaise Pascal's 1642 calculator. To input numbers, an operator used a wooden stylus to rotate the numbered wheels on top. Totals appeared in the rectangular holes at the top of the machine.

Heinrich Rudolf Hertz

Born:
February 22, 1857,
in Hamburg,
Germany

Died:
January 1, 1894,
in Bonn, Germany

In the nineteenth century, technically minded young people had a challenging career decision to make. On one side were the exciting and dynamic technologies such as steam locomotives and machine tools. On the other side were the emerging sciences like electricity and thermodynamics. Both sides had their champions, but that was little comfort for someone who had to choose a career. The situation has not changed much over the years. Today's young people have similarly challenging choices to make.

Some, like American scientist Josiah Willard Gibbs (1839-1903), avoided making a decision until they were almost 30 (See pages 145-147.) . Others, like German communications pioneer Heinrich Hertz, tried their hand at both technology and science. Unsure of his future, Hertz started his college work in engineering but also took advanced science and mathematics courses. Technology and engineering were essentially identical career paths during the nineteenth century. The modern world is a better place because Hertz chose to study science. His discoveries were the first to blend electricity into electronics. Modern wireless electronic communication is used daily by almost everyone. It includes not only radio and television, but also telephones, fax machines, and the Internet. All those communication methods can be traced to an 1886 experiment Hertz conducted in southwestern Germany.

Hertz was the oldest of five children and had a comfortable childhood. His father was a lawyer and politician. His mother came from a family of physicians and religious leaders. His education started in a private school when he was six. The school's headmaster was quite demanding

The Science Museum/Science & Society Picture Library

and Hertz's mother carefully watched over his homework. She wanted him to be first in the class, and usually he achieved that goal. Hertz's interest in technology developed after he received a workbench and hand tools when he was 12. His parents later purchased a lathe for him. Science had been his grandfather's hobby and Hertz received his grandfather's science equipment at an early age.

Graduating from high school in 1875, Hertz was unsure about his future. He worked for a construction organization in Frankfurt for about a year. He then briefly attended a technical school, Dresden Polytechnic. His year of military service was spent with a railway regiment in Berlin. Hertz moved to Munich in 1877 with the intention of continuing his engineering education at the Technische Hochschule, or Technical University. While preparing for his entrance examinations, he studied mathematics and science. He enjoyed those subjects so much that he decided to study physics at the University of Munich.

Hertz's professors were excellent teachers working in the new field of electricity. Hertz was an equally excellent student. He studied advanced mathematics during his first semester and did laboratory work during the second semester. German universities freely exchanged students and after a year of study Hertz was expected to move to another school. People called this arrangement *student migration*. Hertz chose the University of Berlin primarily because Hermann von Helmholtz (1821-1894) taught there. Helmholtz was Germany's leading scientist, with international recognition based on his work in the theory of electricity and magnetism. The Helmholtz coil is named after him. It develops very uniform magnetic fields. A person with several scientific interests, Helmholtz also invented the ophthalmoscope to study the human eye. He drew Hertz into his circle of students and the two developed a strong mutual respect.

To justify Helmhotz's acceptance, Hertz wanted to show his ability in a tangible way. His first technical triumph was winning a medal and a cash award offered by a professional society for the solution to a particularly difficult experimental problem. It dealt with the study of alternating currents. Helmholtz had proposed the problem and was interested in its solution. He provided Hertz with office space and regularly checked his progress. After months of effort, Hertz solved the problem, which was the basis for his first professional article. The solution guided him toward the discovery of radio waves.

Hertz enjoyed his work with electricity and wrote to his parents, "I am now thoroughly happy and could not wish things better." After receiving his doctorate degree in 1880, he stayed in Berlin to do additional work in electricity. He was Helmholtz's salaried assistant at the Berlin Physical Institute, a research facility. Hertz generated 15 publications and was developing a noteworthy reputation. His technical skills touched on both the theoretical and experimental aspects of electricity. Anxious to establish a career on his own, he took a teaching position at the University of Kiel in 1883.

Hertz was an excellent teacher and had more students than he expected. But his main interest was in conducting electrical research and Kiel had limited laboratory facilities. Karlsruhe Technical University had an opening in 1885. The school had a large laboratory and excellent test equipment. Hertz accepted the position when it was offered. At Karlsruhe, he conducted his landmark experiments in radio transmission.

While becoming familiar with the facilities, he met Elisabeth Doll, the daughter of a colleague who taught surveying. They married in 1886 and eventually had two daughters. The same year, Hertz conducted the experiment that would make him world famous. He had noticed that Leyden jars, an early form of capacitor, discharged with a spark across an air gap. A smaller spark would occasionally accompany the main spark across the terminals of a nearby Ley-

A replica of Hertz's 1886 landmark radio transmission experiment is displayed at Munich's Deutsches Museum. An induction coil (far left) produces periodic sparking at contacts in the parabolic metal reflector. Electromagnetic waves are created and received by a similar reflector at the right. The waves pass through an eight-sided polarizer with gratings of wire. Hertz used the device to demonstrate the similarities between light and radio waves. The oscilloscope shows signal patterns to museum visitors.

den jar. Hertz wondered if the main spark caused the other. After countless preliminary experiments, he constructed a high-voltage transmitter circuit that included two metal spheres separated by 1 to 2 cm. They were positioned at the focal point of a metal parabolic mirror. Electrical power came

Some of Hertz's original test equipment is displayed at Munich's Deutsches Museum. The frame-like devices are radio wave polarizers or antennas. His most successful early transmitter included the two wires with different diameter spheres at each end, shown at the bottom of the display case. Hertz called the device a linear oscillator.

from a bank of wet-cell batteries and a can-shaped induction coil about 20 cm in diameter and 52 cm high. Relay contacts provided the make-and-break cycle to generate high voltage from the coil. The rapidly changing current produced sparks that jumped the air gap between the metal spheres. The sparking created electromagnetic, or *radio,* waves. Hertz's receiver was a 70 cm diameter loop of wire with smaller metal spheres at the ends, separated by about 3 mm.

The equipment was crude, but Hertz obtained valuable data from each experiment he conducted. He placed the receiver at different parts of a darkened room, typically 20 meters from the transmitter. Large high-voltage sparking at the transmitter induced weak sparks to jump the air gap at the receiver. Hertz was the first person to send and receive energy through space. Nothing like this had ever been done before.

Hertz noticed a similarity between his electrical signals and light. He conducted a series of complex experiments in 1888 that showed that electrical signals were energy waves, just like light. Hertz reflected and refracted them. He focused them with different devices and calculated wave lengths. His equipment typically generated wave lengths of about 30 cm at a frequency of 100 MHz. He developed efficient antennas and discovered that short waves are better for information transmission than longer ones.

Hertz established the foundation for modern electronic communication. But he had not forgotten engineering and technology. He had a paper published in an engineering journal on a new method for measuring the hardness of metals. Hertz also developed an improved ammeter and a hygrometer for measuring humidity.

Hertz was an excellent teacher with a pleasant personality. He had strong language skills and spoke fluent English, French, and Italian, in addition to his native German. He also had a working knowledge of Arabic. His work resulted in the first long-distance transmission of a wireless communication. The Italian Guglielmo Marconi (1874-1937) sent a Morse code signal across the English Channel in 1899, a distance of about 30 miles. But Hertz did not live to see it happen. He contracted an infection from a decayed tooth and died in Bonn at the age of 37.

Some early technologists were brash and obvious. Examples are the American Isaac Singer (1811-1875) and the Briton Richard Arkwright (1732-1792). (See pages 22-24 and 94-96.) But they were more commonly quiet, low-key persons, like Hertz. In 1890, the British Royal Society awarded Hertz its coveted Rumford Medal, a precursor to the Nobel Prize. Hertz told no one when he traveled to London to accept it at ceremonies with the world's greatest scientists or after returning to Germany. He was modest to the point of never mentioning his discoveries, even when discussing their basic theory with students or colleagues. To honor Hertz's contributions, the international General Conference on Weights and Measures adopted the unit of *hertz* in 1960 as one cycle per second.

References

Pioneers of Electrical Communication by Rollo Appleyard, Books for Libraries Press, 1968.

The Communications Miracle by John Bray, Plenum Press, 1995.

Famous Names in Engineering by James Carvill, Butterworths Publishers, 1981.

Radio's 100 Men of Science by Orrin E. Dunlap, Jr., Books for Libraries Press, 1944.

Frank Julian Sprague

As America began to urbanize in the mid-nineteenth century, city planners saw the need for establishing public mass transportation. Early transit systems relied on horse-drawn streetcars that moved at a walking pace on metal tracks. The world's first trams appeared in 1832 on the New York and Harlem Street Railway. Each carried 30 passengers and was powered by two horses, with the driver on the roof.

However, as power plants, horses had serious disadvantages. They could work only three to six hours per day, were unsanitary, and required huge barns for shelter. It cost about $4 per day to support the horses necessary for a two-horse car. That expense quickly added up for systems like the 212-mile Boston West End Railroad, which had 1,700 cars and 9,000 horses. Cable cars were not the answer because they were expensive and could not cross each other, back up, or easily navigate curved streets. In addition to emitting large amounts of unhealthy smoke, steam locomotives did not work well for making frequent stops and starts. Frank Julian Sprague cut tram operation costs in half and cleaned up both the air and the streets when he introduced the world's first major electric streetcar system. He did it in Richmond, Virginia, in 1888.

Sprague was born in southwestern Connecticut, where his father worked as a supervisor at a hat factory. His mother died when he was eight and from then on he was raised by an aunt in western Massachusetts. He took a competitive examination after completing high school and won an appointment to the U.S. Naval Academy.

Graduating seventh in his class in 1878, Sprague showed a remarkable ability in the new field of electricity. The navy used his

Rendering by Tim Harmon

Born:
July 25, 1857,
in Milford,
Connecticut

Died:
October 25, 1934,
in New York,
New York

services to help develop improved dynamos (dc generators) to power shipboard motors and lights. Sprague continued experimental work in his free time and took courses at the Stevens Institute of Technology in New Jersey when his ship returned to port. In 1881, he was sent to the Mediterranean Sea, where he requested a leave to attend an electrical convention at the Crystal Palace Exposition near London. He was 24 and the only American member on a committee that granted awards for dynamos and lamps. That committee was chaired by Britain's Lord Kelvin, the person for whom the Kelvin temperature scale was named.

The experience struck Sprague like a lightning bolt. He could hardly contain his enthusiasm for investigating electricity as a form of power for transportation. Edward Johnson, a business associate of Thomas Edison, easily persuaded Sprague to resign from the Navy and join Edison's staff in New Jersey. However, this arrangement turned out to be a poor fit.

This 7.5 hp, 500 volt DC electric motor was used in Richmond, Virginia's 1888 pioneering streetcar system. Semi-circular pivots for the wheelbarrow mounting are in front.

An advertisement for Sprague's trucks, the wheel assembly under a streetcar, appeared in the 1888 technical publication *The Electrical World*. Note the single strong coil spring supporting each electric motor. It absorbed road shock while ensuring that the motor and wheel gears stayed in mesh.

For all of his accomplishments, Edison distrusted academic training. He felt that making experiment after experiment was the only proper way to solve a technical problem. Sprague saw staff members make a huge map of a community that was to be electrified. Each house, business, school, and factory had a wooden spool wound with wire proportional to how much electricity it was expected to use. The goal was to determine the required dynamo output. The experiment took weeks to complete. The young Sprague spent only a few days developing the necessary formulas to arrive at the same determination. The company adopted Sprague's method. But he correctly observed that Edison's main interest in electricity was for light. Sprague wanted it used for transportation, so he left Edison after about a year.

With his small savings and some investment capital from Johnson, he established the Sprague Electric Railway and Motor Company in New York City. Sprague's growing company made motors for use in mines, factories, and experimental streetcars. One of his early designs changed the way the motor was mounted. When the motor's output shaft was directly connected to a wheel, parts of the motor often cracked in service. Sprague's method used a "wheelbarrow" technique. The motor was supported by separate strong springs, much as a person's arms support a wheelbarrow. This ap-

proach proved so successful that it found universal use.

In the spring of 1887, Sprague agreed to provide Richmond with a complete 12-mile, 40-car electric streetcar system for $110,000. This was a big contract and an even bigger gamble for Sprague. He had contracted to build as many streetcars as were in use at that time in the entire United States. The system was the first of its kind and Sprague encountered never-ending problems. The gears between the electric motors and drive wheels often locked up, a contractor laid poorly connected rails directly on the clay soil, and Sprague caught a mild case of typhoid fever.

The first 10 cars were more or less ready in early 1888 and the Richmond Union Passenger Railway opened for regular service in February. The line experienced normal start-up problems. Cars jumped the track. Overloaded cars could not climb steep 8 percent grades. Metal motor brushes wore out two or three times a day on each car. Ice formed on the power line. But most of the problems were solved by spring, when all of the 40 streetcars were available for service.

Sprague designed and built not only the motors, controls, and cars, but also the electrical-generating station and the power distribution network. Three 125 hp Armington and Sims single-cylinder steam engines operated six 500-volt dynamos. Steam pressure came from three 16-foot-long coal-fired boilers. Each 15,000-pound streetcar carried up to 40 passengers and traveled up to 12 mph. Two 7.5 hp electric motors on each car received power from an overhead copper line. Electricity was picked up by a rolling wheel called a trolley. Sprague intro-

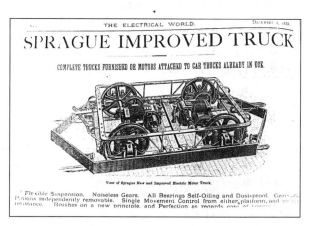

THE ELECTRICAL WORLD.   DECEMBER 1, 1888

SPRAGUE IMPROVED TRUCK

COMPLETE TRUCKS FURNISHED OR MOTORS ATTACHED TO CAR TRUCKS ALREADY IN USE.

View of Sprague New and Improved Electric Motor Truck.

Flexible Suspension.   Noiseless Gears.   All Bearings Self-Oiling and Dust-proof. Gears and Pinions independently removable. Single Movement Control from either platform, and resistance. Brushes on a new principle, and Perfection as regards ease of

duced the spring-loaded under running trolley and it was used on practically all streetcars. The steel rails provided a grounded return path for the electricity. Sprague estimated it would have taken 300 horses to duplicate the work of the electric streetcars.

Before long, the Richmond line began to make a profit and costs dropped to $2 per

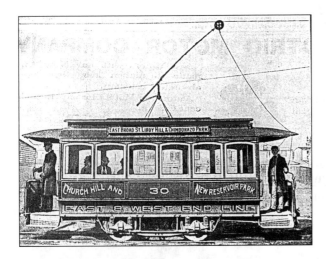

day per car. The project cost Sprague $75,000 more than he received. However, the system was so successful and economical, that it served as the model for more than 200 electric street railways worldwide within the next two years. By 1900, only 1 percent of U.S. streetcar miles still used horses for power.

Perhaps Sprague's most significant and revolutionary invention was the multiple-unit train control. Passenger trains accelerated more quickly, smoothly, and with less wheel slippage—and attained higher speeds—if each car had its own motors.

Sprague invented a mechanism that permitted controlling all of a train's electric motors from a single location. First used on the Chicago South Side Elevated Railway in 1897, the multiple-unit train control made elevated railways and subways possible. Those installations also incorporated a safety feature that Sprague had patented. A button had to be depressed to operate the train. An operator's falling asleep or fainting would cause the release of the button, triggering the automatic shut off of electrical power and application of the brakes.

Sprague's name is not as familiar as some

who accomplished far less. He left the manufacturing business in the early 1890s and started selling his important patents to General Electric, Westinghouse, and Otis Elevator Company. Sprague never insisted that his name be identified with his products and purchasers never credited him with key inventions. Thomas Edison bought Sprague's motor manufacturing business in 1889 and promptly removed Sprague's name from all existing and future products. "Edison-GE" was substituted, giving the false impression that Edison invented some items that he had only purchased.

None of these situations ever bothered Sprague. He was a well-respected and self-assured technologist who had a dazzling array of technical achievements to his credit. He received many awards, including a grand prize at the 1904 world's fair in St. Louis. He died only two years after a nationally broadcast banquet was held in his honor.

References

"The Sprague Electric Road at Richmond, Va.", *Electrical World*, 16 June 1888.

"Horse Power" by John H. White, *American Heritage of Invention and Technology,* Summer 1992.

"The Wrong Track" by George W. Hilton, *American Heritage of Invention and Technology,* Spring 1993.

"The Romance of Invention—VII: A Pioneer of the Electric Motor, the Trolley Car, the Multiple-Unit-Train-Control, the Electric Elevator" by C. H. Claudy, *Scientific American,* 24 January 1920.

One of Sprague's first 1888 streetcars for Richmond, Virginia. It used two 7.5 hp electric motors. Power came from the counter-balanced under running trolley on the roof.

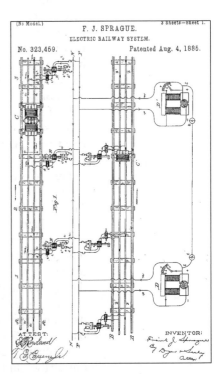

One of Sprague's early streetcar patents used two electric conductors (a-a' and b-b') between the rails (A-A and B-B). This design preceded the more effective overhead power lines.

Konstantin Eduardovich Tsiolkovsky

Born:
September 17, 1857, in Izhevskoye, Russia

Died:
September 19, 1935, in Moscow, Russia

Human flight has fascinated people for centuries and many characters from mythology have been associated with the heavens. The word *mythology* comes from *myth*, which means "fictitious story" and *ology*, which means "a branch of learning." Athena was the Greek goddess of wisdom who lived in the heavens. Mercury was the Roman messenger of the gods who flew on winged feet. Ancient tales from Arabia, Egypt, India, and other countries were written down in the 1500s and called the *Arabian Nights*. The entertaining tales included flying carpets and a flying genie (supernatural spirit). Ordinary nineteenth-century people thought that near-Earth flight by humans would probably happen sometime. But many saw a trip to the moon as a complete impossibility. People commonly belittled those who worked in the field of space flight and seriously evaluated its possibilities. This even happened in the twentieth century. Robert Goddard (1882-1945), American inventor of the liquid-fueled rocket, suggested in 1919 that his invention could be used to fly to the moon. The press immediately labeled him the "Moon Man." In 1969, the *New York Times* newspaper published a belated apology. Although Goddard had died some years before, the article was printed July 17, 1969, with the Apollo XI spacecraft on its way to the first moon landing by a human. Apollo XI used booster rockets of the type originally patented by Goddard.

When the Chinese invented fireworks during the Chin Dynasty (221 B.C.–207 B.C.), no one realized that they would serve as a stepping stone to space flight. The explosive involved was *black powder*, a combination of sulfur, saltpeter, and crushed charcoal. People packed the mixture into bamboo tubes to make firecrackers for colorful

Courtesy Deutsches Museum, München

and noisy celebrations. Black-powder rockets were first used for military purposes by the Chinese in 1206. They were later used for signaling and carrying written messages. Another use involved life saving along sea coasts. Lifelines were fired to boats and ships in trouble. It was such civilian uses that prompted Konstantin Tsiolkovsky to investigate rockets for space flight. He showed that liquid propellants were more effective than solid propellants and became the first technical champion of space flight. Russia honored Tsiolkovsky by launching the world's first artificial satellite on the one-hundredth anniversary of Tsiolkovsky's birth year. Sputnik I went into orbit on October 5, 1957. The Russian word *sputnik* means "traveler."

Tsiolkovsky was born about 600 miles east of Moscow. His father held various jobs as forester, teacher, and government official. He was an adequate provider, but the family often moved to maintain their moderate standard of living. Tsiolkovsky's early education came from his mother and he then attended public school. At the age of 10, he contracted scarlet fever and became nearly deaf. Tsiolkovsky could not continue his school studies and took to reading extensively. Like Thomas Edison (1847-1931), who was also deaf, Tsiolkovsky later credited the affliction with helping him learn to focus on a problem and find a solution. He liked to watch the steam locomotives that were beginning to tie the world together. He took that interest a step further, wondering about steam pressure and then about carriages propelled by jets of steam. Tsiolkovsky read all he could. He was clearly motivated and intelligent, and his father decided to send him to Moscow when he was 16. This was a big step for a young man who looked forward to visiting the libraries and universities of the large city.

While he did not attend college, Tsiolkovsky patterned his studying to match the topics that a college program covered. He was interested in physics and mathematics, and he self-studied through the equivalent of a university major in each. He attended many educational lectures that were open to the public. In his free time, Tsiolkovsky made calculations on the possibility of lifting a vehicle into space. He said his interest in the subject developed after he noticed teenagers jumping from a horse-drawn hay wagon. As each jumped off, the wagon would lurch forward slightly.

This simple action-reaction observation made Tsiolkovsky think about what would be needed to propel a rocket in space.

After three years, Tsiolkovsky returned home and gave private lessons to students who were only slightly younger than he was. He took challenging state examinations in 1879 and obtained certification to teach secondary school. He started teaching arithmetic and geometry in 1880 at a high school 100 miles south of Moscow. His space career was launched.

Tsiolkovsky began an extensive series of theoretical studies. He devoted almost all his free time to space calculation. But Russia was not a world technical leader, so there was little in the way of hardware to support his research. Tsiolkovsky joined the Society of Physics and Chemistry in St. Petersburg to share his thoughts and make new contacts. He did not earn much money and felt he should not ask for expensive equipment to test his theories. Those were difficult days in imperialist Russia. Each new regime was more political than the one before, and the country moved closer and closer to revolution. Tsiolkovsky felt that it was best to lead a quiet life.

Tsiolkovsky also had other aeronautical interests. He designed an all-metal airship that was never constructed. He built a wind tunnel in 1898 to evaluate airplane design. It was the first in

Tsiolkovsky's 1935 book *On The Moon* (image of cover at top) included his design for a liquid-fueled rocket for space travel (bottom).

Tsiolkovsky (left) in the workshop of an unidentified person. The photograph was probably taken about 1930.

Russia and Tsiolkovsky hoped to use data from it to construct an all-metal monoplane. This never happened. But such projects kept Tsiolkovsky mentally active and singly directed toward flight research.

Publication of an article titled "Investigating Space with Reaction Devices" launched Tsiolkovsky's reputation as a rocketry pioneer in 1903. Appearing in the Moscow-based magazine *Survey of Science*, it covered many different space topics, including space sta-

Two nineteenth-century military black-powder rockets of the type in use during Tsiolkovsky's lifetime. The top one was a stick-stabilized life-saving rocket used at coast guard stations from 1859 to 1948. The lower one was spin stabilized by an exhaust gas deflector that caused it to spin like a rifle bullet. It was used by the British army from 1865 to 1919.

tions. Tsiolkovsky also concluded that several rocket stages were better than one large rocket. He proved it was possible to attain orbital and escape velocities, which he correctly calculated at 8 km/sec (18,000 mph). He gave practical instructions on rocket design and evaluated many different propellants. He determined that black powder, a solid propellant, did not have enough energy to carry a payload into space. A liquid hydrogen (LH2) and liquid oxygen (LOX) combination proved the best choice.

Tsiolkovsky was right about both staged rockets and liquid propellants. All large rockets are now boosted into Earth orbit through a series of stages. The Saturn 5 launch vehicle that sent people to the moon in 1969 had three stages. Although its first stage used kerosene and LOX, the upper two stages used LH2 and LOX. The three main central engines on Space Shuttles also use LH2 and LOX. The outer two strap-on solid-propellant engines drop off after use, just like staged rockets.

Tsiolkovsky did not work with hardware. He concentrated on theoretical aspects of rocket engines. The technology of his day could not support the fabrication of sophis-

ticated liquid-fueled rocket engines. They would have required close tolerance machining. Another problem was the low propellant temperatures. LH2 is -425° F and LOX is -300° F. Robert Goddard's first successful firing in Massachusetts came several years later, in 1926.

Tsiolkovsky evaluated propellants based on how much thrust could be expected from one pound of propellant burning for one second. It was a modern approach that rocket technologists have expanded on. They now call it specific impulse (Isp). An Isp of 45 to 70 is common for black powder, and about 400 for an LH2-LOX combination. This means LH2-LOX is about 5.7 to 8.9 times more effective as a propellant than black powder.

The Russian government and other technical professionals slowly began to appreciate Tsiolkovsky's work. He received a government pension in 1919 that allowed him to retire from teaching and continue his rocket investigations. And he remained an active researcher. During his lifetime, Tsiolkovsky wrote more than 500 scientific papers, magazine articles, and books.

The only two countries to develop manned space programs are the U.S.A. and Russia. America's rocket heritage rests on the accomplishments of Robert Goddard, and Russia's rests on those of Tsiolkovsky. No other country has a rocketry pioneer who accomplished as much as either of these two men. Perhaps that is why no other country developed a manned space program. Russia launched the first artificial satellite during the one-hundredth anniversary year of Tsiolkovsky's birth. But there was more than just that. Tsiolkovsky's favored theoretical design used 20 liquid-fueled rocket engines. The Russian Semyorka booster that launched Sputnik I had 20 liquid-fueled rocket engines.

References

K. E. Tsiolkovsky—Selected Works, compiled by V. N. Sokolsky, Mir Publishers, 1968.

Soviet Rocketry by Michael Stoiko, Holt, Rinehart, and Winston Publishers, 1970.

The Rocket—The History and Development of Rocket and Missile Technology by David Baker, New Cavendish Books, 1978.

Rudolf Christian Karl Diesel

The Science Museum/Science & Society Picture Library

Born:
March 18, 1858,
in Paris, France

Died:
September 29,
1913, in the
English Channel

Travel today would be difficult were it not for efficient and powerful internal combustion (IC) engines. Although IC engines have nontransportation applications, such as generating electricity and powering water pumps, their real niche is in transportation. Technologists typically identify five basic types of engines. These include gasoline engines (for automobiles), liquid-propellant rocket engines (for space travel), and gas turbine engines (for airplanes). The other two types are named for their German inventors: Felix Wankel (1902-1988) and Rudolf Diesel. The Wankel engine is only used in some snowmobiles and a few automobiles, but the diesel engine can be found everywhere. It powers trucks, automobiles, small boats, locomotives, off-road construction equipment, air compressors, and many other devices.

Gasoline's explosive nature led some early technologists to seek an IC engine that used a safer and more easily refined fuel. The most successful was Diesel's compression ignition engine, invented in 1897 in Munich, Germany. It's universally known as the *diesel engine.*

Although Diesel was born and died outside Germany, he was a legal resident of that country. He was born in Paris because his family had moved there in search of work for his father, who was a self-employed leather worker. The younger Diesel said he was quickly attracted to technology after his first visit to the technical exhibits at the Conservatoire des Arts et Metiers museum in Paris. He was only 12 when the outbreak of the 1870 Franco-Prussian War forced his family to move from France to London. Diesel enjoyed his time there and furthered his technical interests with visits to London's Science Museum. The cosmopolitan natures of both Paris and London influenced the young Diesel. His experiences there also helped him become a comfortable multilingual world traveler as an adult.

When Diesel's father had trouble earning enough to support his family, other relatives helped Diesel move to Augsburg, Germany, where he attended high school. There he saw a demonstration that made an unforgettable impression on him. A plunger slid inside a glass-walled cylinder, operating like a tire pump. A piece of tinder was placed at the end of the cylinder opposite the plunger. With application of a rapid stroke, the tinder ignited under the heat of the compressed air. The Chinese had long used this concept to start fires, and the device Diesel saw is sometimes called a *Chinese firestick.*

The youngest to graduate from his high school, Diesel finished his studies in just three years. Because of his outstanding academic record, he was admitted to the Technical College of Munich. He was in his late teens and completely self-supporting, living on scholarships and work as a private tutor. Diesel graduated in 1880 with the highest grades in the college's history. Having studied machine design and thermodynamics, he found employment with a refrigeration company in Paris. The ice plant was operated by Carl von Linde (1842-1934), who held many patents in refrigeration technology. Von Linde developed the ammonia compressor and was the first to liquefy air on a commercial scale.

Diesel became plant manager the next year. Two years later, he married Martha Flasche. Working with compressed gases, he developed an interest in the principles of combustion. He first tried replacing steam in steam engines with ammonia vapor. That proved impractical and Diesel

A drawing from page 2 of Diesel's 1895 U.S. patent No. 542,846

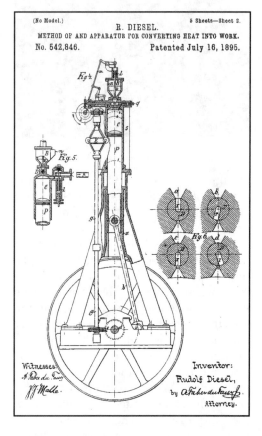

investigated the emerging field of internal combustion. Instead of the spark ignition used in gasoline engines, Diesel envisioned a process that would use heat from compressed air to ignite low-cost fuels. As his piston moved up on the compression

stroke, it would compress the air, greatly increasing its temperature. Fuel injected into the high-temperature air would ignite immediately, pushing the piston down on the power stroke. Diesel calculated that the engine would be four times more efficient than the steam engines popular in his day.

Diesel was the first to design an engine on paper using calculations that applied the new science of thermodynamics. He also constructed many potential pressure-volume charts, which addressed the most important aspect of internal combustion. Diesel was uncertain how to inject the fuel into the cylinder. He decided to use an air tank pressurized to about 500 psi. Automatically opening a small tank valve for a fraction of a second, at just the right time, provided the necessary force to push fuel into the cylinder. Diesel had an experimental engine completed in 1892 that he intended to operate with powdered coal, a plentiful and easily obtained fuel. To get the engine started, he and his assistant, Lucien Vogel (1855-1915), turned a huge flywheel by hand. Pressurized air from the air tank also helped push the piston down. The engine seemed about to start, but soon it suffered a minor explosion, which spewed small parts around the room. Diesel then spent several more years modifying his engine to use a liquid fuel, kerosene.

Diesel encountered many problems designing a fuel-injection system. But he finally had an engine operating for extended periods in 1897. It was a huge single-cylinder, liquid-cooled design about 12 feet tall with a compression ratio of 44:1, twice that of modern diesel engines. Supported by a large A-frame, the engine's bore was 250 mm and its stroke was 400 mm, with a displacement of 19.6 liters. This was the world's first diesel engine. It developed 20 horsepower at 172 rpm at the unheard of thermal efficiency of 26.2 percent. The engine is on display at Munich's Deutsches Museum.

Diesel demonstrated the engine in 1898 and found investment funding to manufacture it at a factory in Augsburg, now called Maschinenfabrik Augsburg-Nurenberg, or M.A.N. Diesel's was not the first patented compression ignition engine design, but it was the first to operate successfully. In the course of defending his patent, Diesel made many court appearances. He never lost a case.

The diesel engine showed immediate promise for powering generators, ships, and locomotives. However, the first production engines had reliability problems caused by the tremendous pressures required for compression ignition engines. Nonetheless, Diesel insisted on having the engines constructed to his original designs. That delayed improvements, which would have made the engine more useful. By 1902, there were 359 diesel engines in service.

A world traveler who was comfortable in any country, Diesel spoke fluent German, English, and French. He met such persons as Charles Parsons, English inventor of the steam turbine, and Thomas Edison. He also met Lord Kelvin, a Scottish scientist who has a temperature scale named after him, and Emanuel Nobel, the brother of the Swedish Alfred Nobel who established the Nobel Prizes. When Diesel journeyed to St. Louis to attend the 1904 world's fair, he met Adolphus Busch (1839-1913). Busch was a German-born American brewer and a heavy investor in M.A.N. Busch had the first American diesel engine constructed in 1898 for use in his brewery. It was a two-cylinder, 30 hp engine constructed by the St. Louis Iron and Machine Company.

Diesel had a pleasant personal life with his wife and three children. But he was sus-

Diesel's 1897 engine on display at Munich's Deutsches Museum was the world's first diesel engine. It includes a pressurized air tank at the right for injecting kerosene at the proper time.

picious of almost everyone else. He suffered from severe headaches all his adult life. Although he was once a millionaire, his inflexible attitude resulted in the loss of almost all his money and caused people to distrust him. He disappeared under mysterious circumstances in 1913, while crossing the English Channel in a ferry.

Few people were as intelligent or as completely technical as Rudolf Diesel. An early advocate of compression ignition, he was unwilling to change his designs to accommodate the wishes of newly arrived technologists. Some criticized his accomplishments by describing Diesel as too theoretical and impractical. Yet he was advancing the state of the art with unheard of cylinder pressures and operating efficiencies. No one but a highly competent professional could have tackled those problems as well as Diesel did. He produced a product that has withstood the test of time.

References

Rudolf Diesel—Pioneer of the Age of Power by W. Robert Nitske and Charles Morrow Wilson, University of Oklahoma Press, 1965.

Diesel's Engine by C. Lyle Cummins, Jr., Carnot Press, 1993.

Internal Fire by C. Lyle Cummins, Jr., Society of Automotive Engineers, 1989.

Charles Edgar Duryea

Born:
December 15, 1861, in Canton, Illinois

Died:
September 28, 1938, in Philadelphia, Pennsylvania

With hundreds of millions of vehicles on the road worldwide, it's not uncommon to hear speculation about who invented the automobile. Identifying such a person would require qualifying words. As with many complex products, the motor car did not have a single inventor. It evolved from the cumbersome 1769 steam-powered carriage built by Nicholas Cugnot (1725-1804) in Paris, France. Although his creation was woefully impractical, Cugnot did invent the first self-propelled vehicle.

Henry Ford Museum and Greenfield Village

During the late nineteenth century, America was primarily a farming country. In 1870, 53 percent of the nation's workforce labored in agriculture. In that field, the most flexible form of power was the horse. Horses were also the only practical method of personal transportation. The combination of having a large number of workers to consider and the heavy reliance on horses kept many early technologists from seriously experimenting with motor cars. Short-sighted people often ridiculed those who suggested that internal combustion power could replace horse power. It took unique personalities to move against the social tide and challenge traditional thinking. Two who did were Charles Duryea and his brother J. Frank Duryea (1869-1967). They organized the Duryea Motor Wagon Company in Springfield, Massachusetts, in 1895. The company manufactured the first pro-duction automobiles in the United States.

The oldest of five children, Charles Duryea was born on a family farm in northwest Illinois. He attended a community elementary school and graduated from high school in La Harpe. People were just starting to use machinery for farmwork, and Duryea enjoyed working with the mechanical devices. He built a high-wheel bicycle from discarded parts when he was 17. He had several different jobs after finishing high school before he settled down for a while to repair bicycles. He worked at bicycle shops in Washington, D.C., and Rockaway, New Jersey, before moving to Chicopee, Massachusetts, where he signed an agreement with the Ames Manufacturing Company. Ames made a wide variety of items from weapons to machine tools, and Duryea wanted them to make bicycle parts to his specifications. He would then assemble them for sale in a bicycle business he had established in Peoria, Illinois. Frank soon joined Charles in Chicopee, working

10-hour days as a machinist for Ames at the standard pay of $2.75 per day.

While displaying bicycles at the 1886 Ohio State Fair, Charles saw his first gasoline engine. He described it as larger than a kitchen stove, and he thought that engines could be made smaller. Returning home inspired, he had long discussions with his brother over possible designs for a horseless carriage. They made many trips to the library looking for information in *Scientific American* magazines. Since Charles was not regularly employed, he had more time to develop detailed designs, primarily for the engine.

Charles found financial backing in Springfield, Massachusetts, in early 1892. He used the money to rent space with belt-driven machine tools from the John W. Russell and Sons machine shop at 47 Taylor Street. The Duryeas built a prototype they called the *buggyaut* and tested it in September 1893. It had a single-cylinder 4 hp engine with a friction drive. The vehicle's design proved unsuccessful. A much better production model followed in early 1895. The 700-pound production car had a specially built body with a 4-foot 9-inch wheelbase. Front wheels measured 38 inches in diameter and the rears were 46 inches. The two-cylinder, liquid-cooled, 1.75 hp engine had a 4-inch bore and 4.5-inch stroke.

The vehicle used a gear-driven three-speed transmission, plus reverse, with gears selected through separate clutches. With a top speed of 20 mph, the driver changed speed with an up-and-down motion of the steering tiller. The Duryeas called this "one hand control." The vehicle was quite advanced and boasted many innovations: pneumatic tires, ball bearings, inclined front-end king pins, bevel gear differential, and external contracting band brakes, with all mechanisms enclosed for safety. But those features came at a high price. At $1,000 to $2,000, the buggyaut was not inexpensive.

At that time, European automobile manufacturers advertised their products through road races. The *Chicago Times-Herald* newspaper decided to sponsor the first one in America. Held on Thanksgiving Day of 1895, the race ran between Chicago and Waukegan, Illinois, a distance of 56 miles. Besides the Duryea driven by Frank, the only other entrants were three Benz and two

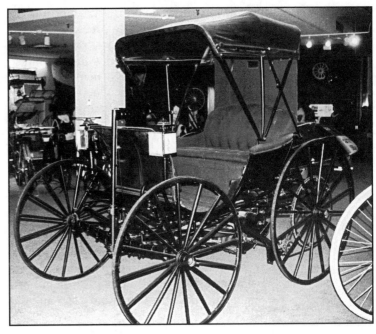

electric cars. The race began at 8:55 A.M. Eighteen inches of snow had fallen two days earlier, leaving the route covered with deep and drifted snow. Duryea crossed the finish line first at 7:18 P.M. He collected the $2,000 first prize. It was an uneventful race at the time and only 50 people saw the finish. (The photograph on the first page of this chapter, taken in 1895 in Springfield, Massachusetts, shows Charles Duryea at the tiller of the car that won America's first automobile road race. The photo has been reproduced more often than any other that shows a particular automobile.)

Great Britain records a Duryea as the first to cross the finish line in the initial London to Brighton Run of 1896. It was not really a race, but a drive meant to commemorate the repeal of a British law that effectively restricted motor cars to a speed of 4 mph. With Frank driving, the Duryea finished first, ahead of 32 other cars. The 52-mile London to Brighton Run still takes place every November for pre-1905 vintage automobiles.

The Duryea Motor Wagon Company was the first American company to manufacture and sell automobiles. The company's production marked the first time that more than one car of the same design was made in the United States. Charles was the major stockholder and Frank was the chief engineer and highest paid employee at $150 per week. The company made 13 cars in 1896,

This 1893 Duryea was the very first of its American breed. It is owned by the Smithsonian Institution's National Museum of American History. With its design based on the high-wheeled horse-drawn carriages of the period, its appearance shows why early motor cars were called horseless carriages.

with its first sale in February. Unfortunately, selling motor cars was a precarious undertaking. Bicycles were far less expensive and many biking organizations had promoted good intercity highways, maps, and local clubs. The 13 cars the Duryea Company initially sold was its total output. The company reorganized under the name of National Motor Carriage Company. The company failed shortly after the brothers left it in 1898.

Frank continued to build Stevens-Duryea automobiles in Chicopee with the J. Stevens Arms and Tool Company. Their first model had a 6 hp engine. Later ones were large and expensive, and were intended for use as chauffeured vehicles. The company remained in business until 1927. Frank had retired from active business at 46 and did

Charles Duryea held several patents. This one for a "Road Vehicle" shows his tiller steering, which doubled as the vehicle's throttle.

some contract work from time to time. He maintained a small machine shop at his home in Madison, Connecticut, and spent much of his time sailing on Long Island Sound.

Charles moved to Reading, Pennsylvania, where he established the Duryea Power Company to manufacture three- and four-wheeled automobiles. The company also made vehicles in Belgium and England, and it remained in business through 1913. Charles made his living as an automobile consultant and writer. His pleasant personality and infectious sense of humor helped him in his new career. He wrote *Roadside Troubles* (1905), *Handbook of the Automobile*

(1906), and *The Automobile Book* (1916). He frequently appeared as an expert witness in legal disputes. Charles was president of America's first automotive organization, the American Motor League.

Charles and Frank had a distant, but not totally unbrotherly, relationship. Some controversy still persists about their respective roles in the creation of the 1895 Duryea. With some justification, Charles saw himself as an innovator—the designer and inventor of a motor car. He held 19 early patents, including the first automobile patent issued to an American manufacturer. Frank was more concerned with day-to-day matters of manufacturing methods and product improvement. Some feel his contributions were pivotal to the company's brief success.

In any case, it was the brothers' teamwork that launched America's first automobile company. Unfortunately, each approached his responsibilities from his own perspective, and they rarely joined forces. Had they done so, their company would have certainly lasted longer. They might be remembered fondly, as people remember other technical brothers' teams— such as the fast-driving twins Francis and Freelan Stanley, known for their steam car, or the even-tempered Orville and Wilbur Wright and their airplane.

References

Carriages Without Horses by Richard P. Scharchburg, Society of Automotive Engineers, 1993.

"Charles E. Duryea's Chicago Winner Made History" by M. J. Duryea, *The Antique Automobile,* July 1945.

Great Inventions, Smithsonian Scientific Series (Volume 12), edited by Charles Greely Abbot, 1934.

The Brighton Run by Lord Montagu, Beaulieu Shire Publications (England), 1990.

Marie Sklodowska Curie

Stories of Scientific Discovery, 1923

Born:
November 7, 1867, in Warsaw, Poland

Died:
July 4, 1934, in Sancellemoz, France

Understanding energy is especially important in technologically advanced countries. But energy can be a confusing topic because it has so many forms. Electrical energy is a flow of electrons. Solar energy is electromagnetic radiation that covers the earth. Gasoline is a liquid and coal a solid. The methods used to compare the energy content of different substances can be perplexing. With that caveat, one method for comparing particular forms of energy involves using the same measure of volume.

One gallon of breakfast cereal, for example, has about 1,300 food calories. A gallon of gasoline has the energy equivalent of about 30,000 food calories and a gallon of coal has about 41,000 food calories. A *food calorie* is 1,000 times greater than a scientific calorie. Uranium is a much more energy dense material. At 158 pounds per gallon, it's the heaviest natural element. That gallon of uranium contains the energy equivalent of about 1.7 trillion food calories. The value is so large, it's all but impossible to visualize. And that's the point: nuclear energy is incredibly powerful.

Nineteenth-century investigators noticed that a substance called *pitchblende* had an unusual characteristic. People did not yet know why, but it exposed photographic film without light. The mineral resembles pitch, a type of asphalt. Pitchblende is a source of uranium and radium, and we now use the word *radioactive* to describe such elements. The word was coined by Marie Sklodowska Curie. Few people have been as closely associated with radioactive materials as Curie. The Polish-born scientist discovered the elements polonium and radium and was the first person awarded two Nobel Prizes.

Although the world knows her as Marie Curie, she was born Maria Sklodowska, the youngest of five children. Curie's father taught mathematics and physics at a high school in Warsaw. Her mother managed a private boarding school for girls. Poland was occupied by Russia and the time around Curie's youth was difficult for Polish nationals. Curie's father lost his job for political reasons and had to move the family several times. Her mother died when she was 10.

A brilliant high school student, Curie graduated at 16. To earn money for her family, she tutored younger students. She also taught herself other languages and could communicate in Polish, Russian, French, and German. Maria and her older sister Bronia studied at a casually organized underground university. Russian officials would not permit advanced education and the two sisters planned to move to France. They agreed that Bronia would go first, while Maria would help support her from home. Bronia left for Paris in 1886 and

Maria became governess for a wealthy family that lived outside Warsaw.

Bronia became a medical doctor and married. The couple insisted that Maria move in with them and she journeyed to Paris in 1891 with her meager savings. She crossed Germany by train, traveling fourth class while sitting on an uncomfortable stool. Curie immediately enrolled at the Sorbonne. Then, she moved into her own quarters to get a quieter study environment.

Stories of Scientific Discovery, 1923

Her academic performance was of the highest caliber. But she lived such a spartan existence that she once passed out in class from lack of food. Curie supported herself with a scholarship and laboratory work. She received a physics degree in 1893 and graduated first in her class. With a scholarship from Poland, she earned a degree in mathematics the following year and graduated second in her class.

Curie in her laboratory sometime after 1906. She often looked tired, which may have been due to exposure to radioactivity

Curie met her future husband, Pierre Curie (1859-1906), while conducting research with the magnetic effects of steel in 1894. The director of laboratories at the School of Industrial Physics and Chemistry, Pierre Curie was an excellent researcher and had made several significant scientific discoveries. The two married the next year and Maria Sklodowska changed her name to Marie Curie. The Curies soon became the greatest husband-and-wife scientific team in history.

Curie had not yet received her doctorate degree and was searching for a research project. Her husband's supervisor suggested she consider working on X-rays, a type of radiation discovered in Germany in 1895 by Wilhelm Roentgen (1845-1923).

Roentgen received the first Nobel Prize in physics in 1901. Curie was enthusiastic about the subject and began work on it almost immediately.

Roentgen's X-rays were created by passing high voltages through gas in a sealed glass tube. The rays appeared to be similar to those emitted by pitchblende ore. Henri Becquerel (1852-1908) of France had discovered in 1896 that it caused photographic plates to blacken. Curie's first step was to determine whether X-rays and pitchblende emissions were related. She also wanted to discover what was in the ore that produced the invisible rays. Curie and her husband had the support of Becquerel. He used his influence in Austria where pitchblende was mined and the government gave the Curies a ton of the material.

Using an old unused shed as a laboratory, the Curies began to process the ore by boiling it down. The method was called *fractional distillation*. Working in the cold, damp, and drafty building between 1898 and 1902, the Curies finally isolated two new elements. Marie named one of them *polonium* for her native country. The other was *radium*, which was a much more powerful element. They obtained only 0.10 gram from more than a ton of pitchblende. The Curies took no precautions to protect themselves from radioactivity because its harmful effects were not yet known. Curie's notebooks are still too radioactive to handle directly.

The Curies' discoveries took the scientific community by storm and opened up the science of nuclear physics. Radium was the most powerful radioactive substance known. It held the promise of saving lives through its potential medical uses. The Curies shared the 1903 Nobel Prize in physics with Becquerel. Marie Curie received her doctorate degree from the Sorbonne the same year. The doctorate was a rare endorsement from such a prestigious university. The Sorbonne was just beginning to acknowledge the scientific ability of women. The Nobel Prize helped Pierre Curie gain appointment as head of the Sorbonne's physics department.

International recognition overwhelmed the Curies with requests and correspondence. They never sought celebrity status and did not enjoy it. It took too much time and drained their strength. They had once taken leisurely bicycle rides in the country, but such activities became less frequent. Their vacations grew shorter. Curie called it "the burden of fame." The husband and wife teamwork continued until 1906 when Pierre was killed in a traffic accident on a Paris street. The devastated Marie had two young children to raise. It took her some time to return to work. When she did, the university appointed her to Pierre's position. Curie became the first female professor at the Sorbonne.

Curie continued her work with radium, trying to process it more easily. She passed an electrical current through molten radium chloride, which was the most important step involved in obtaining pure radium. The discovery was so significant that Curie was named the sole recipient of the 1911 Nobel Prize in chemistry. A most remarkable woman, she was the first person to receive two Nobel Prizes.

During World War I, she helped design

and build X-ray machines. They fit into small trailers for use near the battlefields and were popularly known as "Little Curies." Curie herself was an ambulance driver in France. She wrote a book on the things she learned during the war. Titled *Radiology and War*, it was published in 1921.

Curie received recognition in many forms. The Sorbonne constructed the Radium Institute for her in 1914 and Curie became its director. The *curie* is a common unit of measurement of radioactivity. The Marie Sklodowska Curie University is in Lublin, Poland. Unfortunately, long exposure to radioactive materials caused Curie's health to decline. But she never talked about it and continued working as best she could. Curie died of leukemia at the age of 67.

Marie Curie's name is familiar to many people. She was the first woman scientist with an international reputation. The pantheon memorial in Paris honors France's "great men," and they enshrined Marie and Pierre Curie in April 1995. The Curies were the seventieth and seventy-first persons to receive the honor. Marie Curie was the first woman so recognized by her adopted country. She and her husband had two daughters, Irene (1897-1956) and Eve (1904-1996(?)). Irene also worked with radioactive materials. She and her husband Frederic Joliot (1900-1958) shared the 1935 Nobel Prize in chemistry. Marie Curie almost lived long enough to see her daughter and son-in-law receive the world's highest scientific award—just as she and her husband had received it 32 years earlier.

This lead-lined safe is displayed at the Oak Ridge National Laboratory Historic Site. It shows how much material thickness is needed to shield people from the effects of radioactivity.

The Oak Ridge (Tennessee) National Laboratory is one of the largest energy research centers in America. Its original 1943 nuclear reactor, the first in the world, is a national historic site and open to the public. The loading face features three mannequins to show the reactor's relative size.

References:

Marie Curie—A Life by Susan Quinn, Simon and Schuster Publishers, 1995.

Great Men and Women of Poland by Mary Landon Sague, Kosciuszko Foundation Publishers, 1924.

Guglielmo Marconi

Born:
April 25, 1874,
in Bologna, Italy

Died:
July 20, 1937,
in Rome, Italy

Telegraphy was the only means of rapid long-distance communication in the 1800s. It went underwater in 1865 when a thin wire connected North America with Europe. That was the year Cyrus Field (1819-1892) completed the first permanent transatlantic telegraph cable. The project required seven attempts before it succeeded. So it is not surprising the telegraph company charged $5 to $10 a word to send messages to Europe. The surprise is that there were plenty of customers.

Werner Siemens (1816-1892) spent many years connecting London, England, to Calcutta, India, by telegraph. (See pages 109-111.) Work on the 6,000-mile distance was finished in 1870. Had wireless communication been available, Field and Siemens might not have tried such huge projects. The first person to send and receive a wireless communication signal was Guglielmo Marconi in Italy. His most high-profile technical triumph occurred in 1901, when he sent the first transatlantic wireless signal from England to Newfoundland.

Marconi was born into a wealthy family that enjoyed all the trappings that money could buy. The Marconis had a townhouse in Bologna and an estate outside the city named Villa Grifone. Marconi's father was a landowner. His mother was the daughter of a whiskey distiller in Dublin, Ireland. Between the ages of three and six, Marconi lived in England with his mother and older brother, Alfonso. When he returned to Italy, he could barely speak his native language. Because he spoke Italian poorly and with an English accent, dressed well, and did not enjoy sports, his classmates picked on him. His father decided to educate him with private tutors.

Marconi's mother was 17 years younger

Courtesy Deutsches Museum, München

than his father and she enjoyed social activities. She traveled extensively with her sons, sometimes staying for long periods of time in various European cities. Marconi later became a world traveler. His mother had given him the social skills to move easily among different cultures.

The mother often took her sons to Livorno on the west coast of Italy. When Marconi was 13, he began attending the Technical Institute in that city. He learned that Heinrich Hertz (1857-1894) had recently sent electricity through the air. (See pages 166-168.) Marconi became intrigued by the possibility of wireless telegraphy. He had a weak educational background and failed entrance examinations for the University of Bologna in 1894. His mother asked a neighbor and science professor, Augusto Righi, if her son could unofficially use the university's laboratory facilities. Righi agreed. Marconi's father disapproved and wanted his son to follow a career in the Italian Navy. He gave little encouragement to his son's interest in technical subjects.

Using university resources, Marconi set up a crude laboratory in the attic of the Villa Grifone. Righi cautioned Marconi that he did not have the educational background to succeed at work that had baffled scientists for years. But the young man was undaunted and soon had some success sending signals indoors. He showed his father that he could ring a bell at the opposite end of the house. A wireless signal tripped the bell's relay. His father was so impressed that he gave Marconi $1,000 for additional equipment.

One day in 1895, the Marconi brothers went outside with an oscillator, coherer, meters, switches, antennas, and other pieces of equipment. The oscillator produced sparks that Marconi hoped to receive at a distant point. The receiver included a coherer, a small four-inch glass tube filled with metal filings. The coherer conducted electricity only when it received an electromagnetic signal. The conductivity could be read by the needle of a galvanometer.

Marconi stayed with the transmitter. Alfonso went a mile and a half away with the receiver and a rifle. He would fire the rifle if the receiver picked up a signal. Marconi closed the switch. Alfonso received the signal and fired the rifle. That experiment is often cited as a first in wireless trans-

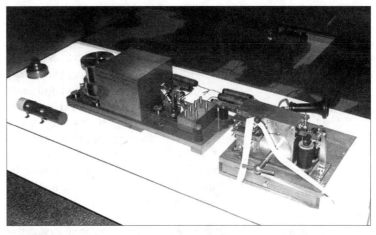

mission.

The Italian Ministry of Posts and Telegraph showed no interest in funding more advanced experiments. Like most others, officials there felt wireless telegraphy would only be useful for ship-to-shore, or ship-to-ship communication. They suggested that Marconi try to find assistance in England, the world's most important seafaring nation. Marconi's mother arranged a meeting with William Preece, chief engineer of the Great Britain Post Office. Marconi was an almost unproven 21-year-old. Nonetheless, Preece was impressed and recommend that the British government support his work.

Using government and family money, Marconi established a research organization and surrounded himself with world-class technical experts. One was John Ambrose Fleming (1849-1945), inventor of the 1904 tube-type diode. (See pages 154-156.) Marconi's goal was to send and receive signals at greater distances. He first worked on a tuning circuit. One small region of the country might have many transmitters. If several signals were sent at the same time, the receivers could not separate the transmissions. The effect would be like a roomful of people all talking at once. Marconi patented his tuning circuit in England in 1900. Numbered 7,777, it allowed for transmitting radio waves at specific frequencies. The effect was like a

Replica of Marconi's 1901 transatlantic receiver. A kite-borne antenna brought the signal into the receiver and Marconi heard it through a telephone earpiece like the one at right center.

Marconi's three-circuit multiple tuner. His installations from 1907 to about 1914 used this unit to couple the antenna to the receiver. Its three tuned circuits, loosely coupled together, gave good selectivity for a pre-electronic device. The three cylinders on top are condensers. The front dial at left varies antenna inductance and the unusual front dial with bars at right is for tuning.

person concentrating on a single speaker.

Marconi established the Wireless Telegraph and Signal Company to market his products. His first major sale was to the British government for use in the 1899-1902 Boer War. Partly for the publicity, Marconi used his equipment to transmit nine miles across the Bristol Channel in 1897. Interrupting high-voltage spark signals sent Morse code. Marconi transmitted at about 15 words per minute. He then succeeded in transmitting 28 miles across the English Channel in 1899. But the greatest challenge involved sending electromagnetic waves

Replica of Marconi's 1901 transatlantic transmitter. Closing the key (right) energized the coil (center) and charged the two enclosed brass spheres (left). The spheres discharged through the antenna, sending an electromagnetic wave. The system was powered by a bank of batteries.

across the Atlantic Ocean.

An unlimited capacity for work and complete faith in his abilities helped Marconi in his efforts. Assisted by Fleming, he set up a transmitter near Lizard Point, England, in the southwest corner of the country. His receiver was in St. John's, Newfoundland, about 2,100 miles away. The transmitter's antenna consisted of two 150-foot poles, placed about 170-feet apart, and strung with 55 copper wires. But the receiver's antenna was simpler: a piece of wire flown from a kite. Both the transmitter and receiver were far more complex than what Marconi had used in Villa Grifone just a few years earlier. The equipment and expenses brought the experiment's cost to $200,000.

Marconi's lifelong friend Luigi Solari operated the transmitter. Solari sent three dots at specific times, Morse code for the letter "S." In Newfoundland, Marconi listened through a telephone earpiece. At 12:30 P.M. on December 12, 1901, he barely heard three clicks. He passed the earpiece to his assistant George Kemp to verify the reception. When the news broke, Marconi's name was suddenly known throughout the world.

Marconi went on to achieve many other technical accomplishments. They were more and more detailed as the complexity of electronic communication became apparent. His business interests assumed international proportions. Marconi's United States branch became the Radio Corporation of America. From 1919, he lived on a large 220-foot-long yacht named *Elettra*. He bought it from the British government who had obtained it from Austria during World War I. Marconi modified the 730-ton yacht to carry tall wireless antennas and outfitted a complete shipboard laboratory. His work was acknowledged with the highest recognition. Marconi received the 1909 Nobel Prize in physics. He shared it with Karl Braun (1850-1918), German inventor of the cathode-ray oscilloscope.

Marconi married twice. First to Beatrice O'Brien from Ireland and then to Maria Cristina Bezzi-Scali from Italy. He had four children. He enjoyed fishing, horseback riding, and traveling. He was a careful dresser and almost all his photographs show his concern for clothing. An automobile accident took his right eye in 1912 and he had an artificial one fitted. It is all but impossible to detect in photographs of him. Marconi died of a heart attack at 63 while preparing for an evening appointment with Italy's leader Benito Mussolini (1883-1945).

As important as Marconi's contributions were, he did not invent radio. He invented point-to-point wireless telegraphy, not voice transmission. Electronic communication progressed rapidly during Marconi's lifetime. Only 27 years after he sent three barely audible clicks across the Atlantic Ocean, John Logie Baird (1888-1946) sent the first television signal from London to New York City. (See pages 199-201.)

References

Marconi by W. P. Jolly, Constable and Co. Publishers, 1972.

Marconi, The Man and His Wireless by Orrin E. Dunlap, Macmillan Publishers, 1937.

Great Lives from History edited by Frank N. Magill, Salem Press, 1987.

The Making of the Modern World edited by Neil Cossons, Science Museum Publications, 1992.

Walter Percy Chrysler

One exciting technology caught the imagination of the American public at the beginning of the twentieth century: automotive design and construction. There have been more than 4,000 brands of automobiles offered for sale worldwide, with about 2,500 of them produced in the U.S.A. Some European manufacturers use acronym names like BMW, for Bayerische Motoren Werke in Germany, or Fiat, for Fabbrica Italiana Automobili Torino, in Italy. But most American companies were named after people. Two examples are David Buick (1855-1929) and Louis Chevrolet (1878-1941). Buick designed the chassis for his first automobile in 1903. He left the business in 1906 to work for gold-mining companies. Chevrolet helped design his company's first 1911 automobile. Then, he worked on racing cars. One he constructed, driven by his brother Gaston, won the 1920 Indianapolis 500 mile race.

Other designers stayed with their companies for many more years. No one controlled an automobile company longer than Henry Ford (1863-1947). He established the Ford Motor Company in 1903 and held a leadership position for more than 40 years. An equally colorful and long-lasting automobile pioneer was Walter Chrysler. Like Ford, he was one of the few industrial leaders who came from the ranks of manual labor.

Chrysler was born into a rough-and-tumble environment in what was then the American West. His hometown was a prairie railroad shop town and Chryler's father drove a locomotive for the Union Pacific Railroad. Father and son occasionally took rides together in a locomotive cab and the father encouraged Chrysler's interest in machinery. The younger Chrysler turned down a chance to attend a nearby college and took a job as a machinist's apprentice for five cents an hour

Courtesy of Chrysler Corporation

Walter Chrysler shown in a 1924 publicity photograph with the first Chrysler automobile, a hand-crafted, five-passenger sedan

Born:
April 2, 1875, in Wamego, Kansas

Died:
August 18, 1940, in Kings Point, Long Island, New York

in the railroad shop. Even as a teenager, he demonstrated an unusual characteristic that served him well his entire life. Chrysler was willing to take risks for the possibility of a distant and ill-defined future reward.

Chrysler loved his work as a machinist and built a 28-inch-long operational locomotive in his spare time. He always spoke fondly of his experiences in the railroad shop, but he left to learn more about machinery. The Atchison, Topeka, and Santa Fe Railroad hired him as a journeyman. He arrived with such outstanding references that the company offered him its highest pay rate. But Chrysler was not satisfied to stay in one place and miss other opportunities for advancement. For eight years, he moved from one place to another, developing a reputation as a capable and gifted machinist-mechanic.

While working in Salt Lake City, Chrysler saw a steam locomotive limp into town one day with one of its two cylinders not working. He repaired it in record time, keeping the train on schedule. He was promptly made roundhouse foreman at the age of 26. Now, with reasonable career possibilities in hand, he returned home to marry his childhood sweetheart, Dell Forker. It was an ideal marriage and she willingly accepted his inclination for risk taking. Chrysler said his wife deserved 70 percent of the credit for any success he may have achieved.

Always successful, Chrysler rose through various management levels at several companies. Because he had worked at so many manual jobs, he also understood the problems of the workers and had their respect. He moved to Chicago and then to Pittsburgh where he served as plant superintendent for the American Locomotive Company.

Chrysler became an automotive enthusiast after he saw a red and white Locomobile at the 1908 Chicago Automobile Show. He persuaded a friend to co-sign a loan and bought it. His main interest was in the car's construction. In 1912, the opportunity came to work as production manager for the Buick Motor Company in Flint, Michigan. Buick offered him $6,000 per year. He was then earning $8,000 at American Locomotive and the company offered an increase to $12,000 if Chrysler would stay. Amazingly, he chose to go to Buick, even though he knew it would forever separate him from the steam locomotives he called "noble mechanisms." Neither his friends nor his wife understood his decision.

Because Chrysler's background was in locomotive production, he brought a different perspective to a company that was still influenced by fine carriage making. He introduced new processes and efficiencies

Courtesy of Chrysler Corporation

Walter Chrysler in a 1932 Plymouth advertisement. Floating power referred to a new way to bond synthetic rubber to steel. It allowed Chrysler to make improved engine mounts that reduced vibration.

more suited to automotive manufacture. Production increased from 45 cars per day to 550 and Chrysler became president of Buick in 1916. Buick was a unit of General Motors, which had William Durant (1861-1947) as chairman. The company's management methods dissatisfied Chrysler and he resigned in 1920.

Chrysler next served as executive vice president of the troubled Willys-Overland Automobile Company in Toledo, Ohio. He had a free hand to oversee the company's two-year reorganization. He successfully reduced the company's debt and placed it on a sound financial footing. While in the middle of that job, the Maxwell Motor Company asked Chrysler to repeat the performance for Maxwell. He agreed and assumed its presidency. He suggested an entirely new car, but others did not want a new car—they wanted a reorganization. Chrysler raised money, bought the company, and began working on an innovative car that would bear his name.

That car would succeed because of the abilities of three outstanding automotive technologists: Carl Breer, Owen Skelton, and Fred Zeder. The industry later called them the "Three Chrysler Musketeers." Breer and Zeder were exceptional engine designers, while Skelton's strength was in transmission and axle design.

Chrysler wanted to build a moderately priced car with features like four-wheel hydraulic brakes and a powerful six-cylinder, high-compression engine. Breer, Skelton, and Zeder delivered the product. The first car to carry a Chrysler nameplate came out in 1924 with a price starting at $1,335. It had a 68 hp 201-cubic-inch displacement engine with a 4.7:1 compression ratio, considered high compression at the time. The engine had innovative features like aluminum pistons, a seven-bearing crankshaft, and pressurized lubrication.

Chrysler had three preproduction automobiles made and displayed them at the 1924 New York Automobile Show. The exhibit brought in 5,000 orders and the company sold more than 32,000 Chrysler cars its first year. Chrysler introduced four models the following year, the 50, 60, 70, and Imperial 80. The numbers referred to the vehicle's maximum miles-per-hour velocity on a level road. At the time, these were almost unbelievable speeds for moderately priced cars.

Six-cylinder Chrysler roadsters took third and fourth places at the 1928 international Le Mans race. A far more expensive Bentley came in first, followed by an equally costly Stutz.

Car sales increased and the company needed more production space. Chrysler purchased the Dodge Brothers Company in 1928 to acquire its nameplate and manufacturing facilities. Shortly afterward, he introduced the Plymouth and De Soto automobiles.

To test vehicle aerodynamics, Breer had a wind tunnel constructed in the late 1920s. The company used his data to design and construct a streamlined unit body Airflow model under Chrysler and De Soto nameplates. The car was introduced in 1934 at $1,345, and only 11,000 sold that year. It was too far ahead of its time. Sales dropped to below 5,000 in 1937, its last year of production. Although not a financial success, the Airflow showed that Chrysler had not abandoned his willingness to take a risk.

To serve as headquarters for corporate offices, the 77-story 1930 Chrysler Building was constructed in New York City. Walter Chrysler initiated the project and personally chose the marble for the corridors and the veneers for the original elevators. The Chrysler Building is a beautiful traditional skyscraper, complete with medieval gargoyles. It was once the world's tallest building.

When Chrysler retired in 1935, the company was out of debt and its production was second in the industry. Chrysler then spent his time boating, fishing, and collecting penny banks. He wrote his biography and helped his son take over the business. He died at his estate on Long Island.

The full force of the Great Depression caused some financial cutbacks at the Chrysler Corporation. One department that never experienced a reduction was the company's research department. Chrysler knew that innovations were important to technology. That heritage continued right into the 1960s with the company's introduction of 50 gas-turbine-powered automobiles for consumer testing. The vehicle never went into production, but it showed the company's continued willingness to test a new technology. The Chrysler Corporation was the only automobile

manufacturer to be a prime NASA contractor for the Apollo moon landing program. It built the first stage of the uprated Saturn I booster. Walter Percy Chrysler's influence lasted well past his lifetime.

References

Birth of the Chrysler Corporation and Its Engineering Legacy by Carl Breer, edited by Antony J. Yanik, Society of Automotive Engineers Publications, 1995.

Life of an American Workman by Walter Chrysler, Dodd Mead, 1937.

Seventy Years of Chrysler by George H. Dammann, Crestline Publishing, 1974.

1934 Chrysler DeSoto Airflow sedan. Though a beautiful automobile, its aerodynamic style never caught on with the public.

1964 Chrysler experimental gas turbine car. A total of 50 were constructed and turned over to randomly selected consumers in a unique two-year market survey. It showed the research heritage left by Walter Chrysler.

Willis Haviland Carrier

Born:
November 26,
1876, in
Angola,
New York

Died:
October 7, 1950,
in New York,
New York

Entering a cool and comfortable building on a sweltering summer afternoon is a great comfort. But buildings have not always had the efficient climate-control systems that they have today. In the nineteenth century, people occasionally used combinations of ice and fans to lower a building's inside temperature. However, this method proved impractical and only worked in small rooms. It took twentieth-century technology to help people achieve the goal of controlling temperature, humidity, circulation, and air purity. Those are the four functions of air-conditioning systems.

The term *air conditioning* was so imprecise that many inventors could argue that they were the first in the field. Factories were the initial targets for air conditioning. Some raw materials used in factories are sensitive to moisture and temperature. These materials include paper, textiles, and celluloid. In early efforts, ventilating fans conditioned stale summer air. Water sprays conditioned dry winter air. But those efforts did not achieve air conditioning in the modern sense. The true developer of interior climate control was Willis Carrier. His first scientifically designed system was used by a printing company in 1902.

Carrier was born into a poor farming family in western New York State. He was an only child and his mother introduced him to mathematics and machines. His mother's death, when Carrier was 11, left him devastated. He decided early on to succeed in engineering and technology and seemed single minded in that pursuit.

Carrier left school at age 17 to work to help his father make payments on the family farm. He wanted to attend college but could not afford the expense. However, he took a statewide competitive examination

Courtesy of Carrier Corporation

and won a scholarship to Cornell University. Carrier studied mechanical and electrical engineering during the day. While not in school, he waited on tables, tended furnaces, and cut grass to earn money for food and rent. After graduating in 1901, he went to work for the Buffalo Forge Company in Buffalo, New York, for $10 per week. The company made industrial air-handling equipment like fans, blowers, and exhausters.

Having focused on studying electricity at Cornell, Carrier had expected to follow a related career path. Later in life, he said, "I had planned to study electricity when still in Angola and chose it because it was then a new art, much as electronics is today." Still, he took on his new work in air handling with enthusiasm. His first significant job involved designing a high-horsepower fan to supply combustion air for a factory boiler. Carrier was dissatisfied with the temperature and humidity data that his company used for its standard calculations. Tests he conducted suggested that the data was not correct. He asked the company to establish a small research laboratory and, to his amazement, the company agreed. Carrier was only 26 and had worked at Buffalo Forge for only one year. He used the lab to obtain the first scientifically collected data for the new field of *psychrometry,* the study of the interaction of heat, moisture, and air.

The assignment that put Carrier on the path to air conditioning came from a Brooklyn print shop in 1902. The Sackett-Wilhelms Lithographing and Publishing Company did color printing by sending paper through presses several times. Each pass added a new color. Changes in humidity altered paper size, which produced fuzzy images. This was particularly troublesome in the summer. Sackett-Wilhelms asked Buffalo Forge to design and install humidity control equipment. The assignment went to Carrier.

While analyzing the problem, Carrier found a solution in recalling that a glass of cold water gets wet on the outside on a hu-

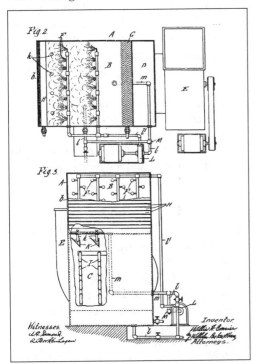

A drawing from one of Carrier's patents for a dehumidifying air washer. The refrigeration coils at "I" cooled water, which was sprayed through the spray bars at "F." Moisture condensed out of the air as it was pulled through the unit by the squirrel-cage fan at "E."

mid day. This happens because some of the moisture in the air condenses onto the glass. Carrier visualized a spray of cold water, in which each droplet would function like the glass of water. A small amount of atmospheric moisture would condense on each droplet, thus reducing the humidity. Carrier designed a cold-water spray system to keep the humidity at the desired 55 percent at Sackett-Wilhelms. Other people thought his method irrational. After all, water sprays were commonly used to *increase* humidity during the dry winter months. It seemed unreasonable that they would also *decrease* summer's dampness. But the system worked exactly as Carrier

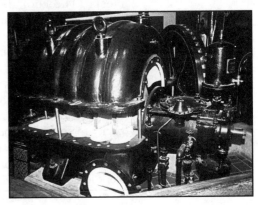

Carrier's 1922 centrifugal compressor was so important to comfort cooling that the Smithsonian Institution displays it at its National Museum of American History in Washington, D.C.

had predicted that it would. Buffalo Forge established a subsidiary company in 1907, naming it the Carrier Air Conditioning Company of America. By the end of that year, the company had installed temperature- and humidity-control systems in several textile mills, a shoe factory, and a pharmaceutical company. All of Carrier's early installations were in industrial plants.

Carrier cooled water and air with mechanical refrigeration. The process used a compressor and evaporator connected in a closed loop with copper tubing. It relied on the cooling effect caused by evaporation, as occurs, for example, when evaporating rubbing alcohol cools a person's skin. A liquefied gas like carbon dioxide, sulfur dioxide, or ammonia served as the working fluid, or *refrigerant.* Liquid refrigerant sprayed inside a metal evaporator changed into a gas, which cooled the evaporator. The compressor then converted the gas back into a liquid, allowing the process to continue indefinitely. That method is still used in modern refrigerators, freezers, air conditioners, dehumidifiers, and heat pumps.

Carrier did not invent mechanical refrigeration. However, he was the first to successfully use it as part of a scientifically designed air-conditioning system. He favored the method of cooling the water with the evaporator in a mechanical refrigeration unit. The cooled water was then used in water sprays to control humidity on the factory floor. Carrier called his devices *air washers*. They were about the size of a large closet. The cooled water also flowed through a heat exchanger that had air blowing over it. Newly cooled air was then ducted to the proper factory area. But that was usually a smaller room, often used for some critical aspects of production. The piston-type compressors used in mechanical refrigeration did not have enough capacity to cool large rooms.

Carrier was a researcher, designer, and builder of air-conditioning equipment. His work led him to reevaluate many practices related to air quality, such as spraying water to reduce humidity or keeping windows shut to improve ventilation. He formally presented his findings in 1911 at the annual meeting of the American Society of Mechanical Engineers (ASME), under the mundane-sounding title of "Rational Psychrometric Formulae." His paper remains the most important document ever prepared on air conditioning. Those who carefully followed its recommendations, could hardly go wrong in designing an effective system.

Carrier had still more to offer. Another major achievement was the centrifugal compressor for cooling large volumes. Piston-type compressors were fine for small applications, but it was the centrifugal units that allowed for comfort cooling. Carrier made his first in 1922 and sold it to the Onondaga Pottery Company in Syracuse. The unit is now displayed in the Smithsonian Institution. It was once powered by a 100-horsepower electric motor. *Comfort cooling* was the term Carrier used to describe nonindustrial applications of his equipment. Some firsts in that area include an installation at the J. L. Hudson Department Store in Detroit in 1924. Then came the Rivoli Theater, a movie house, in New York City in 1925. Next was the Milam Building, an office building, in San Antonio in 1928. Also in 1928, the U.S. Senate and House of Representatives were air conditioned by Car-

rier. Because each building was different, each application required its own unique equipment.

A physically strong, likable man, Carrier was nicknamed "the Chief". He had tremendous powers of concentration, but he could be more than a little forgetful. On one occasion, while visiting a friend, he checked to determine the cause of a painful foot and discovered that he had put three socks on one foot and one on the other. Another time, he lost his ticket on a train and could not remember where he was going. Carrier was also a poor driver whose countless minor accidents amused his staff. Widowed twice, he married three times and adopted two sons. He died of a heart ailment in 1950.

Much as Orville Wright (1871-1948) served as aviation's foremost pioneer and saw his field evolve from wire and fabric to jet power, Carrier saw air conditioning come of age. He held more than 80 patents. His company had some problems after it became independent from Buffalo Forge in 1914, but it soon prospered.

There were other air-conditioning companies. There were other inventors. But no one other than Carrier had contributed so much across the entire range of air conditioning. The ASME regards his 1911 paper as "the outstanding example of our [organization's] commitment to technical excellence." After his achievements, schools began offering courses and programs in air conditioning. Willis Carrier almost single-handedly established a new technology.

References

Willis Haviland Carrier—Father of Air Conditioning by Margaret Ingels, Doubleday and Co., 1952.

Heat and Cold—Mastering the Great Indoors by Barry Donaldson and Bernard Nagengast, American Society of Heating, Refrigerating, and Air Conditioning Engineers, 1994.

Modern Americans in Science and Technology by Edna Yost, Dodd, Mead and Co., 1962.

"Willis H. Carrier—Father of Air Conditioning" by Fritz Hirschfeld, *Mechanical Engineering*, December 1980, pp. 22–23.

Charles Stewart Rolls and Frederick Henry Royce

Turn-of-the-century motorcars gave drivers independence—the ability to go where they wanted, when they wanted. That attracted many purchasers. The large potential market prompted more than 4,000 companies worldwide to build motor cars. Some nameplates still exist from the earliest days of European production. The first car built by Karl Benz (1844-1929) went into production in Germany in 1893. (See pages 148-150.) In France, a youthful Louis Renault (1877-1944) offered his first Renault for sale in 1898.

But few automobiles generate greater appeal than a Rolls-Royce. The Rolls-Royce has always been too expensive for the average person to purchase but never too expensive to appreciate. With a distinctive radiator housing that has remained essentially unchanged since 1904, it has found great popularity among the wealthy. No car denotes handcrafted perfection more than a Rolls-Royce. That was the primary goal established by Charles Stewart Rolls and Frederick Henry Royce. They came from totally different backgrounds to introduce their car to the world at the 1906 London Motor Show.

Rolls was the third son born into an aristocratic London family. His father was a baron, as was his mother's father. Rolls was educated at the prestigious Eton Preparatory School and Cambridge University. In 1898, he earned a degree in "mechanical engineering and applied science." He en-

Courtesy Rolls-Royce Motor Cars Limited

joyed technology, and transportation in particular. Bicycling was among his favorite activities and he served as captain of the university racing team. Rolls was handsome, dark haired, and had an athletic build. He never married. Always looking for adventure, he enjoyed ballooning and automobiles. Rolls made 170 balloon ascents. His Peugeot was the first automobile ever seen at Cambridge. Coming from an aristocratic background, Rolls would not have had to work for a living had he not gotten involved in international motorcar racing. It was an expensive pastime. To supplement income from his family, Rolls opened a car sales and repair business in London. Most of his customers were rich friends.

Royce's early life could not have been more different. The youngest of four children, he was born near Peterborough, about 70 miles north of London. His father was a poor flour miller and his mother was a farmer's daughter. Royce's father died was he was 10, and from then on he had to sup-

Charles Rolls

Born:
August 28, 1877, in London, England

Died:
July 12, 1910, in Bournemouth, England

Frederick Royce

Born:
March 27, 1863, in Alwalton, England

Died:
April 22, 1933, in West Wittering, England

Courtesy Rolls-Royce Motor Cars Limited

An early Rolls-Royce Silver Ghost, built about 1913

This is the second of three engines Royce built just before forming his partnership with Rolls. The 1.8-liter, 2-cylinder engine had a compression ratio of 3:1 and developed 10 hp. It was in a car owned by Ernest Claremont, an earlier partner of Royce.

port himself. He worked as a newsboy, a telegraph messenger, and an apprentice at a locomotive repair facility. He studied mathematics and electricity in his spare time. Unable to complete his apprenticeship due to lack of money, he took work in 1881 at a machine tool factory in Leeds. He worked hard, often putting in 16-hour days. He and Ernest Claremont pooled their finances in 1884, about £70, and established F. H. Royce and Company in Manchester to make electrical devices. Fearful of returning to the poverty of his youth, Royce often worked himself to exhaustion. When faced with a challenging technical problem, he stayed with it continuously, damaging his health by missing sleep and meals. It was an uphill climb, but his company eventually proved successful.

Claremont and Royce married sisters in 1893, the daughters of Alfred Punt. Although not wealthy, Royce had enough income to purchase a house. He invited his mother to live with him and his wife, and she did so until her death in 1904. The house had a beautiful garden that became Royce's favorite form of recreation.

Royce bought a used French Decauville motorcar in 1903. He was totally displeased with it. He spent much of his time repairing the unreliable vehicle. Royce felt sure that he could construct a better car and did so between the autumn of 1903 and April 1904. In that short period, he designed the body that would later define the characteristic Rolls-Royce shape. He powered it with a 2-cylinder, 10-horsepower engine coupled to a 3-speed gearbox. Royce included

no new ideas in his design. He was not an inventor and never claimed to be. However, he did precise work and used only the finest raw materials. He improved the best features of the Decauville and redesigned the ones that needed attention. Anticipating the future, Royce had made enough parts for two more cars. He drove the first, his partner Claremont the second, and shareholder Henry Edmunds bought the third. None still exist, but the engine from Claremont's car is displayed at Manchester's Museum of Science and Industry.

Edmunds was so pleased with his car that he described it in glowing terms to his friend Rolls. Rolls could not find a reliable English car for sale and was eager to meet Royce. Edmunds arranged a meeting in 1904 at the recently constructed Midland Hotel in Manchester. Although separated by social standing and an age difference of 14 years, Rolls and Royce struck up an immediate friendship. Rolls particularly liked the Royce automobile. It was quieter than the competition and operated with less vibration. Rolls purchased 20 to sell in London and the two men took steps to establish Rolls-Royce Motors, Limited. Royce left the operation of his electrical business to others and began to produce motor cars.

The two partners decided to produce one standard, superb model. Royce supplied the technical talent and Rolls marketed the automobiles. Their first prestige motor car was the 1906 Silver Ghost. Most were painted an eye-catching silver. The $9,000 luxury car had a 7.4-liter, 6-cylinder engine that developed 50 horsepower and was extremely durable and smooth running. The car was large enough that wealthy people would have

enough room for their suitcases while traveling. Rolls used his own 1908 model to transport ballooning equipment. A total of 6,000 Silver Ghosts were built before the Phantom series came out in 1925. The *Spirit of Ecstasy* statue on the radiator housings of Rolls-Royce automobiles came into common use after World War I.

Rolls and Royce had symbiotic careers, each needing the other to succeed. Rolls would have had no cars to sell without Royce, and Royce would have had no cars to build without Rolls. Together they created a company that represented the peak of technical excellence. Royce was a lifelong perfectionist and designed the world's most engineering-dominated cars. Given proper maintenance, all Rolls-Royce engine and chassis parts were designed to operate indefinitely without overstressing. Henry Ford once said, "Royce was the only person to put heart into an automobile."

Rolls was somewhat of a daredevil and won many trophies while racing automobiles. Racing was a common way to advertise cars at

This 1904 Rolls-Royce is considered the oldest in existence. Owned by London's Science Museum, it carried F. H. Royce on a ride in 1931.

the time, but companies usually hired professional drivers. Rolls also went to France and learned to fly from Wilbur Wright. He purchased a Wright biplane for his own use. He was the first to fly both ways across the English Channel in June 1910. Rolls took off from Dover, England, and flew to Calais, France. After circling the city for 15 minutes, he returned to his starting point. The entire trip took one and a half hours. He entered an air show in Bournemouth in July. His airplane's

tail section broke apart during a dive and Rolls died in the crash. Rolls became Great Britain's first air fatality.

Royce was the guiding technical genius, but he lacked business skills. Others slowly eased him out of control and gave him an honorary title. The shock of his partner's death, coupled with his poor living habits, combined to cause his complete physical collapse in 1911. Although he lived until 1933, Royce never fully recovered his health.

Royce was unwilling to involve himself with airplane engine manufacture until his patriotism was stirred by World War I. He helped design the liquid-cooled, 360 hp, V-12 Eagle in 1915. It was followed by the Hawk, the Falcon, and others. Rolls-Royce built more than 60 percent of all British airplane engines used during the war. In 1919, John Alcock and Arthur Brown made the first nonstop transatlantic flight in a Vickers Vimy biplane powered by two Eagle engines. Royce died while working on an engine destined to become Great Britain's most famous World War II power plant, the Merlin. Used in British Spitfires, American Mustangs, and many other airplanes, more than 166,000 of the V-12 engines were built. To emphasize their importance to the war effort, Air Chief Lord Tedder stated, "Three factors contributed to . . . victory: the skill and bravery of the pilots, the Rolls-Royce Merlin engine, and the availability of suitable fuel."

Rolls and Royce first met in 1904 at the recently opened Midland Hotel in Manchester, England. It is still in business, now operated by Holiday Inn.

References

The Rolls-Royce by Jonathan Wood, Shire Publications, 1987.

Rolls-Royce in America by John Webb De Campi, Dalton Watson Ltd., 1975.

Complete Encyclopedia of Motor Cars edited by G. N. Georgano, E. P. Dutton Publishers, 1973.

William Edward Boeing

Born:
October 1, 1881,
in Detroit,
Michigan

Died:
September 28,
1956, in Seattle,
Washington

It is neither uncommon nor unexpected for technical projects to be named for non-technical people who spearhead their development. One of America's major manufacturing awards is the Malcolm Baldrige National Quality Award. A secretary of commerce, Baldrige (1922-1987) was a government official, not a technologist. The 6.21-mile-long Moffat Railroad Tunnel in Colorado is the third-longest tunnel in America. It is named for financier David H. Moffat (1839-1911). Steel magnate Andrew Carnegie (1835-1919) and tire manufacturer Harvey Firestone (1868-1938) were investors who possessed limited technical skills.

Not all technologists earned recognition for their personal ability to design or construct new products. Some, like Carnegie and Firestone—and William Boeing— are remembered for their ability to organize groups of technically talented people. Providing them with encouragement and a positive working atmosphere resulted in some superb products. Boeing's company gained a major industrial toehold with its remarkable 1933 Model 247 airplane. It was the first all-metal, twin-engine, low-wing transport monoplane. The 247's general layout still dominates transport aircraft design.

Boeing was an only child born into a wealthy family that had lumbering and mining interests. His father had emigrated from Germany and his mother from Austria. His family had valuable holdings in Michigan, Minnesota's Mesabi Iron Range, and Washington's Pacific Northwest forests. Boeing was educated in private elementary and secondary schools in America and Switzerland. His father had died when Boeing was eight and his mother remarried. Boeing did not get along well with his stepfather and decided to maintain some dis-

Courtesy The Boeing Company

tance. He attended Yale University's Sheffield Scientific School but left in 1902 without receiving a degree. Yet, he credited Yale with teaching him about efficient organization for the manufacture of high-quality products. Boeing moved to the state of Washington to attend to his family's business. He decided in 1908 to settle permanently in the Seattle area.

Boeing enjoyed yachting and purchased a shipyard as an investment. He did not know it at the time, but the shipyard would become his first airplane factory. Boeing met U.S. Navy Lieutenant Conrad Westervelt at Seattle's University Club. Westervelt was a structural engineer who enjoyed the new field of aviation. He and Boeing were both bachelors at the time, had studied engineering, were good at playing bridge, and enjoyed yachting. They became close friends. Westervelt encouraged Boeing's budding interest in flying and both took their first flights in 1914. Early twentieth-century airplanes were rickety, unreliable, and could

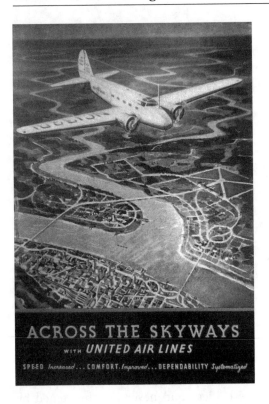

Boeing's 247 shown in a 1930s advertisement for United Airlines.

lift little more than their own weight. Westervelt and Boeing thought they could build a better plane than the pusher-type biplanes that were common.

Boeing went to Los Angeles in 1915 and learned to fly at a school operated by pioneer aviator Glenn Martin (1886-1955). He returned to Seattle with a crated $10,000 seaplane. Boeing wanted to turn his hobby into a business and used the Martin seaplane as a model for his own airplane. Boeing had the necessary money and the building and Westervelt had the technical skill. They hired about 21 workers to build a seaplane. A friend at the Massachusetts Institute of Technology had recommended that Boeing hire Tsu Wong to head the design group. It was good advice. When the first of two B&Ws was completed in 1916, Boeing took it out on its initial flight. The 52-foot, twin-float biplane with a 125 hp radial engine flew without mishap. New Zealand purchased both for mail delivery and pilot training.

World War I was on the horizon and the American government was preparing itself. Westervelt had been transferred to the East Coast. Wong and Boeing used their experience with the B&W to build a navy seaplane trainer they called the Model C. They delivered a prototype to Pensacola and it passed all the necessary tests. Boeing's new company received its first major order when the government contracted for 50 of the trainers.

The years after World War I were difficult for Boeing as orders for airplanes dried up. The company stayed in business by winning contracts based on products from other manufacturers. They modernized 111 de Havilland DH-4B airplanes. They also built 200 Thomas Morse MB-3A pursuit planes. Government contracts had kept the company in business. But Boeing was convinced there would soon be a need for commercial aircraft. That was difficult for many people to understand. There were no large passenger aircraft and many private planes were surplus military trainers, which cost about $700. A new one would have cost about $4,000. Even though the aircraft market experienced a general decline and his company had to build furniture and speedboats to stay in business, Boeing understood the importance of research and development. He hired people who would be assets to the company once the industry passed through its post-war decline. Boeing occasionally used his personal resources to cover his workers' weekly wages.

The company's breakthrough commercial airplane was the 1926 Model 40. Boeing built 24 to carry mail between Chicago and San Francisco. The Model 40 was a biplane with a single 420 hp Pratt & Whitney Aircraft (P&WA) Wasp radial engine. The Model 40 had a top speed of 135 mph and a range of 550 miles. Boeing built it to carry two passengers and 1,200 pounds of mail. It was the first to provide regular passenger service and airmail night flights.

Boeing was now on a firm financial footing and had 800 employees. The company was one of the largest aircraft manufacturers in the country. The 1930 Monomail 200

Boeing's 247 commercial passenger airplane marked a major turning point in worldwide aviation. Introduced in 1933, the basic arrangement of the 247 still dominates modern aircraft design.

The B-17 bomber was the last aircraft Boeing was personally involved with. Among the most important World War II military aircraft, 12,731 were built.

was its next commercial offering. That plane was an outgrowth of the Model 40 and was the first low-wing, all-metal transport monoplane. It had retractable landing gear, an aerodynamically clean exterior, and a single P&WA 575 hp Hornet radial engine. It could carry eight passengers, freight, and mail.

The Monomail led to a revolutionary commercial aircraft, the 1933 Boeing 247. The 247 was a beautiful all-metal monoplane with two wing-mounted 550 hp turbo-supercharged Wasp engines. It also featured an automatic pilot and deicing equipment. The 247 was the first airplane that could maintain altitude on one engine with a full load. With seats for 10 passengers, it was so superior that competitors had to match it to survive. The airplane revolutionized travel with 20-hour coast-to-coast flights.

The 247 paved the way for Boeing's entry into competition for a multiengined military bomber. The term "multiengine" in a proposal request had meant two or three engines. Boeing's designers came up with an unconventional four-engined aircraft they called the Model 299. They flew it to Wright Field in Dayton for testing in 1935. An Army Air Force pilot took off with the flight controls locked and the plane crashed. Still, impressed by earlier flight results, the army ordered more planes for further testing. Boeing, who was most interested in commercial aviation, had built one of the best military bombers in history.

His Model 299 became better known as the B-17 Flying Fortress.

The government decided in the 1920s to allow private companies to deliver airmail. Boeing organized a new corporation in 1929 to provide the service. It included aircraft manufacture (Boeing Company), engine manufacture (P&WA), and an airline (United Airlines). The new corporation was so financially successful that the government conducted an investigation into its business practices. Boeing had done nothing illegal and was so angry at the veiled allegations that he retired from the company in 1933. He had married Bertha Paschall in 1921 and they had one child. The family spent time yachting and fishing in the waters around Seattle. Boeing also raised racehorses and cattle. He died on his yacht, *Taconite,* in Puget Sound at the age of 74.

Boeing attributed much of his success to his philosophy of surrounding himself with enterprising individuals. He supported fresh ideas and never felt threatened by anyone's abilities. He once said, "I've tried to make the [people] around me feel, as I do, that we are embarked as pioneers upon a new science and industry. . . . Our problems are so new and unusual that it behooves no one to dismiss any novel idea." He received the Daniel Guggenheim Medal in 1934 for "successful pioneering and achievement in aircraft manufacture and air transportation." Boeing was the sixth aviation pioneer to receive the award and only the third American. By the time of his death, the company he founded had produced about 86 different airplane types.

References

Biography of William E. Boeing, The Boeing Company, News Bureau Item No. S-7168, undated (circa 1960).

Vision—A Saga of the Sky by Harold Mansfield, Duell, Sloan, and Pearce Publishers, 1956.

Background Information—The Boeing Company, The Boeing Company, Public Relations Item No. C-1114, October 1982.

Boeing Company's Internet homepage [online]. Available: www.boeing.com

John Logie Baird

Rendering by Tim Harmon

Born:
August 13, 1888,
in Helensburgh,
Scotland

Died:
June 14, 1946, in
Bexhill, England

Backpackers are adventuresome but generally stick to established trails. Every once in a while, they strike out on their own on an unmarked trail. Such treks are seldom troublesome because the backpackers carry maps based on information from people who preceded them. The field of technology has no maps and inventors must use their best judgment. Thomas Edison (1847-1931) spent a long time looking for a practical filament for his incandescent lamp. During that trek, he once famously said that he had discovered 2,000 materials that would not work.

Technologists began investigating the concept of television before electronics was even discovered. They first worked at sending a nonmoving image down a telephone wire. This involved breaking the image into small pieces. Current terminology would call that "digitizing the image." At the destination, the pieces would be reassembled, much as they are in a jigsaw puzzle. Fax machines operate that way. In an era before dependable electronics, one inventor developed a transmission and reception theory based on a spinning disk. It came to be called *mechanical television* and John Logie Baird was the first to make a practical system. Baird sent the first live television transmission in 1925. His equipment was the first used in broadcast service in 1929. The British Broadcasting Corporation (BBC) granted him the world's first television license. The BBC may have been influenced by Baird's accomplishment the previous year. In 1928, he sent the first transatlantic television image from England to New York. But his method followed an unmarked trail that had a dead end.

Baird was born about 25 miles west of Glasgow in Scotland. His birth city was across the River Clyde from James Watt's birth city, Greenock. Baird was the youngest of four children. His father was a poorly paid minister and the family was not particularly well off. Baird was a sickly child and suffered from severe respiratory ailments all his life. He attended the local schools and read all the technical books and popular magazines he could find. While in his teens, he wired his house for electric lamps. The lamps were battery powered and recharged with a generator run by flowing water. As a youth, Baird was also interested in telephone operation and photography.

Baird started studying electricity at the Royal Technical College in Glasgow when he was 18. He heard about the early investigations of transmitting images by wire but did not show interest in the subject at that time. Baird was more concerned with gaining an education so he could make a living. He found employment in Glasgow at an automobile company and then at an electrical power company. Both jobs were hard on his health and he often missed work because of chronic illnesses. When World War I broke out in 1914, Baird was rejected for service. He made ends meet by selling socks and soap to department stores. He even moved to the warmer climate of Trinidad in the Caribbean Sea to make jams and jellies for sale. None of these ventures succeeded. Bad luck dogged Baird through his early years. At 34, he was sick, jobless, and had only £200 to his name. He considered himself a failure.

Then, Baird moved in with a childhood friend, Guy Robertson, in Hastings, England, about 60 miles south of London. He took long walks and his health began to improve. Baird felt there was only one way to overcome his failures: he must invent something. He remembered his youthful enthusiasm for electricity and the discussions of television at school. There had been many improvements in electronics over the past few years and Baird began to read technical publications again. He decided to try to make a practical television system, a goal that

The label accompanying this incomplete experimental mechanical television transmitter credited it to Paul Nipkow. Baird used the idea to make a practical system in 1925. The large holes in the spinning wheel reduce its weight. The scanning holes are in a shallow spiral and are too small to be visible in this photograph.

Baird's laboratory was located on an upper floor of this London building at 22 Firth Street. The first floor houses a restaurant named Bar Italia and a circular plaque opposite the clock commemorates the historical location. It states, "In 1926, in this house, John Logie Baird first demonstrated television."

had eluded everyone else.

In the early 1920s, most investigators used a form of the Nipkow disk as an experimental transmitter. The Nipkow disk was a revolving metal disk that had about 24 small holes in a shallow spiral near the edge. The holes scanned a subject and separated the image into smaller sections. In one version, a lens focused reflected light from the subject onto photocells behind the rotating scanning disk. The cells translated the image into a pulsing electric current. The disk was named for Paul Nipkow (1860-1940) of Germany, who developed the idea in 1884. Nipkow's theory was sound, but no one had made such a device work. Sick, almost penniless, and with no laboratory equipment, Baird thought he could.

With his limited savings, Baird purchased some crude equipment and assembled it on a small table. It included a cracker box, a cardboard scanning disk, knitting needles, and army-surplus photocells and diodes. He used a Nipkow disk in his transmitter and receiver. The receiver recombined the scanned image onto a small screen. In 1924, Baird transmitted an image of a Maltese cross over a distance of a few feet. The lines of his picture went from top to bottom, not left and right as with current televisions. And the whole picture was made up of only 30 lines, not the 625 or more used in modern television sets. It flickered at 12 frames per second. His success encouraged Baird to take the apparatus to London, in hopes of finding financial backing.

Baird found a financial backer who offered a small amount of money and found him an upper-flat apartment that he could use for a laboratory. Almost always hungry, often sick, and frequently desperate, Baird spent months working on his crude equipment. His goal was to transmit a three-di-

mensional image and he used a puppet head as a subject. Success came in October 1925, when he transmitted a blurred but recognizable image of the puppet head. In January 1926, in his apartment-laboratory, he showed the transmission and reception of people's faces to an audience of technical people. The faces appeared as flickering pinkish images on a four-inch by two-inch screen. The size of the group may have been as high as 50. Baird became famous overnight. The publicity provided him with investment money and he was never poor again.

Like some others in technology, Baird was part successful inventor and part showman. He was eager to interest the public in televi-

The 1929 Televisor had a rotating disk at the back. Its speed was calibrated with a knob at the left and the image was viewed through the small screen at the right.

sion to create a market and established the Baird Television Development Company. He sent moving images along telephone lines from London to Glasgow in 1927, a distance of 483 miles. And in 1928, he stunned the world with a transatlantic signal from Purley, England, to Hartsdale, New York. The image was of a Mrs. Mia Howe and was received by amateur radio enthusiast Robert M. Hart.

The BBC granted Baird a six-year experimental license to transmit television signals using his mechanical scanning system. The station's call letters were 2TV and it went on the air in 1929. Television sets could be purchased in kit form for £12 or already wired for £20. Called Televisor, 10,000 to 20,000 sets may have been sold. At that time, programming in the traditional sense did not exist. The fuzzy images that people received were often only head and shoulder shots of

people talking during test transmissions.

Baird had put up with many personal hardships and was unwilling to acknowledge the considerable benefits of all-electronic television. Some people thought him quite abrasive. He married Margaret Albu in 1931. She was a concert pianist and the daughter of a diamond merchant. She may have provided funding for some of Baird's later projects. The couple had two children. Baird died at the age of 58.

A major problem with Baird's system was that the rotating disk could only transmit low-definition images. Disks could not spin quickly enough to produce better results. The Televisor operated at only 30 lines of resolution and could not reproduce fine detail. It was for others to develop all-electronic television, which eliminated that problem.

But Baird's work was not a complete dead end. His approach received a new lease on life during the Apollo lunar landing missions of 1969-1972. The National Aeronautics and Space Administration (NASA) wanted to send live color television pictures from the moon. NASA awarded the camera contract to the Westinghouse Corporation. Existing color television cameras were bulky, heavy, and delicate. Westinghouse developed a 3-1/2-inch rotating-disk camera. It used red, green, and yellow filters that rotated at 600 rpm, the same speed used by Baird. The camera was used to televise all lunar landing missions.

References

John Logie Baird and Television by Michael Hallett, Priory Press, 1978.

Electrical Engineers and Workers by P. W. Kingsford, Edward Arnold Publishers, 1969.

The Communications Miracle by John Bray, Plenum Press, 1995.

"John L. Baird Dies; Television Leader," *The New York Times*, 15 June 1946, p. 21.

"The Color War Goes to the Moon" by Stanley Lebar, *American Heritage of Invention and Technology* Summer 1997, pp. 52-54.

This is the first page of Baird's 1924 British patent. He speculated about a television screen with many lamps when he wrote of "small electric lamps arranged in rows to form a screen."

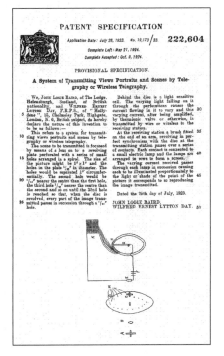

Walter Herschel Beech and Olive Ann Beech

Walter Beech:
Born:
January 30, 1891,
in Pulaski,
Tennessee

Died:
November 29,
1950, in Wichita,
Kansas

Olive Beech:
Born:
September 25,
1903, in Waverly,
Kansas

Died:
July 6, 1993, in
Wichita, Kansas

Courtesy of Beechcraft   Courtesy of Beechcraft

Rock climbing, sky-diving, and acrobatic skiing are examples of physically challenging sports. They require special training and equipment, and each activity involves an element of danger. Enthusiasts often say the danger is part of the attraction. Such high adventure and demanding situations are not a particularly recent phenomenon.

Americans have always participated in dangerous activities, although sometimes not by choice. Pioneers climbed the Rocky Mountains with little technical equipment to assist them. Early settlers and hunters risked the perils of white water while rafting or canoeing on violent rivers. Some encountered threatening animals while carrying only primitive weapons. And early flight enthusiasts purposely put themselves in danger with flimsy contraptions of wood, fabric, and wire. Flying became safer as airplanes became better designed and more reliable. After ordinary people discovered the thrill of piloting an airplane, America became the unofficial seat of the *general* aviation community. That term refers to pleasure and corporate flying, as well as commercial activities with smaller airplanes, like crop dusting and pipeline inspections. Walter and Olive Beech were involved with general aviation from the beginning. They opened the Beech Aircraft Company in 1932 and built the unique Staggerwing biplane.

Walter was born into a farming family and attended local public schools. He worked part time in sawmills and at a municipal power plant. Otto Lilienthal (1848-1896) had brought flying to the world with his gliders in Germany. (See pages 151-153.) Walter Beech must have seen images of Lilienthal in flight, since they appeared in countless publications. At age 14, Beech made a crude glider from a wooden frame and bed sheets. It did not fly, but Beech was not injured in his failed attempt at flight and he never lost his enthusiasm for it.

Beech moved to Minneapolis in 1912 and went to work for an automobile firm. He and a friend purchased a wrecked Curtiss pusher biplane and repaired it during their free time. With very little instruction from the previous owner, Beech easily flew it from a meadow on his first try. He joined the Army Air Corps in 1917 and served as a pilot trainer during World War I. Stationed at Kelly Field, near San Antonio, Texas, his responsibilities included flying airplanes from one base to another. This gave Beech the opportunity to fly several different types of machines.

Flying in an inexpensive military surplus JN4D Jenny airplane, Beech earned his living after the war through daredevil demonstrations. He flew under all weather conditions and gained valuable insights into flight characteristics and aviation equipment. He often went to Wichita, an oil boomtown in south central Kansas. In 1921, Beech accepted a position as test pilot at the Swallow Airplane Company in Wichita. Believing that flight demonstrations were the best way to sell an airplane, Beech flew in competitions all over the country. He won several aerobatic contests, which helped him attain the position of general manager at Swallow in 1923. Airplanes of the 1920s used wooden frames, but Beech felt that metal frames would work better. Disagreements arose within Swallow and both Beech and co-worker Lloyd Stearman resigned in 1924. Stearman started a airplane company two years later.

Beech, Clyde Cessna (another airplane company founder), and others organized the Travel Air Manufacturing Company. Their first plane was the Model 2000, a three-seat, open-cockpit biplane, built around a steel-tube fuselage. Powered by a 90 hp OX-5 engine from the Curtiss-Wright Corporation, it was the first airplane with a faired-in liquid-cooled engine. The Model 4000 was identical except for an air-cooled radial engine made by a different manufacturer. The company sold about 600 Model 2000 and 600 Model 4000 Travel Airs.

One of Beech's major victories came in 1926. It was a 12-day cross-country tour, flying on instruments. His first-place finish demonstrated the safety and practicality of instrument flying. Beech also helped design a special monoplane for the 1929 National Air Races in Cleveland. With another person piloting, Beech's plane won at an average speed of 195 mph. This was the first time a civilian airplane defeated all the military airplanes in a speed competition.

Beech met Olive Mellor at Travel Air, where she worked as office manager. Her father was a building contractor and her family had moved several times. She showed administrative abilities at a young age and was responsible for writing checks and paying the family bills at 11. Olive attended business schools at night, worked as an office manager and bookkeeper, and wound up at Travel

Model 17 Staggerwing with a World War II paint scheme. The lower wing is closer to the front than the upper wing, which prompted its name. The 1932 airplane also had a retractable undercarriage.

Air in 1924. As she told it, of the 12 employees, she was the only woman and the only nonpilot. Walter and Olive married in 1930.

The Travel Air company merged with the Curtiss-Wright Corporation in 1930. Walter became an executive based in New York City. He liked neither the setting nor the job. He preferred smaller cities and enjoyed flying, which was not convenient in his new position. In New York, Olive did not work outside the home. The couple decided to return to Wichita in 1932 and open their own company. In the middle of the Great Depression,

Model 18 Twin Beech, with a Plexiglas nose for its role as a World War II bomber trainer. The plane had two 450 hp radial engines that gave it a top speed of 215 mph and a range of 745 miles.

this was a risky decision.

The Beeches rented a small 900-square-foot factory and hired several people they had worked with at Travel Air. Olive served as secretary-treasurer of the new Beech Aircraft Company and Walter as president. The company's small staff designed and built a new type of general aviation airplane. It was a four-seat, enclosed-cabin biplane with the comfort of a fine automobile and a top speed over 200 mph. Powered by a 450 hp R-985 Pratt & Whitney radial engine, the Beeches Model 17 Staggerwing was introduced in November. The name came about because the lower wing was closer to the front than the upper wing. The plane was moderately priced at $18,000.

The Staggerwing was a fast, comfortable, and popular aircraft. Walter and Olive encouraged its use by women pilots who wanted to attempt new records. Among them was Jacqueline Cochran (1912-1980), who reached a record altitude of 30,052 feet in 1939. Later, flying a jet plane in 1953, Cochran became the first woman to fly faster than the speed of sound. Pilots Louise Thaden and Blanche Noyes broke another record in 1936. They won the Bendix Transcontinental Speed Dash with their Staggerwing, beating the nearest competitor by almost 45 minutes. Thaden and Noyes set a new transcontinental speed record for women. More than 100 of the 780 Staggerwings manufactured are still flying.

The Beeches next production aircraft was a twin-engine, low-wing monoplane. They named it the Model 18 Kansan but it was more often called the Twin Beech. The seven-seat airplane first flew in early 1937 and 1,582 were purchased by the Army during World War II. At the time, more than 90 percent of American navigators and bombardiers trained in the Model 18. Its 32-year production history (1937-1969) was once the

Courtesy of Beechcraft

Walter Beech photographed at the front of a Travel Air 2000 in about 1925. The biplane was the first with a smoothly enclosed liquid-cooled engine.

longest of any airplane in the world. Beech introduced the Bonanza V35 after the war. It had a distinctive V tail instead of a horizontal and vertical fin. The Bonanza first flew in 1945 and about 10,000 were built.

Walter's hobbies were hunting and all types of sports. He put in 10,000 hours as a pilot and particularly enjoyed traveling with Olive and their two daughters. Olive's hobbies were art and music. When Walter unexpectedly died of a heart attack in 1950, Olive took over as president of the company.

The *New York Times* newspaper selected her in 1943 as one of America's 12 most distinguished women. She quickly showed why. The Korean War was in progress and Beech Aircraft had many government contracts. Olive smoothly led the company through the war and right into the space age. The company developed storage and transfer units for the super-cold propellants used during the Apollo moon landing program. Olive served as an advisor to the Smithsonian Institution and *Fortune* magazine honored her twice for her leadership in industry. She served on the company's board of directors, staying involved until she died at 89. Olive Beech saw the construction of more than 49,000 Beech airplanes.

Successful husband and wife technical teams like the Beeches are rare but not unheard of. Washington and Emily Roebling supervised construction of the Brooklyn Bridge through its completion in 1883. Frank and Lillian Gilbreth developed improved factory techniques between 1904 and 1924. Ole and Bess Evinrude worked side by side manufacturing outboard motors during the 1910s and 1920s. The Beeches received many individual awards. But one they shared was membership in the National Aviation Hall of Fame in Dayton, Ohio. The only husband-and-wife team to precede them was Charles and Ann Lindbergh.

References

Walter H. Beech—Air Pioneer, Beechcraft Corporation, undated (circa 1980).

Olive Ann Beech—Chairman Emeritus —Beech Aircraft Corp, Beechcraft Corp., January 1984.

The Vintage and Veteran Aircraft Guide by John W. Underwood, Collinwood Press, 1974.

Frederick McKinley Jones

Courtesy of Thermo King

America is a massive country with cities linked by more than 3 million miles of roads and 300,000 miles of railroad track. Those transportation arteries are used to move perishable products over thousands of miles. Modern refrigeration methods keep farm produce and medicines safe and fresh. Before 1939, trucks had to use ice or an ice and salt combination. August Fruehauf (1868-1930) manufactured the first refrigerated truck trailers for food transportation around 1920. After loading trailers with up to six tons of produce, workers opened a trapdoor in the roof. Crushed ice flowed over the payload and would keep it fresh for a short time. But traffic delays or unusually hot weather overloaded that crude method and shipments sometimes spoiled.

Buildings were cooled by mechanical refrigeration as early as 1902. Willis Carrier (1876-1950) almost single-handedly developed modern air conditioning. (See pages 190-192.) But Carrier's equipment was too large for trucks or railroad cars. And it was not designed to withstand vibration from unpaved and bumpy roads, or jostling during connection of railroad cars. Following a checkered career with many unusual twists and turns, Frederick Jones developed the first practical mechanical refrigeration system for trucks in 1939. It was standard equipment on some vehicles used by the Army Air Corps and the Quartermaster Corps during World War II.

Born across the Ohio River from Cincinnati, Jones was raised by his father until age

Born:
May 17, 1893,
in Covington,
Kentucky

Died:
February 21, 1961,
in Minneapolis,
Minnesota

10. His mother died when he was an infant. Over the next few years, he lived in a parsonage and earned his keep by splitting firewood, scrubbing floors, and cooking. He attended public school for four years—his only formal education. Fascinated by motorcars, Jones left school and talked his way into a job at the R. C. Crothers Garage in Cincinnati. He was about 14 and had an aptitude for engines. He quickly became an apprentice mechanic and three years later, a foreman. Jones's skill with tools stayed with him all his life and he used them as artistically as a concert musician.

Jones loved driving cars and modified some race cars for Crothers. But Crothers considered Jones too young to drive them in races. Jones left the garage and hopped a freight train, eventually winding up in Chicago. Jobs were harder to come by in that unfamiliar city and he decided to return to Cincinnati in 1912. He accidentally jumped onto the wrong freight train and found him-

Jones received more than 40 patents associated with refrigeration. This 1949 patent for a roof-mounted "Air Conditioning Unit" was for use on refrigerated trucks. His patent stated it would be for "trucks used to transport perishables such as . . . citrus fruits, vegetables, and the like."

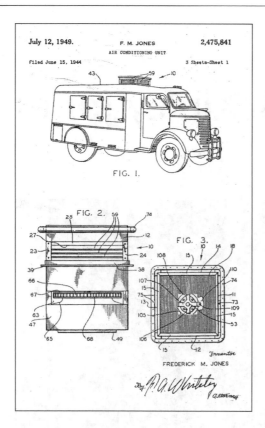

self in Effingham, Illinois. Jones sought work at the Pacific House hotel and immediately began rebuilding the hotel's malfunctioning steam-heating system. It was September and the owner wanted to have the system operating before the onset of winter. One month later, Jones had it working as good as new.

While in Effingham, Jones heard of a large farm in Hallock, Minnesota, that might need someone to maintain equipment. The 30,000-acre Hill Farm was operated by the son of James J. Hill (1838-1916), builder and owner of the Great Northern Railroad. Besides enormous buildings, the farm had its own electrical generating plant, every possible piece of farm machinery, and barns for more than 200 horses. Jones arrived in a snowstorm on Christmas Day 1912 and had no trouble getting a job.

In addition to taking correspondence courses, Jones experimented with electricity, electronics, and automobile engine design. He modified the farm's alternators and increased its electrical power. He also built a portable X-ray machine for a local physician and designed equipment for his small radio station, KGKF. In his spare time Jones played the saxophone in the town band,

fished, and hunted. But his favorite pastime was building a race car with his new friend Oscar Younggren, owner of the local garage. The two men used a Dodge engine on their modified frame. They called their car *Number 15* because they completed it in 1915. Jones enjoyed driving the car in local races and once beat a Curtiss pusher biplane.

During World War I, Jones gave up his job and joined the army. Sergeant Jones served in France, wiring buildings for electric lights. He also maintained motorcycles and trucks. He returned to Hallock, opened a repair shop, and went back to racing *Number 15* on the weekend.

The event that put Jones on the path to truck refrigeration began with motion pictures. The owner of the local movie house could not afford the equipment for talking motion pictures and asked Jones for help. Sound came from a stripe on the film that had to be sensed by a thin beam of light. Jones ground a half-cylinder lens from a glass towel rod, purchased an inexpensive photoelectric cell and exciter lamp, and had a serviceable system. It came to the attention of Joseph Numero at Ultraphone Sound Systems in Minneapolis. He offered Jones a position as product development manager in 1930 and became his lifelong friend.

Jones worked on sound equipment, theater arc lamps, and seats. His first patent was for a ticket-dispensing machine that is the basis for those still used in movie theaters. Theaters were among the first public buildings to be cooled by refrigeration, and Jones also investigated air conditioning.

Competitors brought lawsuits against Ultraphone and the lengthy litigation exhausted Numero. Involved with other businesses, he sold the company and asked Jones to research air conditioning. Not unlike General Electric's treatment of its great researcher Charles Steinmetz (1865-1923), Numero provided Jones with a guaranteed income, housing, and a place to test his ideas.

Numero was a close friend of Harry Werner, head of the Werner Transportation Company. Werner had lost a shipment of poultry because the ice in the trailer melted after the truck broke down on the highway. Numero thought that Jones could do almost anything and offered to build a mechanical refrigeration system for Werner. It was not

The practicality of refrigerated truck trailers and railroad boxcars led to the increased use of frozen and fresh foods. Jones received more than 60 patents, 40 for refrigeration equipment. He was the first African-American member of the American Society of Refrigeration Engineers in 1944.

Jones was quite independent and had erratic living habits. He would sometimes leave work to go fishing for long periods and not tell anyone. Other times, he would work on a project nonstop for several days. He ate and slept irregularly. Jones had a short-lived marriage in the 1920s and married again in 1946. His last 15 years with his wife Lucille were more orderly and the happiest of his life.

Jones often told young people that the three keys to success were to read, have faith in yourself, and don't be afraid to get dirty. He posthumously received the National Medal of Technology in 1991. The medal is the highest honor bestowed by America for technical achievement. Jones's widow accepted the medal from President George Bush in the Rose Garden of the White House. Thermo King did not forget the legacy Jones left. The company had a car in the 1967 Indianapolis 500 Mile Race named simply *15*.

Modern versions of Jones's truck refrigeration units can be seen on the front of some medium-duty trucks. This refrigerated truck transports frozen ice cream.

a new idea. Others had tried and failed. Size restrictions and road vibration presented what seemed insurmountable problems. Nonetheless, Werner delivered a 24-foot trailer for Jones to experiment with.

Jones's first design used a four-cylinder gasoline engine to drive the refrigerator's compressor. He also designed a combination starter-generator-flywheel that would automatically start and stop the engine. It was controlled by an inside temperature sensor. Jones attached the 2,200-pound equipment under the trailer. His race car experience with vibration damping helped him build a compact, sturdy, well-cushioned unit. After some testing, Jones got the weight down to 1,800 pounds, and Numero decided to go into production. Numero borrowed against his life insurance and started the Thermo King Company based on Jones's refrigeration unit. Jones received a patent for the unit in 1940. The company sold 33 of his Model A refrigerators in 1939 and Jones moved into a comfortable penthouse apartment above the factory. Only 10 years later, 5,000 units were in service and the company had annual sales of $3 million.

Dirt, mud, and stones caused maintenance problems and decreased the efficiency of the under-trailer unit. Jones reduced its weight to 950 pounds and mounted the refrigerator at the top front of trailers, where they are located today. He rode countless miles in refrigerated railroad boxcars from 1948 to 1950 to evaluate the company's design for rail transportation.

References

I've Got an Idea!—The Story of Frederick McKinley Jones by Gloria M. Swanson and Margaret V. Ott, Runestone Press, 1994.

"Born Handy" by Steven M. Spencer, *The Saturday Evening Post,* 7 May 1949.

Man With a Million Ideas by Virginia Ott and Gloria Swanson, Lerner Publications, 1977.

Dictionary of American Negro Biography edited by Rayford Logan and Michael Winston, W. W. Norton and Co., 1982.

Juan de la Cierva

Born:
September 21,
1895, in Murcia,
Spain

Died:
December 9,
1936, in Croydon,
England

The world has witnessed many different flight devices from many different inventors in many different countries. There are lighter-than-air craft such as hot-air balloons, blimps, and rigid airships. There are fixed-wing aircraft such as gliders and airplanes. Rotary-wing helicopters are another possibility. The autogiro was an unique aircraft that was briefly popular in the 1920s and 1930s. Combining aspects of fixed- and rotary-wing designs, it was almost crash proof.

During that time, early aircraft designers were cautiously learning about their new technology. But the devil-may-care attitudes of early aviators often thwarted each tentative step they made toward product safety. Such pilots often tried unusual aerobatics that overstressed airframes. They would fly too fast or too slow during maneuvers for air show audiences. Each crash further convinced the general public that flying was an unsafe activity. And there were many crashes. Spaniard Juan de la Cierva saw a test pilot make improper maneuvers with an experimental bomber Cierva had just built. The aircraft stalled and crashed into the ground at slow speed. The incident convinced Cierva to build the most crash-proof airplane he could. His first autogiro took to the air in 1923 and was the safest aircraft of its era.

Cierva was born near the Mediterranean Sea into a family with political connections. His father was a leader of the Conservative Party and twice a member of the Spanish government's cabinet. The younger Cierva received a normal grade school education where he made two lifelong friends, Jose Barcale and Pablo Diaz. At age 14, the three developed an all-consuming interest in aviation. It was 1910 and the golden age of aviation in Europe.

French pilot Jean Mauvais wrecked his small biplane in 1912 while landing in a field

Wings of Tomorrow by Juan de la Cierva

near Cierva's home. The plane was powered by an 80 hp Gnome rotary engine. The engine was an unusual 7-cylinder, air-cooled, radial type. The cylinders rotated at 1,150 rpm while the crankshaft remained stationary. Cierva described the engine as the only undamaged component. The teenagers agreed to purchase the airplane if Mauvais would teach them to fly it. They bought the wrecked airplane and had it operational in only a few months. Using the initials from their family names, they named it the BCD-1. But everyone called it the Red Crab because of the color the young men painted it. They learned to fly in the plane and it never crashed again. Barcale went on to become a civil engineer. Diaz stayed with Cierva and later became foreman of his experimental aircraft shop.

Cierva graduated from the Madrid Technical College in 1918 as an aeronautical engineer. World War I was ending and Spain had done little to keep up with aviation advances. To correct the situation, the government offered prizes of $10,000 for the construction of fighter, bomber, and reconnaissance airplanes. Others showed much interest in two of the designs. So Cierva decided to try his luck in the area that had the least competition: the bomber.

His father's connections helped Cierva locate financial backers for the $32,000 project. Completed when he was 23, his C-3 biplane had an 80-foot wing span and was powered by three Hispano Suiza, 8-cylinder, 150 hp, liquid-cooled engines. The plane could carry the equivalent of 14 passengers. An army air force fighter pilot test flew the experimental bomber. The pilot was more accustomed to fast fighter airplanes and once took the bomber too close to the ground to provide a sufficient margin of safety. The airplane stalled, and the pilot did not have enough altitude to recover and crashed. While the pilot had only bruises and scratches, the bomber was a write-off.

Two autogiros fly over the Hudson River near New York City in 1930. The George Washington Bridge, beneath them, was under construction and was completed the next year.

Cierva was struck by the fact that such a small error had caused so much damage. He decided to build an airplane that would not stall. Stalling is a condition when a wing loses its lifting ability. It is most often caused by a low forward speed. But it can also occur when speeds are too high or at an altitude where the air is thin. Ice buildup causes the airfoil to change its shape, which can also result in a stall. Any airfoil can experience the problem. Compressor blades in jet engines and propellers can also stall, although neither occurs commonly.

The career of an aeronautical designer does not necessarily end with a wrecked airplane. Cierva found additional financial backing and began to consider new flight methods. A heli-

Amelia Earhart held the official altitude record for autogiros. She is shown here before her 1931 flight to 18,415 feet at Willow Grove Naval Air Station near Philadelphia.

Wings of Tomorrow by Juan de la Cierva

copter won't stall but no practical helicopter had been built in 1919. Using the helicopter's rotary wing as a beginning point, Cierva began evaluating a related flying machine. It would use an unpowered, freely spinning rotary wing to provide most of the lift. Cierva called his plane an *autogiro*. *Giro* means "rotation" in Spanish.

An autogiro uses the engine, fuselage, and tail assembly of a conventional aircraft. Cierva's design also included short stubby wings for enhanced stability and aileron control. A large, freely spinning rotor on an overhead mast provides lift. Cierva's autogiro moved along the ground until the wind caused the rotor to spin at about 100 rpm. It then popped into the air and moved forward from the pulling effect of the conventional propeller. Autogiros had a top speed of around 120 mph. If the engine happened to fail, the plane slowly descended to the ground and landed without damage to the aircraft or pilot. Its motion was called *auto rotation* and resembled the action of winged seeds falling from trees.

The most critical design problem facing Cierva was the variation in lift between the left and right sides of the rotor. He explained the effect by assuming that the tip of a clockwise spinning rotor had a velocity of 200 mph. He also assumed a forward flight speed of 100 mph. The advancing rotor blade on the right side of the aircraft traveled through the air at 300 mph. The receding blade on the left side traveled at

Top half of photo: A publicity photo suggests that the autogiro could land anywhere. The police officer is presumably giving pilot James Ray a parking ticket.
Bottom half of photo: The first autogiro takes to the air from Getafe Airdrome near Madrid on January 17, 1923.

Wings of Tomorrow by Juan de la Cierva

An autogiro flies along Miami's shoreline. The photograph was taken in about 1930.

This 1931 PCA-2 was the first American-built autogiro. Manufactured by Pitcairn Aircraft of Pennsylvania under license from Cierva, it had a Wright 300 hp radial engine, a top speed of 118 mph, and a price of $15,000. The aircraft was retired after making 730 flights for the *Detroit News*.

Wings of Tomorrow by Juan de la Cierva

only 100 mph. The unbalanced speed would result in an unbalanced lift, causing the autogiro to flip onto its side. Cierva tried many ways to overcome the problem until he arrived at the flapping hinge, a method still used on modern helicopters. Each rotor blade was individually hinged so it could rise or fall as it rotated. The rotor blades could turn freely on their axis and did so in a type of flapping motion. Cierva patented his autogiro in 1920 and built his first controllable one in 1923. It was based on an old French Hanriot fuselage. The four-blade rotor had a diameter of 32 feet. The power plant was a Le Rhone, 9-cylinder, 110 hp, 1,250 rpm, rotary engine. Test pilot Alejandro Gomez Spencer took off from the Getafe Airdrome near Madrid and kept the aircraft under total control for the four-minute flight. It landed almost straight down. But it took Cierva until 1925 to produce a practical autogiro for sale.

Cierva was a licensed pilot and made the first autogiro flight across the English Channel in 1928. Spain showed little interest in the auto-

Juan de la Cierva photographed with his autogiro

giro. Great Britain and America showed more. Cierva moved to England in 1926 and established the Cierva Autogiro

Company. The country licensed the aircraft for production by the Avro Aircraft Company. The fictional British character James Bond piloted an autogiro named *Little Nellie* in the 1967 motion picture *You Only Live Twice*.

Autogiro manufacture in America was through Harold Pitcairn Aviation in Willow Grove, Pennsylvania. The company built the popular C-8 on an Avro 504-K fuselage. It used a Curtiss-Wright J-5 radial engine, the same type Charles Lindbergh used in his 1927 solo transatlantic flight. Amelia Earhart set an autogiro altitude record of 18,415 feet in 1931.

Cierva died in an airplane crash at the London airport. His death and the emergence of successful helicopters all but extinguished interest in the autogiro.

Cierva's primary concern was flight safety. To assure the strength of the many rotor hub parts, he insisted they all be built 10 times stronger than design requirements. The main rotor was supported by several bearings. The failure of any one bearing would not cause the rotor to stop. No autogiro ever experienced a rotor component failure during flight that resulted in loss of the aircraft. Earhart once said, "The automatic stability of the machine, as well as its peculiar properties of safe vertical descent, make it of immense utility in all types of commercial and sport flying."

References

Wings of Tomorrow by Juan de la Cierva, Brewer, Warren, and Putnam Publishers, 1931.

Aviation–The Pioneer Years edited by Ben Mackworth-Praed, Chartwell Books Publishers, 1990.

The Rotary Aero Engine by Andrew Nahum, Her Majesty's Stationary Office, London, 1987.

Harry Franklin Vickers

Born:
October 10, 1898,
in Red Lodge,
Montana

Died:
January 12, 1977,
in New York,
New York

Electricity is such a dominant power-delivery technology that some might forget its two competitors: mechanical and fluid power. Bicycles, automobile drive trains, string trimmers, and escalators use mechanical transmission methods. Fluid power uses *hydraulic* (liquid) and *pneumatic* (air) power delivery methods. These methods are less obvious to most people because fluid power's emphasis has been on manufacturing machinery and off-road construction equipment. More industrial robots, for example, use hydraulic power than any other type. For safety reasons, many power tools use pneumatics instead of electricity. America produces more than 40 percent of the world's fluid power products, substantially more than any other country. Hydraulic mechanisms are powerful devices that develop large forces. They were used in 1987 to lift a one-mile-long, offshore oil complex in the North Sea. Named Ecofisk, the complex consisted of eight platforms that weighed 44,090 tons each. More than 120 hydraulic jacks followed computer-controlled commands to raise the complex more than 20 feet. Hydraulics is so important to technology that it is one of the four fundamental topics in high school Principles of Technology courses. The other three are force, heat, and electricity.

It is hard to identify the precise date that fluid power entered the technical arena. Some point to the introduction of cargo-handling equipment by William Armstrong (1810-1900) in 1847. Armstrong invented hydraulically operated cranes for loading and unloading British ships in Newcastle. He also invented the weight-loaded accumulator in 1851. It provided a stored reserve of power. Gas-charged accumulators can be found on almost all modern hydraulic equipment. Armstrong used the Newcastle's pressurized water supply for his hydraulic fluid. This common nineteenth-century technique continued in England through the 1940s. By 1858, Armstrong had sold 1,200 cranes, hoists, and elevators. His major customers were shipyards and railroads.

Armstrong saw the tremendous potential of hydraulic power. While just a few years earlier mechanical cranes could move no more than 5 to 10 tons, his cranes could move 160 tons. Only a handful of early technologists worked with fluid power. That exclusive club included such Americans as George Westinghouse (1846-1914), who invented an air brake in 1869. Another was Arthur L. Parker (?-1945), who established a hydraulic company in 1918 that still carries his name. Another important fluid power pioneer was Harry Vickers, who became the president of one of the largest and most diverse corporations in America. Vickers started his career by inventing a balanced vane hydraulic pump in 1925. He

Courtesy Vickers Incorporated

This photograph shows the major parts of Vickers's balanced vane pump. The housing is on the adjustable angle plate. The rotor, with its dozen-or-so movable vanes, is at front center and the pump's cover is at the lower right.

hoped automobile manufacturers would use it in power steering systems.

Vickers was born not far from Yellowstone National Park and attended public schools in Montana and California. He had a youthful interest in the emerging technologies of electricity and wireless telegraphy. He also had machining skills and just before World War I, started an apprenticeship in a machine shop. Vickers enlisted in the army, where his knowledge qualified him for the Signal Corps. While still in his teens, he supervised the construction and operation of several radio stations in France. Following his release from the army in 1919, Vickers worked for a Los Angeles company that repaired hydraulic hoists on trucks. There, he developed an interest in hydraulics. He saw it as an infant industry with much potential.

Two years later, Vickers organized his own hydraulic repair company. It wasn't very large, and the staff comprised only Vickers and one part-time person. Vickers's customers wanted him to make new gear pumps, instead of repairing their old ones. Although gear pumps were inexpensive and easy to manufacture, Vickers saw that they had two major flaws. They had a short life and low efficiency. The short life came from high gear wear. The low efficiency resulted from gear

wear and the poor machining expertise of the time. Worn and imprecisely finished gears allowed oil on the high pressure side of the pump to leak back to the low pressure side.

Another common pump was the unbalanced vane type. Hydraulic oil entered on one side and was forced out on the other. Because the forces were unequal, early bearing failure was a common problem. Vickers used his free time to invent and patent a new high-pressure pump. He called it the Vickers balanced vane pump. It soon became popular for compact industrial equipment, like presses and shears, and wherever hydraulics played a role in delivering power.

The pump had a slotted rotor that turned slightly off center in the housing. The photograph at left shows the rotor to the right of the micrometer. The housing is behind it on the small adjustable table. Rotation caused spring-loaded vanes to move back and forth in the rotor's slots. This drew in hydraulic oil and then forced it out under high pressure. Two inlet and two outlet ports were directly opposite each other. That positioning balanced the pressure forces on the rotor, which prevented bearing side loads and extended pump life. Modern vane pumps are powerful units. They produce pressures up to 3,500 psi and flow rates up to 190 gallons per minute.

Vickers hoped to use his pump for power steering systems in automobiles and trucks. In the meantime, he sold it for pumping oil from the bottom of oil wells. It also found a market in forging presses, hydraulic shears, and other equipment. Vickers fitted an experimental power steering system on a 1929 Cadillac four-door sedan. It included his pump and a special steering assembly. Vickers was an expert machinist and did much of the installation work himself. The Great Depression delayed the use of power steering in passenger cars, but it attracted the attention of automobile body manufacturer Alfred Fisher. He and Vickers organized a company in Detroit to make fluid-power equipment.

Fisher and Vickers manufactured special hydraulic booster units for existing steering systems on heavy vehicles. Their company started producing them in 1931. They also made pumps, valves, and controls for the machine tool and construction industries. They manufactured complete flight control systems for airplanes and variable-speed hydraulic transmissions for interurban railroad cars.

The onset of World War II encouraged Vickers to produce systems for the U.S. Navy. His variable-speed hydraulic transmission elevated and aimed large shipboard guns. It also steered the ships. Other components found uses in winches, cranes, and hoists of all types. The company was the largest one in the power hydraulic transmission field. Vickers held 95 patents and invented many of the products his company manufactured. The first automobile company to offer power steering on passenger cars was the Chrysler Corporation in 1951.

Vickers's company combined with several others over the years. It merged with the Sperry Corporation in 1937 to form Sperry-Vickers Corporation, in which Vickers remained president. The company merged

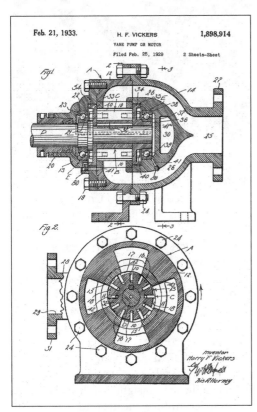

Vickers's patent drawing shows the two inlets and two outlets on his balanced vane hydraulic pump. The inlet ports are numbered "10" and are at the top and bottom. The outlet ports numbered "11" are at the left and right.

again with Remington-Rand Corporation in 1955, becoming the Sperry-Rand Corporation. It was one of the 50 largest industrial companies in the United States. Vickers continued as chief executive officer until his retirement in 1967. By then, his company had become involved with a large variety of products. It manufactured farming equipment, ground probes used during NASA's moon landings, and even the first computer offered for sale: the Universal Automatic Computer, or Univac.

Vickers had a long and enviable record of technical achievement and industrial administration. He received awards from the University of Southern California, the American Society of Mechanical Engineers, and the American Society of Tool Engineers. Vickers married Nell White in 1920 and they had two children. His personal interests included flying, amateur radio operation, fishing, and nature study. He also enjoyed working with his hands in his home workshop. He died at the age of 78.

Harry Vickers established a technological tradition that has extended around the globe, and from the field of instruction to expensive motor cars. He was an important inventor and an able administrator who headed large manufacturing corporations from 1931 to 1967. In the late 1950s and early 1960s, his company expanded its manufacturing sites to Italy, Denmark, Sweden, France, Belgium, Japan, England, Spain, and India. Vickers's name is associated with a hydraulics training program that started in 1945. And during the 1990s, the independent British Vickers Company owned 51 percent of the most prestigious automobile company in the world, Rolls-Royce Motor Cars of Crewe, England. Born in a small Montana community, Vickers left a technical legacy that has continued to expand in almost every country of the world.

References

Our Heritage, Vickers Inc., circa 1991.

Hydraulic Machines, by Adrian Jarvis, Shire Publications, 1985.

"Harry F. Vickers," *Vickers News* (a company newspaper), 5 February 1943.

Howard Hathaway Aiken

Born:
March 9, 1900,
in Hoboken, New
Jersey

Died:
March 14, 1973, in
St. Louis, Missouri

Computers are such a common part of modern technology that it is hard to imagine the world without them. When Charles Babbage (1791-1871) worked on his computer in Britain in the 1840s, no one understood what he was doing. (See pages 67-69.) Very few people could comprehend the concept of a nonhuman computer. Babbage began constructing what he called an *analytical engine.* The high cost of its many complex parts kept him from completing it. The British government had chosen not to get involved with the project.

By the 1930s, the world was considerably different. People flew through the air. Unseen electrons lighted houses with electricity. International news was immediately available through radio broadcasts. Technology's image had progressed to the point where average people generally understood the idea of technical development. But some technological developments did not benefit humanity. Europe and Asia were in political turmoil, with war clearly on the horizon. Howard Aiken had an idea for a computer that could help strengthen America. In the late 1930s, he convinced the International Business Machines Corporation (IBM) to develop his design, which he called the Mark I. Aiken's electro-mechanical computer was the world's first large-scale, program-controlled, general purpose digital computer.

Although born in Hoboken, Aiken grew up mostly in Indianapolis, where he attended Arsenal Technical High School. He worked 12 hours each night at an electrical power plant to help support his family and save money for college. He earned an electrical engineering degree from the University of Wisconsin in 1923. Aiken then took a job with the Madison (Wisconsin) Gas and Electric Company. He worked for two more companies before decid-

Smithsonian Institution Photo No. 81-10040

ing to return to school for a physics degree. He started at the University of Chicago in 1933, then transferred to Harvard University. He received a Ph.D. in 1939 and stayed on at Harvard as a professor.

Aiken often worked with lengthy mathematical equations. The tedious work made him think of using a machine for routine and labor-intensive calculations. He used the writings of Babbage, the nineteenth century computer pioneer, as a basis for his theoretical automatic calculator. On paper, he devised a method in 1937 that used the on/off character of electrical relays to identify numbers in binary code. His insight paralleled that of German computer pioneer Konrad Zuse (1910-1995). (See pages 226-228.) Neither knew about the work of the other. Aiken discussed his idea with colleagues, but they considered work on his hypothetical device too expensive to be sponsored by Harvard. They suggested that Aiken contact IBM. At that time, the company manufactured calculators, accounting machines, printing tabulators, and other office equipment.

In late 1937, Aiken met with James Bryce, one of IBM's most respected inventors, who held 500 patents. His most significant inventions were the multiplying and dividing mechanisms used in calculators. Both would be important components in the Mark I. The clarity of Aiken's proposal and its potential for success impressed Bryce. After several

use normal mathematical sequences. First run in January 1943, the Mark I was huge: more than 50 feet long, 8 feet high, and 5 feet wide. Its 750,000 parts included counters, punches, cam contacts, clutches, and countless rotating shafts. It used four paper-tape readers for input information. One carried the program instructions and three held the

Smithsonian Institution Photo No. 74-7494

meetings, the two men approached company president Thomas Watson (1874-1956). They were surprised at how rapidly Watson agreed to support the expensive venture. Watson often made quick decisions. But he may have also been influenced by the possibility of a second world war, which would begin in 1939. Ultimately, IBM paid two-thirds of the $500,000 project cost and the U.S. government paid the rest.

Work began in 1939 at IBM headquarters in Endicott, New York. Aiken supplied the insights and IBM engineer Clare D. Lake assembled the hardware with a small group of assistants. Lake's many inventions over 30 years with IBM included the company's first printing tabulator. His successes earned him the nickname "Mr. Accounting Machine." Although Aiken's invention would eventually become a general-purpose computer, IBM called it the Automatic Sequence Controlled Calculator. Everyone else called it the Mark I.

Aiken tried to make the machine as simple as possible. It addressed the four operations he considered essential. He thought that the computer should (1) use positive and negative numbers, (2) use mathematical functions like sines and tangents, (3) operate automatically, and (4)

data. Four electric typewriters recorded the output. More than 500 miles of wiring connected everything. The key items were the 3,000 relays that supported the binary mathematics necessary to efficiently carry out calculations. The noisily clicking relays made the operating Mark I sound like a room full of people knitting. The computer was highly secret during the war.

After a brief test run in Endicott, the Mark I was disassembled and shipped to Harvard University near Boston. By this time, Aiken had been drafted into the service for war-time duty. As Lieutenant Aiken, he supervised the Mark I's reassembly and its full-time operation for the Navy. Programming the Mark I required adjusting as many as 1,400 rotary switches on an external panel. Long sections of three-inch-wide paper were punched with programming holes and fed through one of the four tape readers. Lieutenant Grace Hopper (1906-1992) was one of the computer's three programmers. (See pages 220-222.) She and Aiken worked together for six years. Hopper later helped to develop the popular Common Business Oriented Language (COBOL). In 1983, she became the first woman to achieve the rank of Rear Admiral.

Aiken's Mark I patent consists of a large stack of paper: 93 pages of drawings and 138 pages of text. IBM employee Clare Lake was listed as the primary inventor, in part because Aiken was so hard to get along with.

The huge 50-foot-long Mark I computer was slow by modern standards and only had the calculating power of a modern $5 calculator. Four output electric typewriters are at the far right. Four paper-tape input devices are next to them.

Input information was delivered to the Mark I by four punched paper-tape readers.

Smithsonian Institution Photo No. 55754

The Mark I took 0.3 seconds to add or subtract, 4 seconds to multiply, and 12 seconds to divide. Four people once took three weeks to solve a problem that the Mark I finished in only 19 hours. It operated around the clock and was mainly used to calculate trajectories for shells fired from large navy guns. When journalists described the Mark I to the public after the war's end in 1945, they often incorrectly labeled it an "electronic brain." It was electromechanical and had no electronic components. Aiken shared a post-war patent for it with three IBM employees. With the complex nature of the patent and the growth of competing methods, it took more than seven years between the application for and the granting of the patent.

The Mark I operated for more than 15 years until it was retired in 1959. The machine was carefully dismantled with pieces going to IBM, Harvard, and the Smithsonian Institution. Although slow by modern standards, the Mark I was the first of its breed and launched America's leadership in the computer industry.

Aiken completed an improved version, the Mark II, in 1946. The electronic Mark III was operational in 1950, and the Mark IV in 1952. After that, Aiken got out of the business of building computers. He became director of a new computer facility that he founded at

Harvard. His greatest achievement might have been helping universities establish computer science programs. He was a quick-tempered person and difficult to get along with. Due to a conflict with Thomas Watson, Aiken refused to acknowledge the IBM president's role in the Mark I project. But Aiken also had softer edges to his personality. He was not secretive about his work and openly shared his ideas with colleagues. He gave all lecture fees that he received to members of his staff. They often appeared as wedding gifts or loans between paychecks, which never had to be repaid.

Aiken retired to Florida in 1961 to do independent computer consulting, and he became a professor at the University of Miami. He received countless awards from professional organizations, the U.S. Navy and Air Force, foreign governments, and universities. He was a rare person whose insights have had a significant effect on the entire computing profession. Aiken died at the age of 73.

The Mark I showed the way for others as the 1940s brought a rapid increase in computer development. It is not unusual to read that the 1946 Electronic Numerical Integrator and Calculator (ENIAC) was the first computer. The ENIAC may have a claim as the first *electronic* computer, but the electromechanical Mark I had been operational three years earlier. Casually written books sometimes completely neglect the Mark I, but serious references never do. A typical example comes from *Portraits in Silicon* by Robert Slater (1989): "Aiken had managed to build the first program-controlled computer." That simple sentence says it all.

References

Portraits in Silicon by Robert Slater, MIT Press, 1989.

Encyclopedia of Computer Science and Engineering edited by Anthony Ralston & Edwin D. Reilly, Van Nostrand Reinhold, 1993.

The Making of the Micro by Christopher Evans, Van Nostrand Reinhold, 1981.

Timetable of Technology edited by Patrick Harpur, Hearst Books, 1983.

Journey through Inventions by Ron Taylor, Smithmark Publishers, 1991.

Timetable of History—Science and Innovation, Xiphias Software (CD-ROM), 1993 (address: 8758 Venice Blvd., Los Angeles, CA 90034).

Philo Taylor Farnsworth

At my daughter's wedding shower, which included a light lunch, everybody was impressed with the delicious homemade bread. The bride-to-be approached her seventh-grade teacher and proudly told her that she had made the bread. She said she had found the recipe while researching a related topic for a report in seventh grade, and had made it ever since. The teacher became slightly misty-eyed. She said that teachers often don't know how they may have affected their students. That wedding shower conversation was a minor event, but it has an impressive technical heritage. A teacher at Rigby High School in Idaho spent some extra time in 1922 with a student who had an interest in electronics. Partly because of the teacher's additional effort, the student went on to develop a unique television system.

Early in the history of radio and television, it was not obvious that those media would become huge enterprises controlled by only a few networks. Independent investigators who worked alone helped to launch both communications systems. Many early inventors were left behind in the dynamics of such national enterprises. Lee De Forest's (1873-1961) 1907 triode made broadcast radio a reality. But De Forest earned little recognition or profit from his invention. John Logie Baird (1889-1946) developed the first television system used in broadcast service in 1929. (See pages 199-201.) He, too, earned little from his work. Another true pioneer of television is often lost in history's fuzzy memory. Were it not for his work being out of the technical mainstream, Philo Farnsworth might be better remembered as one of the inventors of all-electronic television.

Called Phil by everyone, Farnsworth was born into a south-central Utah farming fam-

Smithsonian Institution Photo No. 64079

Born:
August 19, 1906, in Beaver, Utah

Died:
March 11, 1971, in Salt Lake City, Utah

ily. Farnsworth's family had little money and Farnsworth did not learn much about technology until they moved to Idaho in 1920. The house the Farnsworths moved into had back issues of technical publications left behind by the previous residents. Farnsworth read them all several times. Barely in his teens, with no access to related equipment of any kind, he developed an interest in the emerging electronics field. He was especially fascinated by the prospect of using photoelectric cells and cathode-ray tubes to transmit images through the air. Farnsworth was 15 when he began designing his system.

Some television pioneers like Baird were working on mechanical television. This approach used a spinning disk with holes to break an image into small bits. The theory held that such bits could be reestablished at their destination to produce a real-time image. In practice, even the best of the early mechanical television systems produced barely recognizable images. Farnsworth was thinking about an all-electronic system, one that did not use a spinning disk. For technical advice and counseling, he approached Justin Tolman, Rigby High School's chem-

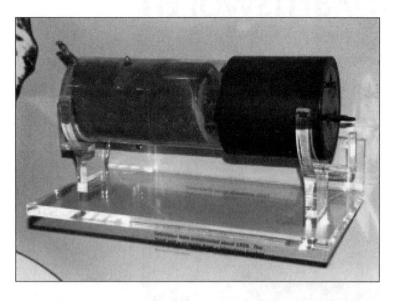

This early Farnsworth dissector was made from a laboratory beaker. Because it had no scanning beam, it instantly converted an image into electrical pulses and was called a "nonstorage" type of camera tube.

istry teacher. Farnsworth was new to the town and only a first-year student. Tolman initially thought the young man had an inflated opinion of his ability. But Tolman's opinion soon changed and he freely met with Farnsworth in a large study hall after school.

Farnsworth sketched complicated circuit diagrams on the chalkboard. Then, he and his teacher would evaluate the diagrams and make changes. It wasn't long before the student's knowledge exceeded the teacher's. But it was the technical discussions that encouraged Farnsworth to expand his views of television transmission. Other than use the school library, there was little he could do to obtain current information. The family lived so far from the seats of technical expertise, they were effectively out of contact with the technical world. Farnsworth was in Rigby for just one year before his parents moved again. However, the time he spent there defined the rest of his life.

After graduating from high school in Provo, Utah, Farnsworth started working his way through Brigham Young University. He wanted to learn all about electronics. He needed an educational foundation if he hoped to invent a practical television system. He completed two years of study before his father's death forced him to drop

out. Farnsworth was the oldest of five children. He felt a moral and a financial obligation to go to work to keep the family together.

Farnsworth found a job as an office boy with the Community Chest in Salt Lake City in 1926, the year he married Elma Gardner. They eventually had four children. Farnsworth would occasionally discuss his thoughts about television with his co-workers. George Everson and Leslie Gorrell showed considerable interest in his ideas. Neither had any technical knowledge, but they liked Farnsworth's integrity, apparent genius, and work ethic. They invested $6,000 in his work and sent him to Los Angeles, where he rented an apartment not far from the Griffith Park planetarium. He used the dining room as his laboratory. Living in Los Angeles allowed him to meet and communicate with others working in the new field of television. He was barely 20 years old and applied for his first patent within one year. He regularly mailed money home to his family.

The key item in Farnsworth's design was the tube used by the television camera. He named his a *dissector*, because it dissected the image into bits. The photograph on the first page of this chapter shows Farnsworth holding a dissector. It was considerably different from the iconoscope being developed by Vladimir Zworykin (1889-1982) in Camden, New Jersey. Zworykin worked for the Radio Corporation of America (RCA), a giant organization in the radio broadcast field. RCA had a large amount of money invested in Zworykin's iconoscope and did not like the idea of potential competition.

The iconoscope had a scanning beam, while the dissector did not. Farnsworth's dissector was a glass tube that resembled a large can with electrical connections on the side. It had a photosensitive mirror at one end, on which the television image was focused. The mirror emitted electrons in response to the different light intensities projected on it. Farnsworth made the first demonstration of his system during a press conference at his laboratory in 1928. At 150 lines, the images were crude—mode rn televisions have 625 lines or more. Also, the dissector's electronic efficiency was low, resulting in a dark picture. But it was good enough to receive positive publicity and encourage financial backers. It had taken $60,000 for Farnsworth to reach that stage, a sizable amount at the time. Television development was clearly an expensive proposition.

Farnsworth established a company named Television, Incorporated, in 1929. It was partly funded by the Philco Corporation. His small company would soon compete with large organizations such as RCA, General Electric, and Westinghouse. All had invested large sums of money in the Zworykin system and offered to buy out Farnsworth's company. An intense, high-strung individual, he routinely refused. Always barely keeping his company solvent, Farnsworth made the first open public demonstration of electronic television in 1934 at Philadelphia's Franklin Institute. The image was on a screen only slightly larger than one square foot.

After spending $250,000, Philco dropped its support. In 1935, Farnsworth established his own broadcasting station,

W3XPF. But, RCA used its considerable political influence to keep the Federal Communications Commission from implementing commercial television on a nationwide scale. Farnsworth had 165 patents and legally controlled several crucial aspects of television broadcasting.

Developing early television was indeed expensive. By 1939, RCA had spent $9.25 million and Farnsworth about $1 million. No end was in sight and the financial hardships of the Great Depression had taken their toll on Farnsworth. He agreed to sell his patent rights to RCA in 1939. The agreement was a financial triumph for Farnsworth, but it removed him from further active research in television.

Farnsworth then shifted his focus to electronic manufacturing with his Farnsworth Radio and Television Company in Fort Wayne, Indiana. The company built radar equipment during World War II and television sets afterward. It became a division of the International Telephone and Telegraph Corporation in 1958. Farnsworth later did research work with radar and nuclear energy, but he spent much of his time at a fishing retreat in Maine. He died at the age of 64.

It would be encouraging to report that America's large communication networks treated all early-twentieth-century inventors with an even hand. Yet, they often did not. Many unpleasant company officials were too concerned about corporate profits or a loss of market share. Technical goals were being controlled by business and financial people. Philo Farnsworth had one of America's most inventive minds, but he often faced heavy odds. In the end, he was forced to give up the invention he first formulated as an excited young teenager in the American west.

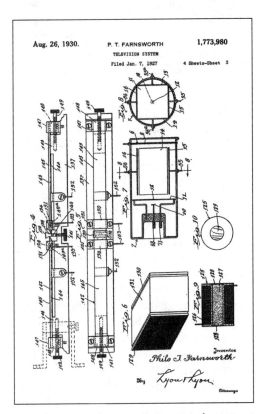

This drawing from a Farnsworth patent shows a cutaway view of his dissector, labeled as "Fig. 7" and "Fig. 8."

References

The Story of Television—The Life of Philo T. Farnsworth by George Everson, W. W. Norton and Co. Publishers, 1949.

Electronic Motion Pictures by Albert Abramson, University of California Press, 1955.

"Transition" (obituaries), *Newsweek*, 22 March 1971, p. 76.

Grace Brewster Murray Hopper

Born:
December 9, 1906, in New York, New York

Died:
January 1, 1992, in Arlington, Virginia

People often say that television has been among the most influential inventions in the world. But that is not precisely true. Television transmitting and receiving equipment does not influence people—programming does. Philo Farnsworth (1906-1971) and Vladimir Zworykin (1889-1982) separately invented different transmission methods. (See pages 217-219.) But they were inventors and technologists, not programmers. They were more concerned with developing practical technical systems than in sports, news, movies, or situation comedies. In a similar way, computers can be compared to television. Without programs, a computer would be about as useful as a television set that had no antenna or input signal.

Modern digital computers began to make their tentative appearances in the late 1930s. Before they arrived on the scene, all mathematical machines were externally controlled. Calculators and tabulating machines required routine inputs from operators. Programming was a new idea that suggested a machine with built-in intelligence. The world's first large-scale, automatic, fully programmable computer was the Mark I, developed for International Business Machines by Howard Aiken (1900-1973). (See pages 214-216.) It was a huge electromechanical device, more than 50 feet long and 8 feet high. During the dark days of World War II, Aiken knew that the Mark I required accurate programs to make calculations for the U.S. Navy. He enlisted Lieutenant Grace Hopper. She served as one of the first three Mark I programmers. Deciding how to accomplish the new task was a challenging technical activity.

Grace Brewster Murray was born into a middle class New York City family. She was named after her mother's best friend, Grace Brewster. Her father was a disabled insurance

Smithsonian Institution Negative Number 83-14876

salesman who walked slowly on two artificial legs and with the aid of two canes. She said he was her inspiration. But it was her grandfather who cultivated her interest in mathematics. He was a surveyor who allowed Grace to help him on the job. She enjoyed using his colored pencils and developed a fondness for geometry that remained with her all her life. Her father made sure his two daughters and one son had comparable educations and sent Grace to Vassar College in Poughkeepsie, New York. She did not disappoint him and graduated with honors in 1928 with a mathematics degree. She later received two advanced degrees from Yale University in New Haven, Connecticut. She married Vincent Hopper in 1930.

Vassar College offered her a teaching position in 1931. She remained on the faculty until World War II broke out and she joined the service. Grace Hopper chose the navy because her great-grandfather had been an admiral. She recalled meeting him when she was three. Lieutenant Hopper was assigned to the Bureau of Computation at Harvard University. The bureau used Aiken's huge Mark I computer to calculate trajectories of shells fired from large navy guns. After a brief orientation the first day, Aiken obviously expected big things from Hopper. He said he "would be delighted to have the coefficients for the interpolation of the arc tangents by next Thursday." Hopper joined younger officers Robert Campbell and Richard Bloch. They were the world's first three modern computer programmers.

Hopper was the first person to compile a manual of subroutines. Several small programs, which had been previously checked for errors, could be combined to make an error-free program. Her work was so uniformly impressive that Aiken asked her to write the operating manual for the Mark I. *The Manual of Operation for the Automatic Sequence Controlled Calculator* was the first of its type. IBM was the sponsoring organization and preferred the ASCC name instead of Mark I.

The Mark I was an experimental handmade device in an experimental technology. So it often malfunctioned. Electrical component failure was a common cause, but the Mark I sometimes stopped for unusual reasons. A handwritten "13" was once encoded in the punched tape input as a "B." A symbol for a delta was read as a "4." A later Mark II also had its problems. A moth that had been crushed inside a relay kept the contacts separated. Hopper removed the bug and the computer resumed operating. That is the origin of the term *debug*, which refers to removing errors from a computer program. The moth was taped into the official logbook with the notation "First actual bug found." From then

on, if the computer happened to be down, Hopper always said they were "debugging" it.

Hopper worked with Aiken for about six years. Then she took a significant career risk and left the navy in 1949. She remained in the navy reserves but went to work as a senior mathematician for a Philadelphia company that later became the Remington-Rand Corporation. Remington-Rand was developing the UNIVAC series of computers, an acronym for Universal Automatic Computer. When unveiled in 1951, the UNIVAC was the first high-capacity computer offered for sale to companies and businesses. The first was purchased by the U.S. Census Bureau. UNIVAC was 14 feet long, had 5,000 vacuum tubes, and was the world's fastest machine. It could use numbers and letters equally well, and it was the first to store data on magnetic tape.

Hopper originated the idea that computer programs should be written in English, instead of complex machine code. The transformation is carried out through a program called a *compiler*. Hopper created the first one in 1952. It was named A-0 and stored on the UNIVAC's magnetic tape. Hopper knew that the key to opening the field of computers was to develop programming languages that nonmathematical people could understand. Her continued work in that area resulted in Flow-Matic, a language aimed at business applications such as automatic billing and payroll calculations. By the end of 1956, Hopper had Remington-Rand's computers understanding 20 near-English statements. It was just one more step to the development of the enormously popular COBOL (Common Business Oriented Language). Hopper was on the six-person organizing committee that began work in 1959. COBOL was based on her

Smithsonian Institution

Hopper stands near a magnetic tape input of a 1960s Remington Rand computer. She helped develop the COBOL book in her hand.

The console of an early UNIVAC computer was far more complex than modern desktop computers. The display included the mannequin at the right.

A moth accidentally flew into this type of relay and caused an early computer to shut down. The relay was "debugged" and the computer returned to operating normally.

Flow-Matic language.

The navy asked Hopper to rejoin in 1967 to standardize business operations. Its payroll program had been rewritten 823 times and confused everyone. Hopper did what was required but recommended against the use of ever larger computers and punched cards. As long ago as the early 1970s, Hopper favored networked computers. She pointed out that when a farmer wanted to move heavy boulders, he didn't grow a larger ox. He simply added another one.

Hopper's duties often cast her in the role of teacher. Her favorite teaching tool was a wire nanosecond (billionth of a second). A 12-inch piece is about the distance electricity traveled in a nanosecond. Hopper compared it to a 984-foot-long wire microsecond (millionth of a second). She said every programmer should have a wire microsecond so they would know exactly what they were wasting with each extra microsecond.

Hopper's honorary degrees and awards took up eight single-spaced pages in her official navy biography. The most unusual was her election in 1969 as the first ever "Computer Science Man of the Year" from the Data Processing Management Association. And the most significant was probably receiving the 1991 National Medal of Technology from President George Bush. Hopper was the first woman to receive the award for individual accomplishment. A special appointment from President Ronald Reagan in 1983 advanced her to Rear Admiral, making

Hopper the first woman to hold the rank. As an active-duty officer or reservist, Hopper served in the navy for 43 years before retiring in 1986. Her retirement ceremony was held on the USS Constitution, the navy's oldest ship, in Boston Harbor. It was not far from Harvard University where her career had begun more than four decades earlier. Affectionately called Amazing Grace by her subordinates, she never retired from technology. Hopper died quietly at the age of 85, while working as a senior consultant for the Digital Equipment Corporation.

Hopper's wire nanosecond has become something of an icon. But she is also known for a clock that ran backward while still indicating the correct time. She wanted to encourage young people to think in unconventional ways. She felt her greatest accomplishment was the training of countless new computer professionals. Even while in her 80s, Hopper was a charismatic and valued speaker who identified with young people. She often made insightful comments for them to think about. "Go ahead and do it, you can apologize later," was one. Another was, "A ship in port is safe, but that is not what ships are built for." Navy destroyer USS Hopper (DDG 70) was commissioned in her honor on September 6, 1997, in San Francisco.

References

Portraits in Silicon by Robert Slater, MIT Press, 1989.

The Book of Women's Firsts by Phyllis J. Read and Bernard L. Witlieb, Random House Publishers, 1992.

Encyclopedia of Computer Science and Engineering, 2nd Ed., edited by Anthony Ralston, Van Nostrand Reinhold Co., 1983.

USS Hopper's website. Available: http://www.navsea. navy.mil/hopper/

The Mark I computer used four paper-tape inputs like this one. Hopper hand wrote coding commands. An operator used a typewriter-like device to punch holes in paper tape that represented the commands.

Frank Whittle

No engine has helped unite countries more than the gas turbine, or *jet,* engine. By powering long-distance airliners, jet engines help connect international cities with rapid nonstop flights. The first gas turbine engines of the early 1900s provided only rotational power for pumps and alternators. They were inefficient and not suited for transportation. They were internal combustion engines that included three main sections: compressor, combustion chamber, and turbine. The compressor worked much like a household fan, compressing air to about 3:1. Fuel burned continuously in containers the size and shape of large fruit-juice cans. Exhaust gases passed through a turbine, which also resembled a fan. The high-speed gases caused the turbine to spin, and that rotation powered pumps and alternators. The turbine connected to the compressor, which kept the process self-sustaining. Few people considered the gas turbine a potential thrust-generating powerplant for airplanes. But it held the promise of high power-to-weight ratios. A major stumbling point was that no known metal could withstand the stress and high combustion temperatures. Manufacturers and government officials were all but convinced that a jet engine would never produce enough thrust to propel an airplane. Through the turmoil, one person rose to the top as a champion of the jet engine: British-born Frank Whittle. He applied for his first patent when he was only 22.

Whittle was the eldest son born into a family of modest means in the industrial city of Coventry. Whittle's father developed several minor inventions in his spare time, while operating a small machine shop. His father's shop provided countless opportunities for the younger Whittle to learn the fundamentals of machine tool operation. Growing up

Born:
June 1, 1907, in Coventry, England

Died:
August 8, 1996, in Columbia, Maryland

during World War I, he was fascinated with wood-and-fabric biplanes. Whittle's academic performance in public school was adequate but not outstanding. However, he read every book he could find on aviation.

Whittle joined Britain's Royal Air Force (RAF) in 1923, becoming one of 600 apprentices learning the new specialty of maintaining and repairing metal airplanes. His supervisors noticed his excellent work and sent him to flight school at the RAF College in Cranwell. Whittle learned to fly and became an excellent pilot. But he was sometimes overenthusiastic. One of his records states, "He gives aerobatics too much value . . . and flies too low." Whittle made 47 risky experimental catapult launches in one month while serving on an aircraft carrier. Whittle graduated second in his class. He had become interested in jet engines while writing a technical paper for a final-year class. Whittle determined that a properly built engine could develop enough thrust to push an airplane through the air. But this was a theoretical effort by an unproven new officer and the Air Ministry showed no interest. Whittle applied for a patent anyway. It was January 1930 and he was 22.

The same year, Whittle married Dorothy Lee. They eventually had two sons. As a military liaison during the early 1930s, Whittle often visited aircraft and engine manufacturers. He hoped company leaders might show interest in his engine. But the Great Depression was in progress and no company had extra money to invest in such a long-term research project. Most also thought Whittle's design was too advanced for the available metals. Highly stressed jet engine parts operate at temperatures of 1,800° Fahrenheit or higher.

In 1941, the Gloster E28/39 was the first British jet aircraft to fly

No one showed any serious interest until two former RAF friends teaching at prestigious Cambridge University came to Whittle's assistance. They were Dudley Williams and J. C. B. Tinling. Through their influence, the British government approved a complex legal agreement with a newly established research company named Power Jets. Whittle, Williams, and Tinling organized the company in 1935 to build a jet engine. Whittle kept his military rank and was stationed at Power Jets in Farnborough near London.

Spurred on by looming World War II and faith in their project, the men worked night and day. The required performance for the compressor, combustion, and turbine sections far exceeded anything that had been achieved. Staffing and financing for the project were barely adequate. An old 10 hp BSA automobile engine rotated the test parts. The researchers's efforts were rewarded in April 1937 when a complete jet engine briefly ran at 2,500 rpm. The high-pitched sound of the WU (Whittle Unit) caused everyone to run for cover. Whittle said he was paralyzed with fright as he continued to open

the control valve. The engine oversped to 8,000 rpm and then dropped to 0. A higher speed could have torn it apart and killed Whittle. That minor success earned the company a development contract from the British government. Power Jets was to design and construct an engine for an airplane being built by the Gloster Aircraft Company.

While supervising the project, the strain of intense wartime work took its toll on Whittle's health. He was hospitalized several times for nervous exhaustion and severe headaches. Like many technical pioneers, Whittle drove himself relentlessly. He was also regarded as eccentric and many high-ranking officers did not understand him. That is partly why he did not attain high military rank. He retired in 1948 as a temporary air commodore. Although Whittle was devoted to military service, his superiors often treated him in indifferently. He had to fight for everything associated with the jet engine project.

Power Jets completed its first operational engine in 1941 and aptly named it the W.1. The engine had a one-stage centrifugal flow compressor. The 14-inch-diameter axial flow turbine had 72 blades. The turbine and compressor were connected and rotated at the same speed, 16,500 rpm. The engine had 10 combustion chambers, or *burner cans*. The W.1 developed 860 pounds of thrust and was installed in an experimental airplane called the Gloster E.28/39. Whittle personally conducted several taxi trials. The airplane's initial 17-minute flight during May by test pilot G. E. R. Sayer was a complete success.

Power Jets was not a production company, and the Rolls-Royce Company assumed

The WU (Whittle Unit) was the first British gas turbine engine to run, in 1937. It was extensively modified during development. This is its third and final form. It features a double-sided centrifugal flow compressor, 10 reverse-flow combustion chambers, and a single-stage axial flow turbine.

manufacturing responsibility for the W.1. Whittle was disappointed because it removed him from work on engine development. He also lost a personal investment that had grown to £47,000, but he never publicly complained. Britain's first production jet-powered aircraft was the 1943 Gloster Meteor. It used two W.2 engines rated at 1,700 pounds thrust each. Rolls-Royce in Derby remains Britain's major manufacturer of jet engines.

Having served its purpose, Power Jets was dissolved after the war and Whittle retired from the RAF. The British government awarded him £100,000 in partial payment for the economic problems it had caused. Whittle and his wife separated in 1952. They dissolved their marriage in 1976. Whittle then married American Hazel Hall and accepted a position on the faculty of the United States Naval Academy in Annapolis, Maryland. He wrote two books: *Jet: The Story of a Pioneer* came out in 1953 and *Gas Turbine Aero-Thermodynamics* in 1981. Finally realizing Whittle's important technical contributions, the world showered him with accolades. These included a knighthood, the Rumford Medal of the Royal Society, and the American Legion of Honor. In January 1998, General Electric of Great Britain opened a £20 million power-generation research center, the Whittle Centre, on the old Power Jets site. Whittle died at 89 at his home in a Baltimore suburb.

Many people have taken issue with the various individuals who ignored Whittle's invention during the 1930s. To his great credit, Whittle rarely did. He realized that people in decision-making positions were flooded with countless new ideas that failed to pass the common-sense test. To manufacturers dealing with 1930s technology, the jet engine would have required new and expensive metal alloys. The world was in the grip of the Great Depression and economic resources were not available for such an advanced project. But perhaps most important, Britain was preparing for war with a powerful adversary. Government officials were more concerned with survival than developing a new engine. Only after the removal of that imminent threat did they begin to properly support Whittle's pioneering efforts. Whittle appreciated the situation and never expressed concern that he was not properly recognized. Because of his

great technical contribution and his understanding nature, Britons are particularly proud of Sir Frank Whittle.

References

Whittle–The True Story by John Golley, Air-life Publishers, 1987.

Making of the Modern World edited by Neil Cossons, John Murray Publishers, 1992.

"Air Commodore Sir Frank Whittle," Obituary, *The Times* [of London], 10 August 1996.

"Jet Engine's Inventor, Who Changed Aviation, Dies," Obituary, *Reuters New Media,* 9 August 1996.

"They Created the Jet Age" by Noel Vietmeyer, *The Reader's Digest,* May 1987.

Fundamentals of Gas Turbines by William W. Bathie, John Wiley and Sons, 1984.

Aircraft Aircraft by John W. R. Taylor, Paul Hamlyn Publishing Group, 1967.

The Modern Gas Turbine Engine by R. Tom Sawyer, Prentice-Hall Inc, 1947.

The 1,700-pound-thrust Whittle W.2 engine was manufactured by Rolls-Royce. It powered Britain's first production jet aircraft in 1943, the twin-engine Meteor.

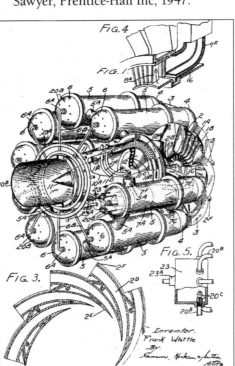

This drawing from an American patent shows the general layout of early Whittle engines. Item 1A/1B is the compressor. Item 5 includes the burner cans, or combustion chambers. Item 12/12A is the turbine. Item 10B is the exhaust case, which provides the engine thrust. This patent is one of Whittle's total of 27.

Konrad Zuse

Born:
June 22, 1910, in
Berlin, Germany

Died:
December 19,
1995, in Huenfeld,
Germany

In the early 1900s, people typically performed multiplication and division with a slide rule. The slide rule is an *analog* device, from the Greek word *analogos*, which means "ratio". Electrical meters with needles and wristwatches with hands are also analog devices. They use the position of needles or hands to indicate numerical values. A slide rule has two identical logarithmic scales that slide over each other. To divide 6 by 3, for example, one number is placed over the other by sliding the scales. The answer, 2, appears where the end of one scale meets the other. Like most calculating devices, the slide rule takes practice, but it's generally easy to use. British mathematician William Oughtred (1575-1660) invented the slide rule in about 1620. American William Burroughs (1855-1898) patented the first keyboard-operated calculator in 1892. (See pages 163-165.) Burroughs's device was actually an adding machine. Slide rules were simple, inexpensive, easy to use, and were used to multiply and divide for many years.

However, large and intricate technical projects often required many interim calculations with slide rules. The results were written on pieces of paper for later use. Keeping track of the information often proved a demanding and detailed responsibility. Final answers had to be checked and rechecked. Konrad Zuse, a young German working at his first job in an aircraft factory, had to make many lengthy calculations. He though about the possibility of creating a digital computer. Zuse began building an experimental unit in his parents' kitchen in the late 1930s. He was unaware that others were working on similar projects in America and Great Britain. The world was preparing for war and commu-

Courtesy Deutsches Museum, München

nication among technologists was often strained.

Mechanical devices fascinated Zuse when he was a youngster. Early on, he hoped to become an inventor. His father worked for the post office and did what he could to support his son's interest in technology. The younger Zuse studied civil engineering during his college years in Berlin. His class work often required lengthy calculations and he spent time developing a form to keep up with the detailed work. While Zuse didn't know it at the time, he was working on a rudimentary computer program.

Zuse went to work for the Henschel Aircraft Company after graduation in 1935. In his free time, he began building an experimental computer in the apartment he shared with his parents. He named it the Z1. Zuse was proficient with Erector set parts and used them to make his all-mechanical computer. It took three years to complete and was about the size of a four-foot cube. Zuse brilliantly decided to use calculation methods based on binary numbers, 0 and 1. He was the first to determine that using only two characters with his computer would be faster than using the 10 characters in the base 10 number system. That approach is now used in computers throughout the world. Zuse's binary system used rod positions, with one direction indicating "on" and the other "off."

The Z1 had a keyboard input and an output of electric lights. Calculation instructions were punched into discarded 35 mm film and passed through a program reader. Zuse used all these methods in his subsequent computers, as well. He received a patent in 1936. Because the Z1 was made with many lower-quality toy parts, it operated only a few minutes at a time before jamming.

The German army drafted Zuse in 1939 and he served as an ordinary infantryman. Managers at Henschel Aircraft thought his abilities could be put to better use at their factory and they got him back in six months. Zuse returned to the important job of analyzing aircraft wings. The company gave him space and financing to work on his new Z2 computer. Zuse was the only German permitted to develop computers during World War II. He had a proven track record in the field. Government leaders felt that Germany would quickly win the war, and they did not want to make heavy investments in an unnecessary technology. Zuse's was a lonely position. Few friends or colleagues understood what he was doing. Like the Z1, his Z2 was a concept computer and not intended for production use. But his Z3 was meant to be a full-scale operational computer.

Binary numbers require on and off switching. Vacuum-tube diodes worked well for the purpose because they can switch

The replica Z3 computer has three major sections. First, the console contains the input keyboard and the output bank of lights on the vertical section. Second, the program reader at the right passes 35 mm film, which contains the calculating commands. Third, the electromagnetic relays at the back wall provide the on and off switching for binary arithmetic.

very rapidly. Zuse suggested a computer with 2,000 tubes, but Henschel was unwilling to pay for such an expensive machine. Instead of tubes, Zuse used 2,600 electromagnetic relays for his 1939 Z3. Their switching rate was a few hundred times per minute. His method paralleled one under development in America by Howard Aiken (1900-1973) with his Mark I computer. Zuse and Aiken did not know of each other's work. (See pages 214-216.)

When completed in 1941, the Z3 had cost only $6,500 and it went into service with the German aircraft industry. The three major components included the operator's console, a film reader, and the relays in three cabinets. The Z3's first use was for calculating airplane vibrations during the high stresses of battle. Like the Z1, it used programs punched in 35 mm film. Data was reentered at the console for every new calculation. Operators viewed a series of small lamps to read the binary number output. The Z1, Z2, and Z3 were destroyed in a 1944 air raid.

Zuse supervised the 1962 construction of a Z3 replica for the Deutsches Museum in Munich. Entering data was a precise procedure using numbers and exponents. It followed the form of "A times 2 to the B power." The operator input the proper val-

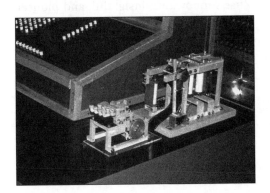

The 35 mm film would normally have punched holes in the center to provide the calculating commands. The film in the reader is for demonstration purposes only.

ues for A and B, both of which have to be binary numbers. It was a job requiring great care and attention to detail. But it released the operator from the drudgery of calculations. A 35 mm filmstrip in the reader at the right of the console had the program commands. The 2,600 relays behind the console provided the on and off switching for the binary numbers used in the calculations. The Z3 could add, subtract, multiply, divide, and take square roots. Multiplication took three to five seconds. Data was input through the keypad on the sloped front of the console. The final answer was read from the lamps on the vertical section of the console. There was no paper or screen output.

The replica Z3 computer is often demonstrated to student groups visiting the Deutsches Museum in Munich. The demonstration is only a simulation because the computer would require too much maintenance to keep it fully operational.

Zuse quickly completed a slightly upgraded version that he called the Z4. As the war drew to a close in 1945, he loaded the one-ton computer in an old truck. He hid it in the cellar of a farmhouse near the Swiss border. Zuse moved to Zurich in 1945 after the war ended, and the Z4 followed him five years later. It was used by the Federal Polytechnic Institute until 1955.

Zuse established a small computer company in Germany in 1950 and received about 50 patents. He developed one of the first computer languages and named it Plankalkül. It aimed at simplifying mathematical calculations. He sold his company to the massive Siemens corporation in the 1960s and became a consultant to the firm. In his later years, Zuse worked with inventions but spent most of his time doing oil painting. He and his wife, Gisela, had five children. Zuse died at 85 in the western German state of Hesse.

It is not unreasonable to ask who invented the modern digital computer. Most historians bypass Zuse and point to Howard Aiken and his Mark I computer. Reference books have been almost unwavering on this detail. A typical example comes from *The Making of the Micro* by Christopher Evans

(1981, page 72): "Though Zuse undoubtedly had a working system operating before the war and Aiken's giant system didn't start working in earnest until 1943, it is the latter system that is generally accepted as having been the first to fulfill [Charles] Babbage's dream."

A related question is, why don't historians credit Zuse's invention as the first modern computer? The answer involves trust. Aiken's computer was clearly operational, had a well-documented history, and survived the war. Zuse's Z1, Z2, and Z3 were presumably destroyed in a 1944 Berlin

bombing raid. It is possible they were dedicated calculators and not general-purpose computers. There is little definitive evidence to support Zuse's claim of priority, not even scrap parts following the bombing raid. Unless additional information becomes available, most of the world will credit Aiken's Mark I as the first general-purpose large-scale program-controlled digital computer. However, that hardly detracts from Zuse's innovative, insightful, and pioneering role in computer development.

References

The Making of the Micro by Christopher Evans, Van Nostrand Reinhold Publishers, 1981.

Pioneers of Computing by F. Gareth Ashurst, Frederick Muller Publishers, 1983.

Portraits in Silicon by Robert Slater, MIT Press, 1987.

Special Section:
Technology's Past Overviews

Technical Fairs

Introduction

For many Americans, visiting a county or state fair is a pleasant summertime activity. The exhibits typically appeal to agricultural professionals, but they attract people from cities and suburbs as well. Local agricultural fairs have a heritage that extends back many centuries. The earliest was regularly held in Hangchow, China, in the 1200s. It was mostly a large market where people bought meats, vegetables, jewelry, and other items. One of the first argricultural fairs in America was the Berkshire Cattle Show. It was held in Pittsfield, Massachusetts, in 1810.

Technical fairs are a more recent event because they came after the Industrial Revolution, which began in Great Britain around 1760. Near the middle of the 1800s, England was the world's major manufacturing country. The country's ruler was the well-liked Queen Victoria (1819-1901). Her husband was German-born Prince Albert (1819-1861). Albert thought that a world fair would publicize the accomplishments of all countries in agriculture, art, and industry. Although there had been successful national exhibitions in France as early as 1798, there had never been an international one.

Many of Albert's advisors thought his idea of a world fair was ill-conceived. But he prevailed. In May 1851, Queen Victoria opened the Great Exhibition of Industry of All Nations in a huge wrought-iron and glass building in London's Hyde Park. Because of the building's striking use of glass, the event was called the Crystal Palace Exposition. It was the first world fair and the first to emphasize technical accomplishment.

Crystal Palace Exposition— London 1851

An insightful member of the royal family, Albert was popular with the public and used his influence to help expand the British presence. One reason he wanted a world fair was to emphasize the industrial accomplishments of his adopted country. Since such a complicated project had never before been tried, others worried about whether people would accept it.

Even more unsettling, the man Albert selected to serve as the building's designer had previously constructed only greenhouses for flowers and plants. Joseph Paxton (1803-1865) had built the largest greenhouse in the world in 1828 for the Duke of Devonshire. Called the Great Conservatory, it was 277 feet long, 123 feet wide, and had a six-story double-curving roof. His work caught Albert's eye and Albert selected him to design the building that would house all the exhibits for the proposed exposition. It turned out to be a strikingly beautiful structure, 1,848 feet long and 408 feet wide.

The Story of Exhibitions

One main entrance to London's 1851 Crystal Palace Exposition was from the Carriage Road along the south perimeter of Hyde Park.

It had a mile of galleries above the main floor. Paxton built the Crystal Palace in just six months.

The exposition had four major divisions:

The Story of Exhibitions

The Crystal Palace was moved in 1854 to this site in Sydenham, in Greater London about four miles south of the River Thames. At the time of this photograph, the building was smaller than it was originally because of an 1866 fire. The two towers provided water pressure for fountains and other uses.

raw materials, manufacturing, inventions, and works of art. Unsure of this new approach to a fair, many countries sent displays of sculptures, paintings, furniture, and similar items. There were almost 14,000 exhibitors, including 6,861 from Great Britain and 520 from the United States. America was one of the few countries that sent major evidence of industrial achievement. Its exhibits included such unusual items as an eight-ton block of New Jersey zinc and a full-sized bridge. The exposition's highest prize was a Council Medal, and America received five of them. More than 150 other American exhibits also received awards. The British government was so impressed that it sent a commission to study manufacturing methods in the United States.

At the exposition, Cyrus McCormick (1809-1884) entered contests with his reaper and tests with his reaper and

defeated all competitors. The reaper won a Council Medal. London's *Times* newspaper wrote that the chance to see the virtues of McCormick's Virginia Reaper was worth more to England than the entire cost of the exposition. Charles Goodyear (1800-1860) displayed some of his rubber products. Samuel Robbins and Richard Lawrence brought rifle samples from their Vermont factory. Robbins and Lawrence were the first in the world to achieve interchangeability of parts on a practical level.

During the few months it was open between May and October 1851, the Crystal Palace Exposition had more than 6 million visitors. It was more popular and profitable than anyone had anticipated, and it set the stage for future world fairs. The large iron-and-glass building was originally scheduled to be torn down. But the British decided to move it in 1854 from Hyde Park to the London suburb of Sydenham for use as an exhibition center. It had permanent art and industry exhibits, a natural history museum, and a concert hall that seated 4,000.

A fire destroyed the building in 1936 and no artifacts from it remain. But the eye-catching arched iron-and-glass design of the Crystal Palace continues to influence modern architects. Two examples are the United Airlines terminal at Chicago's O'Hare International Airport and the eight-story International Information Processing Market Center (Infomart) in Dallas, Texas.

Courtesy of Infomart

The modern Dallas Infomart Center on the Stemmons Freeway was patterned after London's 1851 Crystal Palace. It is made of aluminum and glass.

Crystal Palace Exposition—New York 1853

The year was 1853. The place was New York City. People paid 50¢ to attend the first world fair held in the United States. Similar to the 1851 London fair, it had the official name of Exhibition of the Industry of All Nations. But, it was commonly called the New York Crystal Palace, a reference to the successful London Crystal Palace Exposition.

The London fair marked the first international recognition of American technology. Many U.S. citizens wanted to establish a similar exhibition in America. New York City was the chosen location because it was a port city that would attract European exhibitors and the country's financial center. Private investors sponsored the venture. No public money went into it because people thought the Constitution did not permit financial involvement by the government.

A design that purposely resembled

The tall wooden frame is a replica of Elisha Otis's safety elevator, which he demonstrated at the 1853 New York City Crystal Palace Exposition. It was powered by a twin-cylinder steam engine at the bottom. Otis stood on a platform that is about 10 feet above the floor.

London's Crystal Palace was accepted in August 1852. The first support column went up in October on a 13-acre Manhattan site now occupied by the city library and Bryant Park. The octagonal building had a floor area of 173,000 square feet and a 100-foot-diameter dome that was the largest in the country. The building stood 71 feet tall and had a colorful enameled-glass exterior. The external framework was cast iron. Translucent glass panels fit into wooden frames attached to the metal structure. Eighteen staircases connected the main floor with the second-level gallery.

President Franklin Pierce opened the fair on July 14, long after the scheduled May 2 opening. Horse-drawn streetcars filled with people arrived at one per minute on opening day. The building's three 147-foot-wide entrances were on 6th Avenue, 40th Street, and 42nd Street.

To show visiting Europeans that America was not a backward nation, many of the 4,383 exhibits featured paintings or sculptures. But the industrial exhibits were the main thrust of the fair. They included Elias Howe's (1819-1867) sewing machine, Samuel Colt's (1814-1862) revolving firearms, Erastus Bigelow's (1814-1879) power loom, and Linus Yale's (1821-1868) locks. (See pages 100-102 and 103-105.) Elisha Otis (1811-1861) gave the first public demonstration of his safety elevator.

The fair was scheduled to close in November 1853. It continually lost money, but the board of directors decided to keep it open. They hoped to turn the building into

This original Virginia Reaper built by Cyrus McCormick in 1850 is almost identical to the one he operated at London's 1851 Crystal Palace Exposition. It won the exposition's highest award.

a museum and recover some of the losses. Circus owner Phineas T. Barnum (1810-1891) took over but only stayed four months. He soon realized he could not make the fair profitable. New York's Crystal Palace Exposition cost $500,000 and ultimately suffered a $300,000 loss. It remains the smallest world fair in terms of attendance and building size. It had a total attendance of 1.2 million compared with London's 6 million.

The five-acre building was occasionally used for small exhibitions, charity balls, and other celebrations, much like today's convention centers. Like London's Crystal Palace, it was destroyed by fire. It burned in October 1858 after an accidental fire started in a storage area.

United States Centennial— Philadelphia 1876

America's Centennial Exposition celebrated the 100th anniversary of the signing of the Declaration of Independence. The six-month Philadelphia exposition opened in May 1876. Before a crowd of 200,000, President Ulysses Grant opened the throttle of a huge Corliss steam engine that powered all the mechanical displays. More than 8 million people eventually visited the seven main buildings in Fairmount Park. There were 30,864 exhibits from 50 countries.

The exposition grounds, encircled by three miles of board fencing, had seven miles of walkways and more than five miles

The Story of Exhibitions

A huge Corliss two-cylinder steam engine powered all the exhibits in Machinery Hall at the 1876 Centennial Exposition. This drawing represents President Ulysses Grant starting the engine during the opening ceremonies.

of narrow-gauge railway. The 14-acre Machinery Hall had 8,000 operating machines and included demonstrations of rifling gun barrels, operating printing presses, veneering furniture, and other industrial operations. Pieces of electrical equipment received only limited notice because they had no rotating gears or wheels to capture people's attention. All the machines were powered by a central power unit. It was a huge two-cylinder steam engine built by George Corliss (1817-1888) in Providence, Rhode Island. (See pages 112-114.)

The industrial exhibits were located in Machinery Hall. One unique invention was Alexander Graham Bell's (1847-1922) 1876 telephone. Bell received special recognition at the exposition. Another important invention was George Westinghouse's (1846-1914) 1869 air brake. Thomas Edison (1847-1931) had a display for his 1875 electric pen. It vibrated and made small holes while a person wrote on a mimeograph master that could make multiple copies. It was the first appliance to use an electric

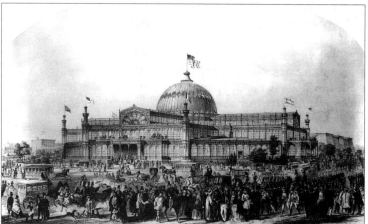

The Library of Congress

New York City's 1853 Crystal Palace Exposition building was similar to the one built in London two years earlier. It was in central Manhattan, approximately at the intersection of 42nd Street and the Avenue of the Americas.

motor. A bridge over the Mississippi River at St. Louis had a special display. The world's first major steel bridge, it was designed and built by James Buchanan Eads (1820-1887). It opened in 1874.

The Corliss steam engine was the largest in the world and the exposition's most impressive exhibit. Almost four stories tall, the 700 ton, 1,500 hp, 36 rpm engine was transported to its site in 65 railroad cars. A geared flywheel extended under the floor. It turned a shaft that powered all the exhibits. The engine was purchased by George Pullman (1831-1897) for use in his railroad-car factory near Chicago. The Centennial Exposition closed in November 1876 and was the first American world fair to make a profit. It cost about $8 million and took in more than $8.5 million.

World's Columbian Exposition—Chicago 1893

One U.S. world fair holds the record for

The Centennial Exposition, Described and Illustrated

A huge Corliss two-cylinder steam engine powered all the exhibits in Machinery Hall at the 1876 Centennial Exposition. Its 56-ton, 30-foot-diameter geared flywheel, at the left rear, was the heaviest machined wheel ever made.

The Story of Exhibitions

having had more words written about it than any other in history. Hundreds of catalogs, photo albums and essays, as well as four full-length histories were printed. Spread along two miles of Chicago's lake front, the World's Columbian Exposition of 1893 encompassed nearly three times the land area of any previous exposition. It had twice the amount of area under roof. The fair commemorated the 400th anniversary of Christopher Columbus's landing in the New World. Its 150 plaster-finished, marble-like buildings led people to call it the White City.

Chicago was selected to host the exposition because of its central location, excellent transportation, and availability of local financing. Americans also wanted to show foreign visitors how their country had developed west of the Allegheny Mountains. The exposition was designed and constructed on a grand scale. Pumps supplied 24 million gallons of fresh water per day. The fair consumed the greatest amount of electrical energy used in the nineteenth century. As many as 14,000 workers used construction materials delivered by more than 36,000 railroad cars.

The fair had four features of major interest, which included the Renaissance architecture of many of its buildings, exhibits by 86 foreign nations, and American exhibits of art and industry. The fourth major feature was a mile-long carnival of sideshows and rides called the Midway Plaisance, French for "pleasure." George Ferris's (1859-1896) Big Wheel made its debut along the Midway Plaisance.

The striking architecture and displays of

This image shows people unpacking crates inside the enormous main building at Philadelphia's 1876 Centennial Exposition. The building was the only one larger than Machinery Hall.

Some unique exhibits were displayed at the 1876 Centennial Exposition. This specially made plow has items from at least 21 historic settings. Examples include a piece of metal from Thomas Edison's first incandescent lamp, a door hinge from Abraham Lincoln's house, and wood from an apple tree carried to America on the Mayflower.

artwork were purposely selected to demonstrate the civilized nature of the United States. But the stars of the show were the wide variety of technical exhibits. The Renaissance-styled Manufactures Building attracted most of the visitors. With 44 acres under roof, it was the largest covered structure ever built. Its four elevators took people to the six-story roof for an excellent view.

Some visitors arrived on the grounds through the largest train depot in the world, which had 35 tracks. They saw such marvels as the linotype, power textile looms, and the largest searchlight in the world. They also saw a 100 mph 130-ton locomotive, the phonograph, and Pullman railroad cars for overnight traveling. Some areas of the fair were connected by a 6 mph moving sidewalk. Sixteen Westinghouse generators, each driven by an engine of almost 1,000 hp, produced more electrical power than the entire city of Chicago.

Somewhere around 140,000 people witnessed the dedication ceremony in October 1892. The exposition opened the following May. Before it closed in October, more than 27 million people had paid admission to this third U.S. world fair. In spite of its huge scale and short duration, the exposition made a profit of $1.85 million and set the style for expositions for the next 20

years. An arsonist's fire in early 1894 destroyed many buildings. Another fire the following summer destroyed still more. The Palace of Fine Arts survived and is now Chicago's Museum of Science and Industry. Many people unknowingly acknowledge the World's Columbian Exposition when they recite the Pledge of Allegiance to the American flag. Francis Bellamy (1855-1931) wrote it for the exposition's dedication.

Other Fairs

There were 22 international exhibitions between 1851 and 1939. One was the Paris Exposition of 1889, which commemorated the 100th anniversary of the French Revolution. Organizers wanted a dramatic structure to symbolize the event and conducted a design competition won by Gustave Eiffel (1832-1923). His Eiffel Tower was 984 feet tall. It remained the tallest structure in the world until Walter Chrysler (1875-1940) completed the 1,046-feet-tall Chrysler Building in New York City in 1930. (See pages 136-138 and 187-189.)

The first Ferris Wheel at the 1893 World's Columbian Exposition in Chicago was a huge amusement ride. It was 264 feet tall and carried up to 1,440 passengers in 36 cars. Each ride lasted two revolutions and took 20 minutes.

Other world fairs were held in Brussels, Buffalo, Glasgow, San Francisco, and Vienna. One popular one was the 1904 Louisiana Purchase Exposition in St. Louis. It attracted more than 14 million visitors. Chicago's Century of Progress in 1933 and 1934 hosted almost 40 million visitors. New York City's 1939 World of Tomorrow covered more than 1,200 acres and attracted about 60 million visitors. World fairs before World War II often had impressive buildings, lagoons, fountains, green space, and a wide variety of displays and entertainment. Going to a fair was an exciting experience unlike any other available then—or now.

References

The Story of Exhibitions by Kenneth W. Luckhurst, The Studio Publications, 1951.

The Crystal Palace by Patrick Beaver, Hugh Evelyn Publishers, 1970.

The World of Science, Art, and Industry Illustrated From Examples in the New York Exhibition, 1853-54 edited by B. Silliman Jr., and C. R. Goodrich, G. P. Putnam and Co., 1854.

The Centennial Exposition, Described and Illustrated by J. S. Ingram, Hubbard Brothers Publications, 1876.

Chicago's White City of 1893 by David Burg, University Press of Kentucky, 1976.

All the World's a Fair by Robert W. Rydell, University of Chicago Press, 1984.

Fairs, The Travelers, Hartford Connecticut, circa 1939.

"The Exhibition in the Palace" by Earle E. Coleman, *Bulletin of the New York Public Library*, September 1960.

"Crystal Palace," *New York Daily Times*, 15 July 1853, page 1.

Popular Culture in America by James William Buel, Arno Press, 1974.

Chicago History, Chicago Historical Society, Fall 1977.

"United's Crystal Palace" by Frank Getlein, *Air and Space*, November 1988.

This firehouse at Harpers Ferry National Historical Park, West Virginia, is the only original building left from the nineteenth-century armory. It was disassembled in 1893, shipped to Chicago, reassembled for the World's Columbian Exposition, and then returned to Harpers Ferry.

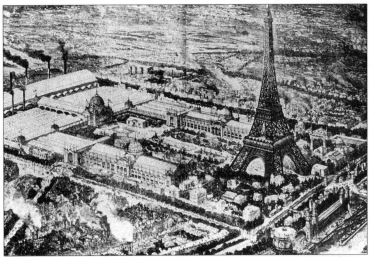

The Story of Exhibitions

The Paris Exposition of 1889 introduced the Eiffel Tower to the world.

The Story of Exhibitions

One of the largest buildings at the 1893 World's Columbian Exposition in Chicago was the Manufactures Building. Partly surrounded by a peaceful lagoon, it was designed to suggest ancient architecture.

Women in Nineteenth-Century Technology

Introduction

Long before there was technology, there was *engineering*. The word *engineer* is related to *engine*, which came from *ingenuity*. To early people, the word "engine" did not refer to a steam- or gasoline-driven power-plant. It meant a *mechanical device*, like Eli Whitney's cotton gin (engine).

During the Middle Ages, trained people designed and constructed mechanisms to capture or defend castles during battle. There were battering rams, catapults, and towers. The job title *engineer* described a person who built such devices. Leonardo da Vinci (1452-1519) designed castle defense equipment for the Duke of Milan in Italy during the 1480s. (See pages 4-6.) Until the mid-twentieth century, there was no significant difference between the work of engineers and technologists.

Early American technologies like bridge building, blacksmithing, and steam engine construction required great strength. Men were physically stronger than women, which was one reason they dominated the field. Since technologists were almost all male, technology was considered a man's occupation. But changes began to appear in the nineteenth century. Technology expanded into areas where strength was not an important consideration. These areas included drafting and design, electricity, and photography. The use of powered machine tools also helped open the field to more people. But there were still social and economic obstacles to deal with. It took particularly confident and intelligent women to venture into technical fields. If they did not succeed, they had no government-sponsored safety nets like health care, food stamps, or welfare assistance to fall back on.

Scientific American

Women were frequently involved with factory production during the 1800s. This 1895 drawing shows the various steps in manufacturing light bulbs at the Swan Lamp Manufacturing Company in Cleveland. Sir Joseph Swan was a British inventor of a practical incandescent lamp.

Science to Engineering to Technology

Throughout history, women have made countless contributions to science, engineering, and technology. But only since the nineteenth century have large numbers of women had the opportunity to study those subjects. The world's first technical area was science. It was broken down into smaller topics such as biology, chemistry, physics, and astronomy. Engineering evolved from science and had specializations like mechanical, chemical, and civil. Rensselaer

Polytechnic Institute in Troy, New York, awarded America's first engineering degree in 1835.

In a similar manner, technology evolved from engineering. The relationship between engineering and technology is a close one. Consider the name of one major United States higher education accrediting board in New York City: the Accreditation Board for Engineering and Technology. Also, note that the American Society for Engineering Education includes the classification of "Technology College Member."

Early on, it was not unusual for women to enter technical fields through science. America's first important woman scientist was the astronomer Maria Mitchell (1818-1888). (See pages 115-117.) Like many professionals—both male and female—during her era, Mitchell was largely self-educated. She earned her living as a librarian in Massachusetts. Her father introduced her to astronomy and she was the first person to locate a comet with a telescope. That 1847 discovery brought her international fame. She became a member of the faculty at Vassar College in Poughkeepsie, New York, and taught there for 23 years. Mitchell was the first woman member of the American Academy of Arts and Sciences.

Ellen Swallow Richards (1842-1911) was a student of Mitchell's who graduated from Vassar in 1870. The Massachusetts Institute of Technology (MIT) then accepted her as its first woman entrant. Richards graduated with a degree in chemistry in 1873. She was the first woman to graduate from MIT and the first woman to receive a technical degree in America.

Richards was a remarkable woman with many successes. Her background in chemistry led to specialized work in the mining industry. She became a member of the American Institute of Mining Engineers in 1879 and was the first female member of any U.S. engineering organization. In dealing with Boston's public water supply, she organized America's first sanitation

lab. Richards established the field of *domestic science,* or *home economics.* She had a demonstration of an efficient kitchen at the 1893 World's Columbian Exposition in Chicago. It included a Rumford stove, named for its inventor, Count Rumford (1753-1814). (See pages 34-36.)

America's first woman technologist worked to make homes more efficient. Catharine Beecher (1800-1878) thought of herself as an educator who wanted to improve the workplace for female homemakers. (See pages 73-75.) Beecher developed a series of innovative house designs in the 1840s. Existing houses were often poorly designed and caused people to waste time or energy during the completion of various tasks. An architect, designer, drafter, and efficiency analyst, Beecher made important contributions toward establishing a place for women in the field of technology.

By 1876, women were making so many technical contributions to American life that the organizers of the centennial celebration in Philadelphia built a Woman's Pavilion. It was in the shape of a Maltese cross, 208 feet by 208 feet. It featured displays of the products of women's art and industry. The 1893 World's Columbian Exposition Fair in Chicago included a three-story 200-

MIT Museum

Ellen Richards was the first woman to graduate from an American college with a technical degree. She was active with laboratory instruction for women, city water systems, and the field of Domestic Science.

EUTHENICS

THE SCIENCE OF CONTROLLABLE ENVIRONMENT

A PLEA FOR BETTER LIVING CONDITIONS AS A FIRST STEP TOWARD HIGHER HUMAN EFFICIENCY

The national annual unnecessary loss of capitalized net earnings is about $1,000,000,000.
Report on National Vitality

By ELLEN H. RICHARDS
Author of Cost of Living Series, Art of Right Living, etc.

WHITCOMB & BARROWS
BOSTON, 1910

Ellen Richards wrote a book on euthenics, which aimed to improve lives by controlling environmental factors.

Elizabeth Keckley wrote a book in 1868 about her experiences while she was dressmaker to Mary Todd Lincoln. She made the First Lady's Inaugural Ball gown and worked in the White House between 1861 and 1865.

foot by 400-foot Woman's Building. Sophia G. Hayden of Boston designed the building in the Italian Renaissance style. Exhibits included artwork and library cases filled with literary works by women. Women gave talks every day on such topics as science, philanthropy, and literature. More than 330 women read papers to a total audience of more than 150,000.

Technical Employment

Women were an important part of the workforce during the nineteenth century. Many worked in textile mills. The first American factories were textile mills and many were concentrated around Providence, Rhode Island. There were about 140 factories in the area in 1815. Another center of textile manufacturing was Lowell, Massachusetts. In 1816, the Committee on Commerce reported to Congress that 66,000 women were employed in the industry. Many lived in boarding houses built by the companies.

In 1829, the 62 shoe factories in Lynn, Massachusetts, employed 1,500 women as binders and trimmers. Another 1,500 had the job title "mechanic." By 1900, more than 25,000 worked in an area called "metals and metal products other than iron and steel." More than 225,000 women worked in American manufacturing in 1850. The number rose to more

BEHIND THE SCENES.

BY

ELIZABETH KECKLEY,

FORMERLY A SLAVE, BUT MORE RECENTLY MODISTE, AND FRIEND TO MRS. ABRAHAM LINCOLN.

OR,

THIRTY YEARS A SLAVE, AND FOUR YEARS IN THE WHITE HOUSE.

NEW YORK:

G. W. Carleton & Co., Publishers.

M DCCC LXVIII.

than a million by 1900. In 1890, the government classified all industries in 369 groups. Women worked in 360 of them.

Philadelphia was the nineteenth-century center of African-American life. The city boasted black newspapers and a black-owned and -operated hospital. It had several of the largest black churches in the country and the foremost black entertainers in the world. Black women flourished in that atmosphere. As early as 1795, 36 percent of the free black women of Philadelphia were retailers or proprietors. Another 4.5 percent were listed as professionals and an additional 4.5 percent as artisans.

Elizabeth Keckley (1818-1907) was a skilled African-American seamstress who saved enough money to purchase her freedom from slavery. She had earned extra income by doing sewing work for the friends of her owner in North Carolina. She moved to Washington, D.C., and worked as the personal dressmaker to Mary Todd Lincoln, while Lincoln's husband, Abraham Lincoln, served as president. Keckley developed an improved system for cutting and fitting dresses. She taught her methods to professional dressmakers in the Washington area.

A self-educated woman, Keckley wrote a book titled *Behind the Scenes*. It was about life in the White House during Lincoln's presidency. She also once briefly worked for Varina Howell Davis, wife of Confederate president Jefferson Davis.

Education

In the nineteenth-century, very few Americans of either sex went beyond high school in their education. In fact, most left after the sixth or eighth grade. Education was not perceived as particularly important. But women knew that it was the keystone to success in the workplace. They started educating themselves by attending public lectures and visiting museums. Many popular books written in the early nineteenth

Behind the Scenes, Carlton Publishers, 1868

Elizabeth Keckley was an African-American who developed improved methods for making dresses. She taught her technique to professional dressmakers in Washington DC.

century were intended for female readers.

America responded more rapidly to women's pleas for education than most other countries. Public education began to expand in the 1820s. Within a few decades, dozens of academies and colleges were established for women. By mid-century, America led the rest of the world in the amount of public and private education available for women. By 1873, there were 223 schools for women in America. Those new schools, like schools for men, placed a strong emphasis on mathematics, science, and other technical subjects. The two most important early coeducational colleges were Cornell University and the University of Michigan.

Maria Mitchell was an important astronomer and she also had influence in a much wider area. Successful brewer Matthew Vassar (1792-1868) had established Vassar College for women on 1,000 acres of land in 1865. He invited Mitchell to join the faculty. She accepted and influenced hundreds of students during her many years of teaching. Vassar College was the first college for

Vassar College, about 50 miles north of New York City, was the first college for women. It is set on a lovely, wooded 1,000-acre campus. This is its original 1865 administration building.

women that had resources equal to those found at men's colleges. *Who's Who in America*, a book listing important national figures, once included 25 of Mitchell's graduates.

There were two primary female pioneers in the field of engineering. Kate Gleason (1865-1933) was the first woman to formally study engineering in the United States. She began an apprenticeship under her father at the age of 11 and entered Cornell University in Ithaca, New York, in 1884 to study mechanical engineering. The school had been founded in 1865 by Ezra Cornell (1807-1874), a poor-born wealthy philanthropist who helped establish the Western Union Telegraph Company.

Although she did not graduate, Gleason worked in the engineering field. She and her father jointly designed a machine that rapidly and inexpensively produced beveled gears. Henry Ford credited her, rather than her father, for the invention when he said it was "the most remarkable machine work ever done by a woman."

Not only was Gleason the first female member of the American Society of Mechanical Engineers in 1918, she was also America's first woman bank president. From 1917 to 1919, she served as president of the First National Bank of Rochester, New York. The previous president had resigned to join the armed forces fighting in World War I. While at the bank, Gleason's techni-

Maria Mitchell used several telescopes during her lifetime. Many resembled this 5.6 inch refracting telescope made by Henry Fritz in 1849.

cal roots became apparent. She promoted financing for large-scale development of low-cost housing.

The first woman to earn an engineering degree was Elmina Wilson. She earned a civil engineering degree in 1892 from Iowa State College in Ames, Iowa. She was also the first woman to earn an advanced engineering degree in 1894. Her degree was des-

A. Zeese and Company, 1892

The Woman's Building at Chicago's 1893 World's Columbian Exposition was the site of the World's Congress of Representative Women. It was a meeting ground for the exchange of ideas and information.

ignated "Civil Engineering, Professional." Wilson remained at Iowa State as a professor and may have been the first woman to teach engineering. She joined a New York City architectural firm in 1904 and then another in 1907.

Wilson collaborated with Carrie Chapman Catt (1859-1947) on issues concerning women. Catt was a leader in the campaign for women's right to vote in America, Canada, and Europe. She served as president of the International Woman Suffrage Alliance. Catt's husband was an engineer.

Bertha Lamme was another nineteenth-century engineering graduate. She earned a degree in electrical engineering from Ohio State University in 1893. Marion Sara Parker received a civil engineering degree from the University of Michigan in 1895.

Invention

The most direct path to technical recognition is to invent a worthwhile product. The first American woman who had her name

on a patent was Sybilla Masters (1670(?)-1720) in 1715. It was British Patent No. 401, for a machine to prepare corn meal. The patent was actually issued to her husband, Thomas Masters, who had been the mayor of Philadelphia some years earlier. But the document clearly credits Sybilla with the statement, "a new Invencon [sic] found out by Sybilla, his wife." The patent application also bears her signature. Masters received a second patent for weaving straw hats. In 1717, the state Provincial Council approved her patents for use in Pennsylvania. The U.S. Patent Office was not established until 1790.

The nineteenth century saw at least two American women who could be considered professional inventors. The first was Margaret Knight (1838-1914) of Boston. Her first invention, not patented, came to her at age 12 during a visit to a textile mill where her brothers worked. Knight saw a young girl get hurt when a rotating spindle flew out from a machine. She designed and constructed a guard on the spot to prevent future accidents. Its exact design is lost to history, but it set Knight on the path to invention.

Although Knight received 22 patents, her unpatented inventions increase the total to 89. And they were certainly a diverse lot. Knight invented machines for cutting shoe soles and for planing concave surfaces. She patented sleeve valve rotary engines. But she is best known for her paper bag machine. It was the first to make bags with flat bottoms and Knight received a patent for it in 1870. The basic design is still used today. Knight's local newspaper called her a "Woman Edison."

The second important nineteenth-century female inventor was Helen Blanchard (1840-1922). She held 28 patents, with her first issued in 1873. Almost all dealt with improvements to sewing machines.

Blanchard established her own company in Philadelphia in 1876. Named the Blanchard Overseaming Company, it was intended to exploit some of her sewing machine inventions. The record is unclear, but the company may have existed until 1885.

Blanchard was apparently successful. Her father had lost their Portland, Maine, homestead through some financial reversals. Blanchard gained enough wealth to buy it back, along with some business property. Much more remains to be discovered about the life of Helen Blanchard.

Although Knight and Blanchard may have been the best of the best, there were other notable nineteenth-century women inventors. Maria Beasley (1847(?)-1904(?)) had 14 patents, most dealing with making barrels. Harriet Tracy had at least 16 patents. Many were mechanical improvements for sewing machines, elevators, and fire escapes. Harriet Hosmer (1830-1908) invented an artificial marble and a page-turning device for musicians. She had five patents all together. A mysterious "Mary S." (1851(?)-1880) created more patented nineteenth-century inventions than any other American woman. She had 53, but none were in her name. The evidence is unclear, but she may have lacked confidence when approached by male attorneys and patent agents.

Without early role models and viewing technology as primarily a man's profession, women slowly and cautiously entered the field. Today, though, you don't have to look far to find examples of their many technical accomplishments.

References

Mothers and Daughters of Invention by Autumn Stanley, Scarecrow Press, 1993.

Technology and Woman's Work by Elizabeth Faulkner Baker, Columbia University Press, 1964.

Women Scientists in America by Margaret W. Rossiter, Johns Hopkins University Press, 1982.

Black Women in Nineteenth Century American Life, edited by Bert James Loewenberg and Ruth Bogin, Pennsylvania State University Press, 1976.

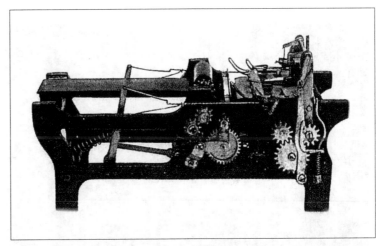

This patent model accompanied Margaret Knight's application for a paper bag machine, patented in 1879. The machine manufactured flat-bottomed paper bags, which were far stronger and more convenient than those with folded bottoms. She was called the "Woman Edison" by her hometown newspaper.

Mothers of Invention, by Ethlie Ann Vare and Greg Ptacek Quill, William Morrow Publications, 1987.

The Real McCoy: African-American Invention and Innovation by Portia P. James, Smithsonian Institution Press, 1989.

The Engineer by C.C. Furnas and Joe McCarthy, Time-Life Books, 1966.

American Society for Engineering Education web site. Available:www.asee.org/pubs3/html/women_engineers.htm

"Women in Science in Nineteenth-Century America" (pamphlet) by Deborah Warner, Smithsonian Institution Press, 1978.

World Heritage Sites

The University of Virginia in Charlottesville was declared a World Heritage Site in 1987. This central rotunda housed the original library and classrooms.

Introduction

Every country has structures in which its citizens take particular pride. Great Britain has Westminster Abbey (named a World Heritage Site in 1987) and France has the Palace at Versailles (1979). China has the Great Wall (1987) and Egypt has its pyramid fields (1979). A United Nations (UN) program allows countries to declare such treasures World Heritage Sites (WHS). Through that program, more than 500 throughout the world are officially acknowledged as having universal significance. Each must be recognized as having played an important role in its country's history. The agency that evaluates WHS proposals is UNESCO, the United Nations Educational, Scientific, and Cultural Organization.

Many WHS are natural geographic features, such as the Everglades (1979) in Florida and the Monte Alban archaeological site (1987) near Oaxaca, Mexico. Two others are the Djoudj Bird Sanctuary (1981)

near Dakar, Senegal, in Africa and the Great Barrier Reef (1981) in northeast Australia. Some sites are ancient ruins, like the impressive Roman aqueduct (1985) at Segovia, Spain, and the beautiful Acropolis (1987) in Athens, Greece. Others have an artistic nature, like the monastery of Santa Maria delle Grazie (1980) in Milan, Italy, with its painting of *The Last Supper* by Leonardo da Vinci (1452-1519). Another is the site of the rock drawings of Alta (1985) in extreme northern Norway. Some sites fall under a cultural category that deals with relatively recent structures of national importance. These are often called technical sites.

The United States' World Heritage Sites include Independence Hall in Philadelphia, the Statue of Liberty in New York City, and Monticello and the University of Virginia (combined) in Charlottesville. Such sites in Europe are often cathedrals, castles, and palaces. But one in England is the world's first metal bridge and one in France is a long nineteenth-century canal.

The United Nations and UNESCO

Chartered in 1945, the UN moved into its New York City headquarters in 1952. The European headquarters of the UN is in Geneva, Switzerland, just down the street from the headquarters of the International Red Cross. Like the UN, UNESCO was born from the ashes of World War II. It came into existence in 1946. The United States was among the first 20 countries to ratify UNESCO's constitution. The organization was established to work on five areas of international concern: illiteracy, primary education, social tensions, mutual appreciation of cultures, and living conditions. Headquartered in Paris, France, UNESCO has about 186 member countries.

The U.S. established the Yellowstone area of northwestern Wyoming as its first national park in 1872. Exactly 100 years later, the United States introduced the idea of a World Heritage Convention. It was based on the concept of a national park. The convention would oversee the establishment of historically important sites. It aimed to promote cooperation among all nations to help protect significant cultural and natural sites.

The United States was the first country to sign the agreement. Participating nations pledge to identify and protect natural and cultural sites within their boundaries. Governments nominate specific sites and acknowledge their obligation to care for them as part of their national heritage. The World Heritage Committee in Paris decides which sites should be added to the ever-growing list. The committee has representatives from 21 nations. They are elected from the 150+ countries that ratified the WHS agreement. On its twenty-fifth anniversary in 1997, more than 500 World Heritage Sites were listed. All represent irreplaceable testimonies of past civilizations and natural landscapes of great beauty and significance. Their preservation should be the concern of everyone in the world.

American World Heritage Sites

The United States has a total of 18 WHS. The first technical one was Independence Hall, which received its designation in 1979. Andrew Hamilton (1676-1741) designed the beautiful but modest building in an architectural style known as Georgian. Construction began in 1732 and finished in 1753. It took so many years because Pennsylvania paid for the building's construction as it was built. The steeple was intended to hold a 2,080 pound bell. It is called the Liberty Bell partly because an inscription cast into the bell reads, "Proclaim Liberty throughout all the land. . . ." The bell cracked in 1835 and it is now displayed in a special shelter not far from Independence Hall. A

The Iron Bridge in west-central England was the world's first all-metal bridge. It was built in 1779 and declared a WHS in 1986. The bridge is in a historic area known as Ironbridge Gorge.

Independence Hall in Philadelphia's historic district was America's first WHS in 1979. Busy Chestnut Street passes in front and carries horse-drawn carriages and automobiles.

reproduction bell hangs in the steeple.

Independence Hall was built to house the Pennsylvania legislature, but it is better remembered as the birthplace of the United States of America. It was the site where all important early national agreements were adopted. The Declaration of Independence was signed there in 1776. The Articles of Confederation, which united the 13 colonies, were ratified inside the building in

1781. And the Constitution was adopted there in 1787. A separate building just to Independence Hall's left in the photo on the preceding page housed the Supreme Court between 1791 and 1800. The one at the building's right was the meeting place of the United States Congress between 1790 and 1800. In 1950, the National Park Service restored Independence Hall to its 1776 appearance.

Many World Heritage Sites are natural formations like the Smoky Mountains on the North Carolina-Tennessee border, which received the status in 1983.

The Statue of Liberty became a World Heritage Site in 1984. France presented it as a gift to commemorate the 100th anniversary of American independence. The disassembled statue arrived in the U.S. in 1885 in 214 huge crates. President Grover Cleveland dedicated the monument in 1886. Some of the money for building a foundation for the statue came from the contributions of American school children.

Gustave Eiffel (1832-1923) designed the internal supporting frame and the 151-foot-tall statue was made by Frederic Auguste Bartholdi (1834-1904). (See pages 136-138.) Bartholdi molded 300 sheets of copper and riveted them over Eiffel's iron framework. Bartholdi intended the statue to be an immense symbol of human liberty. He originally called it Liberty Enlightening the World. The statue underwent a major restoration that was completed in 1986, its centennial year.

At that time, workers replaced much of the iron framework and made a new torch,

Many European World Heritage Sites are cathedrals like Durham Cathedral in northeastern England. The cathedral and nearby castle received WHS status in 1986.

because the original was beyond repair. They installed new elevators and fitted the crown with new stairs and windows. An instantly recognized symbol of America, the Statue of Liberty continues to inspire people all over the world. In 1989, Chinese students at Tianammen Square made a model of the statue to symbolize their revolution. They called it the Goddess of Democracy.

Thomas Jefferson (1743-1826) designed the original buildings at the University of Virginia and his nearby home, Monticello. The buildings combined became a World Heritage Site in 1987. Jefferson built his low red-brick home with its white dome on top of a hill. It was in an architectural style Jefferson had seen in France during his years as ambassador. He worked on the large house between 1784 and 1809. Jefferson had an endless string of guests who often traveled great distances to meet with him. The large house could accommodate about 50 overnight visitors.

Jefferson's heirs were forced to sell Monticello in 1833. It fell into disrepair until its acquisition in 1923 by a nonprofit organization. By 1954, the organization had restored the house and today it appears much as it did in Jefferson's time.

Jefferson also founded the nearby University of Virginia in 1819. He modeled its buildings after those of the Roman republic. Sometimes described as neoclassical, its original 28-acre layout included a central domed rotunda that served as the library and classroom building. It had two rows of pavilions with student and faculty rooms. The rooms are small and dark by modern standards, but they have a historical heritage and are coveted as housing units by today's university students. Jefferson felt unusually close to the university. He supervised its construction, organized its curriculum, selected its library books, and hired its faculty. The University of Virginia opened in 1825 with 40 students.

Other World Heritage Sites

The world's first all-metal bridge was built in England in 1779. It crossed the River Severn a few miles northwest of Birmingham. It became a World Heritage Site in 1981. The bridge was also the world's first all-metal structure. It was made from cast iron produced in a nearby city with the colorful name of Coalbrookdale. There were coal and iron ore in the region, which led to its development as a prime source of cast iron. Because of the bridge, the area soon attained the name of Ironbridge Gorge.

The presence of useful raw materials brought commerce to the region. But it was cut by the river, which produced transportation difficulties. Ironmaster Abraham Darby III (1750-1789) received the contract to build a metal bridge over the river. He assembled it the only way he knew, with pins and slots as barns were constructed at the time. The bridge originally had no threaded fasteners or welded parts. Darby melted the iron and cast everything at his furnace, about a mile away. The four main

Trier, Germany, is the oldest European city north of the Alps. Established by the Romans, it has many ruins like this one called the Porta Nigra, the most magnificent monument in the ancient city. All the Roman monuments in Trier were collectively declared a WHS in 1986.

half-ribs were 70 feet long and weighed almost six tons. Each rested partly on the ground for support and the bridge span totaled 100 feet. Darby constructed a wooden framework to support the metal bridge as it was being erected. The lovely bridge opened in 1779 and quickly acquired the name of Iron Bridge. It was restored in the 1970s and still carries foot traffic.

Southern France has a lovely 180-mile canal called the Canal du Midi. Its peaceful character disguises the efforts of 8,000 people who constructed it by hand between 1666 and 1681. Tunneling through hills and floating over rivers and valleys on viaducts, it was a complicated project. The canal has more than 100 locks, including a staircase of eight locks. A 27-mile feeder canal brought water from the River Sor to maintain the proper level. The Canal du Midi extends from Toulouse near the Spanish border to

The ramparts, or cliffs, of Luxembourg City, Luxembourg, offered a natural defense many centuries ago. With a network of caves handcut into the rock, they now add to the city's attraction as a popular tourist center. They were declared a WHS in 1994.

the Mediterranean Sea at Sete and was declared a World Heritage Site in 1996.

The canal was designed and construction was first supervised by Paul Riquet (1604-1680), a tax official who hoped to connect the Atlantic Ocean with the Mediterranean Sea. By linking up with other canals and the River Garonne, the Canal du Midi completed the route. The route began at Bordeaux and ended at Sete. It ran a distance of 310 miles—far shorter than that involved in sailing around Spain and through the Strait of Gibraltar.

The cities along the Canal du Midi profited enormously and the region is peppered with prosperous farms, grand homes, and hotels. Its success encouraged other European countries to develop ambitious canals. The railroads eventually took over the canal's commerce in the mid-nineteenth century. Today, the restored canal mostly carries tourists who enjoy the scenery and eat in restaurants that once served as the homes of lock keepers.

The James Buchanan Eads Bridge is just a short distance up the Mississippi River from the Gateway Arch in St. Louis. The world's first major all-steel bridge, it opened in 1874 and is on America's current list of WHS applicants.

Some American Applicants for WHS Status

The Secretary of the Interior, through the National Park Service, has responsibility for nominating American sites. Proposed sites are usually federal property or those designated as national historic landmarks. Private properties can be nominated if their owners guarantee to protect them forever.

The United States has more than 20 technical applicants waiting for WHS sta-

This four-step group of hand-operated canal locks is in Birmingham, England's second largest city. A narrow pleasure boat is locking itself through to a higher water level. These locks are similar to ones found on the Canal du Midi in southern France. The Canal du Midi locks were declared a WHS in 1996.

tus. One is the Brooklyn Bridge, which crosses the East River and connects Brooklyn with Manhattan in New York City. It was designed by John Roebling (1806-1869), who died following an accident during the early stages of construction. (See pages 88-90.) The bridge was completed by his son Washington Roebling (1837-1926) and his son's wife Emily Roebling (? -1903). The bridge opened for traffic in 1883.

Another bridge on the list is the Eads Bridge, which crosses the Mississippi River at St. Louis, Missouri. Designed and built by James Eads (1820-1887), it was the world's first all-steel bridge. It opened in 1874. The bridge was the only bridge that Eads built and it has withstood more than 100 years of spanning the most powerful river in the United States.

Two building complexes are being proposed. One is the Michael Pupin Physics

Laboratory at Columbia University in New York City. Pupin (1858-1935) worked with electrical inventions. One of his most significant was a telephone loading coil. It allowed people to engage in understandable conversation over distances as great as 1,000 miles. Pupin also won a 1924 Pulitzer Prize for his autobiography, *From Immigrant to Inventor*.

Another notable building complex is the General Electric Research Laboratories in Schenectady, New York. It was the first company-sponsored industrial research facility in America and succeeded because of Charles Steinmetz (1865-1923). Starting in 1892, Steinmetz did theoretical work with alternating current electric motors for GE. He was the person most responsible for establishing America's early leadership in that field.

You can find the complete list of United States WHS nominees from the National Park Service Internet Web site. It is <www.nps.gov/worldheritage/list1.htm>. Another worthwhile Web site is sponsored by the World Heritage Committee. Go to <www.unesco.org/whc> to find the list of all the World Heritage Sites. For more information, you can also write to:

UNESCO World Heritage Sites
7 Place de Fontenoy
75700 Paris
France

References

The World Heritage, UNESCO, 1991.

UNESCO web sites. Available: <www.unesco.org> and <www.unesco.org/whc>

The Iron Bridge and Town, Ironbridge Gorge Museum Trust, Jarrold Publishing, 1995.

National Park Service Web sites. Available: <www.nps.gov/worldheritage/list1.htm> and <www.nps.gov/worldheritage/list2.htm>

Biographical Dictionary of the History of Technology edited by Lance Day and Ian McNeil, Routledge Publishers, 1996.

Works of Man by Ronald W. Clark, Viking Penguin Publishers, 1985.

Independence (a National Park Service brochure), Government Printing Office, 1996.

Industrial Archaeology

Introduction

People who live in Connersville, Indiana, have an advantage over other Americans. So do the folks who live in Springfield, Massachusetts; Poughkeepsie, New York; and Diamond, Missouri. They all live near sites of significant technical accomplishment and can easily take advantage of their presence. Francis Roots (1824-1889) and his brother established a world-renown blower factory in Connersville. The factory and

The modern entrance area of the Business Design Centre in London was influenced by the design of the 1851 Crystal Palace. The top of the restored 1862 Royal Agricultural Hall can be seen above and just behind the entrance addition.

other aspects of Roots's life are still there. Charles Duryea (1861-1938) and his brother built America's first automobile factory in Springfield. Although the building no longer exists, the city commemorates its location with a small, well-kept urban park. An astronomer, Maria Mitchell (1818-1889), was America's first important woman sci-

entist. (See pages 115-117, 127-129, and 178-180.) Her 1865 observatory on the Vassar College campus in Poughkeepsie is open to the public. George Washington Carver (1861-1943) was America's first biotechnologist. He discovered many uses for farm crops. Carver's home in southwest Missouri is a national monument.

Visiting those sites, and many others, is a very rewarding experience. They combine aspects of technology with American history and provide a sense of place. It allows a visitor to focus on one person and on that person's contribution to our modern way of life. To almost literally walk in the footsteps of such giants is an exciting sensation that is hard to describe.

People who live near one of America's technical museums have similar advantages. Some companies have private museums that are open to visitors on request. Linus Yale, Jr. (1821-1868) improved his father's pin-tumbler lock in 1851. The Yale Lock Company near Charlotte, North Carolina, has an extensive display of locks extending back to Egyptian times. Starrett Tools in Athol, Massachusetts, was established by Laroy Starrett (1836-1922) in 1877. The family-operated business has a small museum centered around measuring tools and the company's founder.

America produces more than 25 percent of the world's manufactured goods and has a technical legacy that stretches back over two centuries. Historic locations and museums help preserve that birthright for future generations. Acknowledging and publicizing America's technical heritage also helps to give young people a firm foundation on which to establish their careers.

Not everyone lives near a technical site of national or international proportion. But some people might live near one of local significance. Perhaps a now-defunct factory down the road made bricks for many of a city's buildings. Maybe an area machine shop once produced parts for the venerated Ford Model T automobile. A derelict power plant might have been the first in the region to produce electricity for public use. Recording information about the physical remains of irreplaceable technical treasures is called *industrial archaeology*.

Origins of Industrial Archaeology

Industrial archaeology started in England in the 1950s. Its purpose is to discover, record, and study the tangible remains of past industries. It is not specifically intended to restore such facilities, which is often an expensive undertaking. Industrial archaeology is merely an observation and recording procedure, including taking photographs. It is not as wide sweeping as the field of *technical history*, which began in 1919 with the founding of the Newcomen Society. The words "industrial archaeology" first appeared in print in the autumn of 1955. Michael Rix of the University of Birmingham in England, wrote an article for *The Amateur Historian*.

Business Design Centre

Sam Morris (1917-1991) is shown inside London's Royal Agricultural Hall in about 1982, before the restoration began.

He was concerned that modernization was unnecessarily destroying Britain's technical icons.

The title of the magazine that published Rix's article shows that industrial archae-

ology is an activity for everyone. A person need not be an educator, a technologist, a scientist, or an engineer to see the value in saving local technical history. No better example can be given than Sam Morris (1917-1991). A successful construction company owner, Morris saw the need to save the Royal Agricultural Hall in Islington, on the north side of London. The huge building, shaped like a Quonset hut, is 312 feet long by 123 feet wide by 73 feet high. It opened in 1862 and was used for agricultural shows. It also housed exhibitions of general inventions, leather goods, furniture, and other items. England's first motor car show was held there in 1899. Its last use during the 1960s was with the British Royal Mail service. After its abandonment, it rapidly deteriorated.

Morris wanted to save the historic building, but the local government had scheduled it for demolition. So in 1982, Morris bought the building from the Islington Council for £1 million. He then worked tirelessly to locate the funding for restoration. The Victorian structure reopened in 1986 as the Business Design Centre. With a new entrance influenced by the 1851 Crystal Palace, it is Britain's only integrated exhibition and trade center.

Industrial archaeology projects do not

The interior of the restored Royal Agricultural Hall in London maintained the effect of a glass roof with safer, stronger plastic. Its Victorian heritage is obvious in the embellished cast-iron trusses.

typically require so much effort. Most are aimed at photographing remnants and recording information about a particular site. Photographs can capture the stark beauty of an abandoned railroad building or a derelict factory. If a restoration seems possible, the coordination is often handled by a committee of citizens on behalf of a local government.

The Saugus Iron Works in Massachusetts was America's first major industry. The forge building had four waterwheels to operate bellows for the furnace and trip hammers. The shed roof at the immediate left is part of the restored furnace area.

Saugus Iron Works

One government-sponsored restoration from seventeenth-century America is the Saugus Iron Works in Saugus, Massachusetts. The first major American industrial enterprise, it began operating in 1646, not long after settlers tamed the region. Iron ore was readily available in the Saugus area and nearby streams were used for transportation. Plentiful trees provided wood for charcoal and lumber for buildings.

The specific site was selected partly because it had a suitable hill. Early iron furnaces were built next to hills because that allowed for taking raw materials to the hilltop for easier loading. Iron ore, charcoal, and other materials were dumped into the furnace through a large hole at the top. Fire inside the slate-lined furnace melted the materials from the high heat generated by the burning charcoal. There were no outside flames. Two huge bellows operated by a waterwheel forced air into the furnace and showered sparks from the loading hole. Once or twice every 24 hours, a clay plug

was removed and molten iron ran into crude forms dug in the sand floor. The cooled cast-iron ingots weighed 200 pounds or more.

The forge building had trip hammers powered by four waterwheels. The heavy ingots were reheated and placed between the hammer and anvil. The hammer rose and fell 30 to 50 times a minute. An ingot could lose up to half its weight as additional impurities were beaten out or burned in the reheating process. Workers were constantly exposed to suffocating heat, flying streams of hot metal from the hammer blows, deafening noise, and physical exhaustion. During an era when industrial safety programs were not even thought of, injuries commonly occurred.

The Saugus Iron Works closed in 1678. The company had experienced raw material shortages, conflicts with local authorities, and difficulty keeping skilled workers when settlement land was cheap. The skilled workers who left Saugus helped expand the colonial iron industry. The National Park Service credits the Saugus Iron Works with providing the foundation for the American iron and steel industry. The region was excavated and reconstructed by the U.S. government in the 1950s.

Baltimore and Ohio Railroad Roundhouse

Most reconstructed industrial buildings have found a rebirth as museums. The 1884 B&O Railroad roundhouse in Baltimore, Maryland, is not only a museum, but has continued its connection with American rolling stock. It is America's oldest train shop and the birthplace of much railroading technology. A wagon repair facility was built on the site in 1829 and a full-time railroad shop was operating in 1832. It was a place where locomotives were constructed and passenger cars repaired. At one time, it employed more than 3,000 people.

The beautifully restored roundhouse opened in 1964 as the B&O Railroad Museum. It has original locomotives, replicas, and engines. The collection includes one of only two remaining Allegheny locomotives. At 778,000 pounds and 7,500 hp, they were among the heaviest and most powerful steam locomotives ever built in the United States. One replica is the *Tom Thumb*, the first American-built locomotive. Con-

Trip hammers at the Saugus Iron Works in Massachusetts were lifted by four wooden cams. In the photo, one points up and another is positioned to begin lifting the trip hammer. The hammer's own weight caused it to fall onto the workpiece.

structed by Peter Cooper (1791-1883), it made various runs in 1830 from the B&O site. The museum has more than 120 pieces of full-size equipment and a large collection of artifacts.

The building has a historical niche that is not shared by any other. Samuel Morse (1791-1872) sent the first telegraphed intercity message from the Supreme Court chambers in Washington, D.C. "What hath God wrought?" was received at the attached Mount Clare Depot on May 24, 1844. The first American railroad ticket was sold at the original depot in 1830.

Robbins and Lawrence Armory and Machine Shop

America has more than 100 smaller technical or industrial museums. Few are as impressive as the American Precision Museum (APM) in Windsor, Vermont, a city with a population of about 3,500. It is housed in the 1846 Robbins and Lawrence Armory and emphasizes nineteenth-century machine tools. It was the first International Mechanical Engineering Heritage Site of the American Society of Mechanical Engineers

(ASME). The 1987 commemorative bronze plaque states, in part, that the museum "contains the largest collection of historically significant machine tools in the nation."

Samuel Robbins and Richard Lawrence received a contract for 25,000 U.S. Army rifles, and a similar order from the British government. The cost was $10.90 per rifle. They fulfilled their contract in 1847 and were the first to achieve practical interchangeability of parts. Their operation was so efficient that the orders were completed barely a year after they opened the factory.

Ed Battison (1916-) was born in Windsor. He worked as a curator at the Smithsonian Institution in Washington, D.C., in 1964. The old Robbins and Lawrence building was scheduled for demolition, but Battison would have none of it. The original rifle company had gone bankrupt in 1856. The local electric company purchased it in 1898 and converted it to a generating plant. During the late 1950s and early 1960s, it was used as a warehouse. The electric company no longer needed the facility and considered razing it. Battison used his skills as a curator to save the structure and fill it with historically significant machine tools.

Opening in 1965, the museum is near the Connecticut

The Baltimore and Ohio Railroad Museum houses the largest rail exhibit in the United States. The building is a restored nineteenth-century roundhouse about 10 blocks from Baltimore's popular Inner Harbor.

Richard Arkwright built the world's first factory in the isolated community of Cromford, England. He constructed a series of canals to transport materials for his cotton factory. This restored building marks the start of the canal to Manchester and Nottingham.

River in a well-restored building that acknowledges its past with nicely finished wooden floors. Large wooden upper-floor-support posts, clapboard interior siding, and traditional windows have been maintained. As the ASME plaque attests, the museum boasts an impressive array of machine tools. It has the original Bridgeport milling machine and a replica Simeon North (1765-1852) milling machine made for the APM by the United Technologies Corporation. It has a Thomas Blanchard (1788-1864) lathe for making irregular wooden shapes like gunstocks. There are also many other specialty tools and displays.

Another historically significant structure is near the APM. The longest covered bridge in America crosses the Connecticut River at Windsor. Named the Cornish-Windsor Bridge, the 460-foot structure is also the longest two-span covered bridge in the world.

Eli Whitney Factory Site

Not all industrial archaeology results in restored buildings. The Eli Whitney factory site near Hamden, Connecticut, consists simply of flat land. Yet it regularly attracts industrial archaeologists from Yale University, about 10 miles away. It has been the subject of archaeological inquiry since 1972.

In recent years, one group of students excavated a region they presumed to be a case-hardening shop. Another group excavated the domestic area, where factory workers lived. Over the course of three months, they found materials ranging from shoe leather to slag, from ceramic dolls to brick walls. The area was once known as Whitneyville. Eli Whitney left no written record of his factory and the reason is easy to understand.

Whitney invented and patented the cotton gin in 1794. It was such a simple, straightforward machine that once people saw a cotton gin, they could copy it. In 1794, the newly enacted patent law was quite weak. So many infringing machines were built that Whitney exhausted himself and his money fighting them in court. By 1798, he was almost penniless and his patent, important as it was to the economy of the South, was worthless to him. He turned his back on the cotton gin forever.

The new federal government was looking for a manufacturer to fill a large order

Thomas Edison's industrial research laboratory was originally located in Orange, New Jersey. To help preserve it for posterity, Henry Ford moved the complex to Greenfield Village in Dearborn, Michigan, in 1928. Steam-powered machine tools operated in the brick building at the rear.

for flintlock muskets. Whitney had no background in making firearms, no factory, no raw materials, and no workers. He won the contract by offering the possibility of producing interchangeable parts. He never came close to that goal.

Using money from 10 local investors, Whitney built a water-powered factory. It did not have any particularly unique machinery, but Whitney hoped his idea of a division of labor would improve production. Unlike his competitors, he did not have each worker make one musket from start to finish. He had 50 workers, each making or assembling an individual part. Each job was simple and could be completed quickly and accurately. Stung by his experience with the cotton gin, Whitney never revealed the details of his factory's operation. No descriptions or drawings have ever been found and the factory slowly disintegrated. Piecing everything back together has presented a great challenge.

Some British Restorations

The Industrial Revolution began in Great Britain, so it seems logical that Britain would start the field of industrial archaeology. Through laws aimed at historical preservation, it has protected countless buildings and structures. Some are fairly well known, like St. Katherine Dock near the Tower Bridge in London. It was built in 1828 by Thomas Telford (1757-1834) to provide convenient ship loading. It is now a series of apartments and shops. Other structures, like the 1831 Middletop Engine House near Nottingham, are less well known. This building is the sole surviving steam-engine house of those that stood at the top of inclines and pulled up rail cars.

Well known or not, all Britain's technical artifacts played a role in making it the mightiest manufacturing country of the nineteenth century. At one point, Britain made more than half the world's manufactured products.

Richard Arkwright (1732-1792) built a factory town in the isolated community of Cromford, England, in the 1770s. (See pages 22-24.) He wanted to take advantage of the power offered by the River Derwent, which never froze in the winter. He built a canal to connect to the industrial cities of Nottingham and Manchester. The canal's beginning point has been restored, near Arkwright's large home.

British immigrant Samuel Slater (1768-1835) established America's first factory in Pawtucket, Rhode Island, in 1790. He served part of his apprenticeship in Belper, England, under Jedediah Strutt (1726-1797). While keeping its exterior appearance, Strutt's factory has been remodeled into an office complex.

Eli Whitney's factory site near Hamden, Connecticut, has no original production buildings. Industrial archaeologists carefully investigate the remains of his rifle factory. It obtained water power from the Quinnipiac River.

The American Precision Museum in Windsor, Vermont, is housed in the Robbins and Lawrence Armory. Built with handmade bricks, it was the most modern rifle factory in the world when it opened in 1846.

Industrial Archaeology in Your Community

Industrial archaeology can be an academic discipline or an enjoyable hobby. In either case, interested people can apply it to the history of any nearby industry. The subject is concerned with all physical relics, from factories to warehouses, to engines, to worker housing, to transportation—and everything in between.

See one of the references listed below for information on how to begin an investigation. The Society for Industrial Archaeology can also provide assistance. Established in 1971, the society's mailing address is:

> Society for Industrial Archaeology
> Michigan Technological University
> 1400 Townsend Drive
> Houghton MI 49931-1295.

The internet address is <http://www.ss.mtu.edu>. Click on the <Industrial Archaeology Program>.

References

The Archaeology of the Industrial Revolution by Brian Bracegirdle, Fairleigh Dickinson University Press, 1973.

Industrial Archaeology–A New Look at the American Heritage by Theodore Anton Sande, Stephen Greene Press, 1976.

Dictionary of Industrial Archaeology by William Jones, Sutton Alan Publishers, 1997.

Historic American Engineering Record Catalog compiled by Donald E. Sackheim, National Park Service, 1976.

World Industrial Archaeology by Kenneth Hudson, Cambridge University Press, 1979.

Industrial Archaeology by Arthur Raistrick, Eyre Methuen Publishers, 1972.

Industrial Archaeology by Kenneth Hudson, John Baker Publishers, 1963.

The *Building That Would Not Go Away–The Story of How Sam Morris Rescued the Royal Agricultural Hall* by Tadeusz Grajewski, Gresham Press, 1989.

Baltimore and Ohio Railroad Museum Web site. Available: http://www.borail.org

"The American Precision Museum" by Dave Goska, *Vermont Life,* Spring 1985.

Recommended Reference Sources

The following particularly helpful technical dictionaries and encyclopedias are listed in the general order of their usefulness, from the author's experience, to researching material of the sort found in *Technology's Past:*

Biographical Dictionary of the History of Technology edited by Lance Day and Ian McNeil, Routledge Publishers, 1996.

Dictionary of American Biography, Charles Scribner's Sons Publishers, 1932; with updating supplements.

National Cyclopedia of American Biography, James T. White & Co. Publishers, 1891; with updating supplements.

Dictionary of National Biography (British) edited by Leslie Stephens and Sidney Lee, Oxford University Press, 1917; with updating supplements.

American Science and Invention by Mitchell Wilson, Bonanza Books, 1960.

Timetables of Technology by Bryan Bunch and Alexander Hellemans, Simon and Schuster Publishers, 1993.

International Dictionary of Women's Biography edited by Jennifer S. Uglow, Macmillan Publishers, 1982.

Dictionary of Scientific Biography, Charles Scribner's Sons, 1976.

Chambers Concise Dictionary of Scientists by David Millar and others, W & R Chambers Ltd. Publishers, 1990.

Asimov's Biographical Encyclopedia of Science and Technology by Isaac Asimov, Doubleday & Co. Publishers, 1964.

World of Invention edited by Bridget Travers, Gale Research Inc., 1994.

Biographical Dictionary of American Science by Clark A. Elliott, Greenwood Press, 1979.

Scientists and Inventors by Anthony Feldman and Peter Ford, Facts on File, 1979.

Encyclopedia of World Biography, Gale Research Publications, 1998.

Famous Names in Engineering by James Carvill, Butterworth Publishers, 1981.

Great Lives from History edited by Frank N. Magill, Salem Press, 1987.

Heroes of American Invention by L. Sprague de Camp, Barnes and Noble Books, 1993.

New Illustrated Science and Invention Encyclopedia, H. S. Stuttman Publishers, 1987.

These additional references proved helpful. They are listed alphabetically by title:

Encyclopedia of the History of Technology by Ian McNeil, Routledge Publishers, 1990.

Famous First Facts by Joseph Nathan Kane, H. W. Wilson Publishers, 1981.

Great Events from History, Science, and Technology Series edited by Frank N. Magill, Salem Press, 1991.

Great Inventions Through History by Gerald Messadie, W & R Chambers Publishers, 1991.

Great Modern Inventions by Gerald Messadie, W & R Chambers Publishers, 1991.

Making of the Modern World edited by Neil Cossons, John Murray Publishers Ltd., 1992.

Mothers of Invention, by Ethlie Ann Vare and Greg Ptacek Quill, William Morrow Publishers, 1987.

Nuts and Bolts of the Past by David Freeman Hawke, Harper and Row Publishers, 1988.

Pioneers of Computing by F. Gareth Ashurst, Frederick Muller Ltd., 1983.

The Real McCoy by Portia P. James, Smithsonian Institution Press, 1989.

Smithsonian Book of Invention, Smithsonian Institution, 1978.

Those Inventive Americans, National Geographic Society, 1971.

Women Scientists in America, by Margaret W. Rossiter, Johns Hopkins University Press, 1982.

Works of Man by Ronald W. Clark, Viking Penguin Inc., 1985.

Index